Principles and Methods of Applied Mathematics

Principles and Methods of Applied Mathematics

Michael (Misha) Chertkov
University of Arizona, USA

World Scientific

NEW JERSEY • LONDON • SINGAPORE • BEIJING • SHANGHAI • TAIPEI • CHENNAI

Published by

World Scientific Publishing Co. Pte. Ltd.
5 Toh Tuck Link, Singapore 596224
USA office: 27 Warren Street, Suite 401-402, Hackensack, NJ 07601
UK office: 57 Shelton Street, Covent Garden, London WC2H 9HE

Library of Congress Control Number: 2025010517

British Library Cataloguing-in-Publication Data
A catalogue record for this book is available from the British Library.

PRINCIPLES AND METHODS OF APPLIED MATHEMATICS

ISBN 978-981-98-0824-3 (hardcover)
ISBN 978-981-98-0928-8 (paperback)
ISBN 978-981-98-0825-0 (ebook for institutions)
ISBN 978-981-98-0826-7 (ebook for individuals)

For any available supplementary material, please visit
https://www.worldscientific.com/worldscibooks/10.1142/14184#t=suppl

Cover art by Ilya Margolin
Desk Editors: Nambirajan Karuppiah/Cian Sacker Ooi

Typeset by Stallion Press
Email: enquiries@stallionpress.com

About the Author

Dr. Michael "Misha" Chertkov is a professor of mathematics and the chair of the Graduate Interdisciplinary Program in Applied Mathematics at the University of Arizona. His research addresses foundational challenges in mathematics, statistics, machine learning and artificial intelligence, particularly as they apply to and are inspired by physical systems such as fluid mechanics. He also works on applications in the control of engineered systems, such as energy grids, and bio-social systems. Dr. Chertkov received his PhD in physics from the Weizmann Institute of Science in 1996. After obtaining his PhD, he spent three years as an R. H. Dicke Fellow at the Department of Physics at Princeton University. In 1999, he joined the Los Alamos National Laboratory, first as a J. R. Oppenheimer Fellow and later becoming a technical staff member in the Theory Division. He transitioned to the University of Arizona in 2019. Throughout his career, Dr. Chertkov has contributed about 300 research papers. He holds the title of fellow at both the American Association for the Advancement of Science and the American Physical Society and is a senior member of Institute of Electrical and Electronics Engineers.

Contents

Introduction

This living book began in 2019 when the author moved to the University of Arizona to chair the Graduate Interdisciplinary Program in Applied Mathematics. His challenge was to update the program's applied math curriculum, consisting of three courses, to align with what we now call (in 2025) Applied Mathematics of the Artificial Intelligence Era. This encompasses natural and artificial applied mathematics, serving various scientific disciplines in the physical, engineering, biological and social sciences.

Since 2021, Math 581A and 581B, taught every fall and spring semester, have become the backbone of the program's core courses, and this book developed from the lecture notes of this course. Many of the author's colleagues from the University's Mathematics Department and across the campus have contributed with inspiration, encouragement and support.

Colin Clark, a postdoctoral researcher and recitation instructor for this and other core classes of the program from 2019 to 2023, has made significant contributions. Colin meticulously reviewed, edited and suggested updates for nearly all the book's pages and sections over the years. Homework exercises and midterm exams for this course were designed in collaboration with Colin and teaching assistants of the course, especially Craig Thomas, who served in this capacity from 2020 to 2022.

The book contains two appendices. Appendix A features final exams prepared by the graduate qualification committee for the Math 581 course. From 2019 to 2023, this committee included Professors Ibrahim Fatkullin, Ildar Gabitov, Kevin Lin, Laura Miller,

Marek Rychlik and Charles Wolgemuth. The author extends sincere gratitude to the committee members for their invaluable contributions, advice and guidance. Appendix B presents material on convex and non-convex optimization, prepared by Yury Maximov from Los Alamos National Laboratory and edited by the author. This material was used in guest lectures for the Math 581B course, delivered by Yury in 2020–2021. The author is thankful to Yury for granting permission to include this content in the book.

This book is accompanied by Jupyter/Julia notebooks, available on the author's living-book website at https://sites.google.com/site/mchertkov/living-books/applied-math-book, where periodic updates, improvements and corrections will also be posted.

Chapter 1

Applied Math Core Courses

Every student in the Program for Applied Mathematics at the University of Arizona takes three core courses during their first year: Methods (Math 581), Theory (Math 584) and Algorithms (Math 589). Each course provides a distinct set of tools and expertise for tackling problems in modern applied mathematics. The courses often cover overlapping topics, sometimes in parallel (Fig. 1.1),[a] which helps demonstrate the unique contributions of each approach and fosters a deeper understanding of applied mathematics. The curriculum includes traditional subjects, such as differential equations, alongside more modern topics, such as optimization, control and elements of computer science and statistics. Every aspect of the material is carefully selected to reflect the most relevant concepts for both current and future applications.

The core courses aim to equip students with a variety of toolboxes in applied mathematics. Occasionally, exact solutions can be found by applying powerful – though often highly specialized – techniques or methods. However, more commonly, solutions must be formulated algorithmically, with approximate answers obtained through numerical simulations and computational methods. A deep understanding of the theoretical foundations of a problem enhances the design

[a] All figures in this book are available in color in the online version on the author's website. Every effort has been made to ensure that the figures remain clear and legible in both black-and-white print and the printed version.

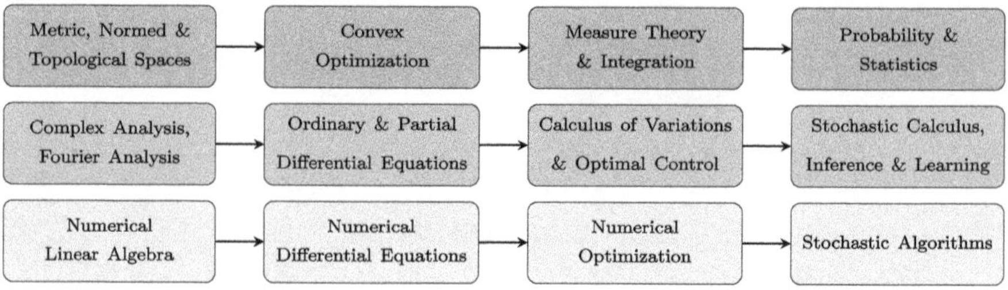

Figure 1.1. Topics covered in Theory (top column), Methods (middle column) and Algorithms (bottom column) during the fall semester (columns 1 and 2) and spring semester (columns 3 and 4).

and implementation of these methods and algorithms, enabling us to make precise predictions about when and how they will be effective.

The core courses cover a broad range of mathematical topics, highlighting some of the most intriguing and significant areas in applied mathematics. This wide exposure often helps students identify specific fields they wish to explore further within the program. While the core courses provide a strong foundation, they do not – and cannot – meet the in-depth requirements needed for dissertation work. Students are expected to pursue more specialized courses and engage in independent study to deepen their expertise in their chosen areas.

Moreover, the courses do not – and cannot – cover every topic within applied mathematics. Instead, they offer a carefully curated, cohesive and necessarily selective overview of material that, based on the instructors' expertise and perspectives, is most relevant for students during and after their graduate studies. In this introductory chapter, we aim to share our view of what defines modern applied mathematics, presenting the material in a way that unifies concepts that may initially seem unrelated.

What is Applied Mathematics?

We study and develop mathematics to model, optimize and control various physical, biological, engineering and social systems. Applied mathematics is an interdisciplinary field, combining mathematical science, domain-specific knowledge and understanding and, often, insights from related fields, or "math-adjacent" disciplines (Fig. 1.2).

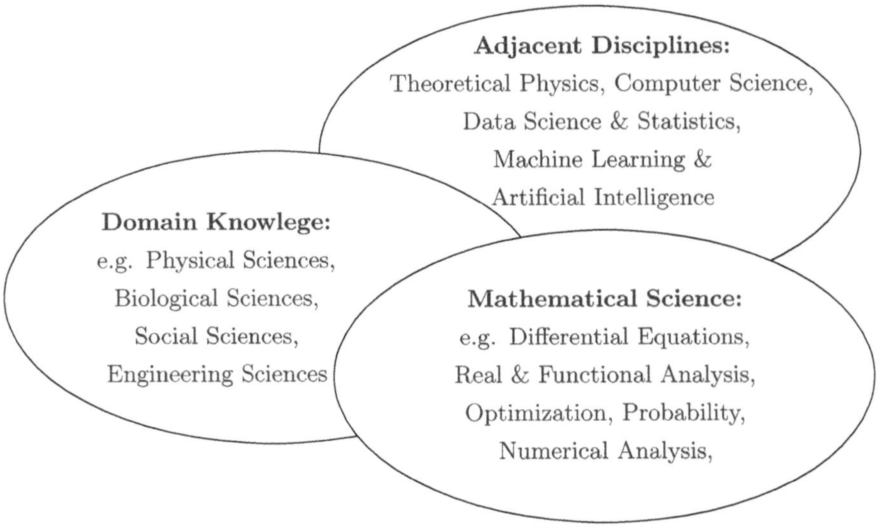

Figure 1.2. The key components studied under the umbrella of applied mathematics: (1) mathematical science, (2) domain-specific knowledge, and (3) a few 'math-adjacent' disciplines.

In our program, the core courses emphasize the mathematical foundations of applied mathematics, while more specialized topics and domain knowledge are explored through advanced coursework, independent research and internships.

Applying mathematics to real-world problems demands mathematical approaches that have evolved to meet the complexity and challenges of these problems. In some cases, relatively simple governing equations can effectively describe the phenomena. In such situations, the challenge often lies in developing methods that are both highly accurate and efficient, sometimes requiring adaptability to varying data without compromising precision. However, in other cases, there may be no established governing equations, either because we do not know them or because they may not exist. In these instances, the task is to create better mathematical models by interpreting and synthesizing incomplete or imperfect data. Throughout the core courses, we emphasize a general methodology that includes:

1. formulating the problem, first informally, using common scientific and engineering terms, before refining it into a formal mathematical representation;

2. analyzing the problem using every available tool, including theoretical, methodological and algorithmic approaches developed within applied mathematics;
3. determining the type of solution required and implementing the most suitable method to achieve that solution.

Contributing meaningfully to a specific domain requires more than mere mathematical expertise; it demands a deep understanding of the domain itself. This domain-specific knowledge can shift our perspective on what constitutes an appropriate solution. For instance, when system parameters are no longer well-defined and must be estimated from measurements or experimental data, interpreting the solutions becomes more challenging, and it becomes critical to quantify the uncertainty in these solutions. Likewise, in complex systems that couple many subsystems, exact expressions may no longer be feasible to compute or interpret. In such cases, approximate or "effective" solutions may offer more insight. In every domain-specific context, it is crucial to identify the most pressing problems and determine which solutions will be most valuable.

Mathematics is not the only discipline that provides valuable insights to other fields. Physics, statistics and computer science have each contributed distinct frameworks, models and intuitions for solving complex problems. This is particularly evident in the rise of data science, machine learning and artificial intelligence (AI). The explosion of available data has created new opportunities in engineering, as well as in the physical, natural and social sciences, solving problems that once could only be addressed empirically. Machine learning, largely driven by artificial neural networks inspired by biological ones, has accelerated many classical numerical methods, while AI now aids in automating mathematical derivations, analyses and even proofs. These "math-adjacent" fields – statistics, theoretical physics and computer science – are foundational to data science, machine learning and AI because they offer powerful tools for analyzing, modeling and interpreting data, as well as for guiding decision-making. Despite the immense progress, significant challenges remain, and we believe that integrating mathematical rigor with the intuition and methodologies from these disciplines will be essential for addressing them.

Problem formulation

We will use a diverse range of instructional examples from various fields of science and engineering to demonstrate how to transform a loosely defined scientific or engineering phenomenon into a well-formulated mathematical problem. Some of these challenges will be resolved, while others will remain open for further research. Examples include the Kirchoff equations for power systems, the Navier–Stokes equations for fluid dynamics, network flow equations, the Fokker–Planck equation from statistical mechanics and constrained regression from data science.

Problem analysis

We approach problems extracted from real-world applications using every available method, drawing on both domain-specific intuition and mathematical expertise. Often, we can make precise statements about a problem's solutions without fully solving it in the mathematical sense. For example, dimensional analysis from physics offers a valuable form of preliminary analysis. Additionally, we can uncover key properties of solutions by examining underlying symmetries and determining the expected principal behaviors. Important examples include oscillatory behavior (waves), diffusive behavior and the distinction between dissipative/decaying and conservative behaviors. A great deal can also be learned by analyzing the problem in different asymptotic regimes, such as when a parameter becomes small, simplifying the analysis. By matching different asymptotic solutions, we can construct a detailed – though ultimately incomplete – understanding of the problem.

Solution construction

As noted earlier, a key aspect of applied mathematics is the use of specialized techniques to find analytic solutions. However, these techniques are not always practical. In such cases, developing strong computational intuition becomes crucial for selecting the most appropriate methods for numerical or hybrid analytic–numerical analysis. This computational toolbox is essential for breaking down and solving complex problems effectively.

Part I
Applied Analysis

Chapter 2

Complex Analysis

The real number system, while powerful, has certain limitations: Not all operations are permissible for real numbers. For instance, it is impossible to take arbitrary roots of negative numbers within the real number system. This limitation is overcome by introducing the concept of the *imaginary unit*, denoted by $i := \sqrt{-1}$. A number that is a real multiple of the imaginary unit, such as $3i$, $i/2$, or $-\pi i$, is called an *imaginary number*. A number composed of both a real and an imaginary part is known as a *complex number*.

Complex analysis is the branch of mathematics that explores functions involving complex variables. A central idea in this field is that many operations between real numbers can be naturally extended to complex numbers and that many real-valued functions have counterparts in the complex domain. Remarkably, these extensions often reveal a deeper richness, unlocking new techniques for problem-solving that go beyond what is possible with real numbers alone.

Complex analysis serves as an indispensable tool in various branches of both pure and applied mathematics, as well as in diverse fields of physics – ranging from hydrodynamics to quantum mechanics – and engineering disciplines, such as aerospace, mechanical, and electrical engineering.

Our treatment of complex analysis is focused on topics that will be essential for later chapters. For readers seeking further foundational material on complex analysis, we recommend *Methods of the Theory of Functions of a Complex Variable* by Lavrentiev and Shabat [1], which presents a rigorous and comprehensive approach to complex analysis. Another excellent reference is *Functions of One Complex Variable* by Conway [2], which offers a detailed introduction, with a focus on theoretical aspects. Finally, *Complex Analysis* by Ahlfors [3] remains a classic reference, providing a thorough introduction to the theory of analytic functions of one complex variable, with a balance of theory and applications.

2.1 Complex Variables and Complex-Valued Functions

2.1.1 *The Cartesian representation of complex variables*

For two complex numbers $z_1 = a_1 + ib_1$ and $z_2 = a_2 + ib_2$, we have the following operations:

- **Addition:** $z_1 + z_2 = (a_1 + ib_1) + (a_2 + ib_2) = (a_1 + a_2) + i(b_1 + b_2)$,
- **Multiplication:** $z_1 z_2 = (a_1 + ib_1)(a_2 + ib_2) = a_1a_2 + i(a_1b_2 + b_1a_2) + i^2 b_1b_2 = (a_1a_2 - b_1b_2) + i(a_1b_2 + b_1a_2)$.

The addition and subtraction of complex numbers are straightforward extensions of their real-valued counterparts.

Example 2.1.1. Let $z_1 = 1 + 2i$ and $z_2 = 4 - i$. Compute (a) $z_1 + z_2$ and (b) $z_1 - z_2$.

Solution.

(a) $z_1 + z_2 = (1 + 2i) + (4 - i) = (1 + 4) + (2 - 1)i = 5 + i$,
(b) $z_1 - z_2 = (1 + 2i) - (4 - i) = (1 - 4) + (2 + 1)i = -3 + 3i$.

Since the behavior of addition and subtraction of complex numbers resembles the translation of vectors in $\mathbb{R}^2$, we often visualize complex numbers as points in the Cartesian plane by associating the real and imaginary parts of the complex number with the x- and y-coordinates, respectively (a) is also illustrated in Fig. 2.1.

Definition 2.1.2. The *complex conjugate* of a complex number z, denoted by z^* or $\bar{z}$, is the complex number with the same real part but with an imaginary part equal in magnitude and opposite in sign. That is, if $z = x + iy$, then $z^* := x - iy$.

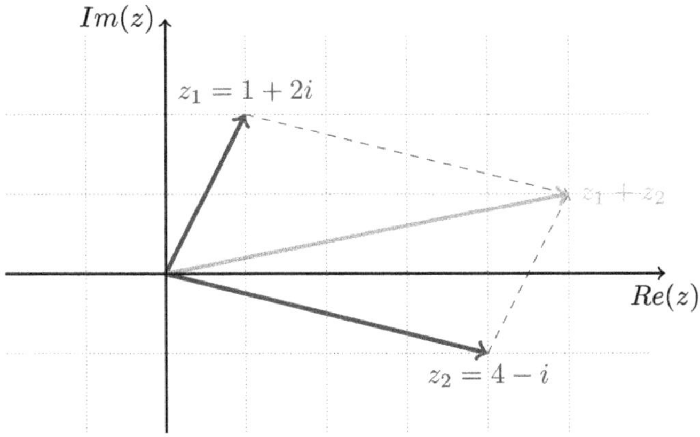

Figure 2.1. Complex numbers can be visualized as vectors in $\mathbb{R}^2$. By convention, the real component is plotted on the horizontal axis, and the imaginary component is plotted on the vertical axis. The addition of two complex numbers is analogous to vector addition in $\mathbb{R}^2$.

The multiplication and division of complex numbers follow directly from their real-valued counterparts, with the additional rule that $i^2 = -1$.

Example 2.1.3. Let $z_1 = -1 + 2i$ and $z_2 = 4 - 3i$. Compute (a) $z_1 z_2$ and (b) z_1/z_2.

Solution.

(a) $z_1 z_2 = (-1+2i)(4-3i) = -4 + 3i + 8i - 6i^2 = 2 + 11i$,
(b) To compute z_1/z_2, we first multiply by z_2^*/z_2^* so that the denominator $z_2 z_2^*$ becomes a real number:

$$\begin{aligned}\frac{z_1}{z_2} &= \frac{-1+2i}{4-3i} \cdot \frac{4+3i}{4+3i} = \frac{(-1+2i)(4+3i)}{(4-3i)(4+3i)} \\ &= \frac{-4-3i+8i+6i^2}{16+9} = \frac{-10+5i}{25} = -\frac{2}{5} + \frac{1}{5}i.\end{aligned}$$

Complex conjugates

Theorem 2.1.4. *For algebraic operations such as addition, multiplication, division, and exponentiation, consider a sequence of operations on n complex numbers $z_1, \ldots, z_n$ resulting in a complex number w. If the same operations are applied in the same order to* *the conjugates $z_1^*, \ldots, z_n^*$, then the resulting number will be the complex conjugate of w, denoted w^*.*

Example 2.1.5. Let us illustrate Theorem 2.1.4 using the example of a quadratic equation, $az^2 + bz + c = 0$, where the coefficients a, b, and c are real. By direct application of Theorem 2.1.4, if the equation has a complex root, then its conjugate must also be a root, which is consistent with the well-known quadratic formula:

$$z_{1,2} = \frac{-b \pm \sqrt{b^2 - 4ac}}{2a}.$$

Exercise 2.1. Use Theorem 2.1.4 to show that the roots of a polynomial with real-valued coefficients of *any* order must occur in complex conjugate pairs.[a]

Example 2.1.6. Find all the roots of the polynomial $p(z) = z^4 - 6z^3 + 11z^2 - 2z - 10$, given that one of its roots is $2 - i$.

Solution. Since $p(z)$ has real-valued coefficients, its roots must occur in conjugate pairs. Given that $z_1 = 2 - i$ is a root, $z_2 = 2 + i$ must also be a root. We verify this by direct substitution. We can factorize $p(z)$ as

$$p(z) = (z - z_1)(z - z_2)r(z),$$

where $r(z)$ is found by polynomial division, giving $r(z) = z^2 - 2z - 2$. Thus, the four roots of $p(z)$ are found by solving

$$0 = z^4 - 6z^3 + 11z^2 - 2z - 10 = (z - z_1)(z - z_2)(z^2 - 2z - 2).$$

Solving $z^2 - 2z - 2 = 0$ using the quadratic formula gives $z_{3,4} = 1 \pm \sqrt{3}$. Therefore, the four roots of $p(z)$ are

$$z_1 = 2 - i, \quad z_2 = 2 + i, \quad z_3 = 1 + \sqrt{3}, \quad z_4 = 1 - \sqrt{3}.$$

Example 2.1.7. Let $z_1 = x_1 + iy_1$ and $z_2 = x_2 + iy_2$. Show that if $\omega = z_1/z_2$, then $\omega^* = z_1^*/z_2^*$.

---

[a]The exercises in this book are intended as part of assigned homework. No solutions are published to encourage independent problem-solving and allow students taking the course to work through the exercises on their own.

Solution. From the definition of a complex conjugate, we have

$$z_1^* = x_1 - iy_1, \quad z_2^* = x_2 - iy_2.$$

We need to compute ω^* and verify that it equals z_1^*/z_2^*. First, compute ω:

$$\omega = \frac{x_1 + iy_1}{x_2 + iy_2} \cdot \frac{x_2 - iy_2}{x_2 - iy_2} = \frac{x_1x_2 + y_1y_2}{x_2^2 + y_2^2} + i\frac{x_2y_1 - x_1y_2}{x_2^2 + y_2^2}.$$

Now, compute ω^*:

$$\omega^* = \frac{x_1x_2 + y_1y_2}{x_2^2 + y_2^2} - i\frac{x_2y_1 - x_1y_2}{x_2^2 + y_2^2} = \frac{(x_1 - iy_1)(x_2 + iy_2)}{(x_2 - iy_2)(x_2 + iy_2)} = \frac{x_1 - iy_1}{x_2 - iy_2}.$$

Thus, $\omega^* = z_1^*/z_2^*$, as required.

2.1.2 *The polar representation of complex variables*

In addition to their Cartesian representation, complex numbers can also be represented in polar form using the components r and θ. The value r, known as the modulus of z, satisfies $r^2 = |z|^2 := zz^* = x^2 + y^2 \geq 0$, and θ, called the argument or polar angle of z, is defined as $\theta = \arg(z)$ for $|z| > 0$, modulo 2π:

$$x + iy = r\cos\theta + ir\sin\theta, \quad \text{where } r = \sqrt{x^2 + y^2}, \quad \theta = \arctan(y, x).$$

Polar representation is illustrated in Fig. 2.2. Using trigonometric identities, the product of two complex numbers is another complex number whose modulus is the product of their moduli and whose argument is the sum of their arguments. For example, if $z_1 = r_1\cos\theta_1 + ir_1\sin\theta_1$ and $z_2 = r_2\cos\theta_2 + ir_2\sin\theta_2$, then

$$z_1z_2 = r_1r_2\cos(\theta_1 + \theta_2) + ir_1r_2\sin(\theta_1 + \theta_2).$$

This summation of arguments when multiplying complex numbers is reminiscent of multiplying exponential functions. In fact, polar representation simplifies complex multiplication, as the modulus of the product is the product of the moduli, $|z_1z_2| = |z_1||z_2|$, and the argument of the product is the sum of their arguments, $\arg(z_1z_2) = \arg(z_1) + \arg(z_2)$. These properties are key motivations for defining the complex-valued exponential function.

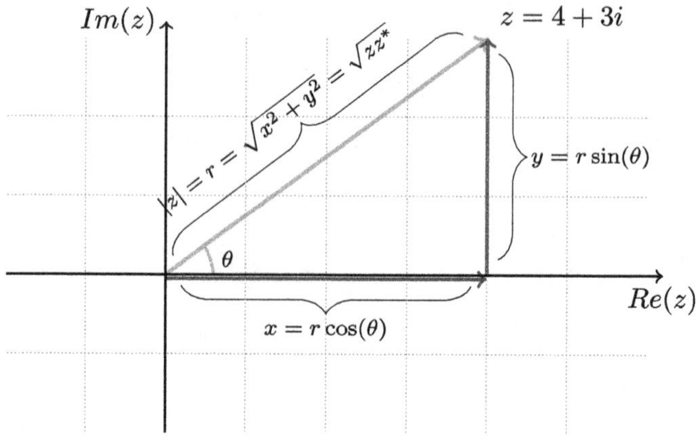

Figure 2.2. A complex number, z, has both a Cartesian representation and a polar representation. Its modulus, denoted by $|z|$ or r, is non-negative and satisfies $|z|^2 = r^2 := zz^* = x^2 + y^2$. Its argument, θ, is the angle measured modulo 2π, counterclockwise from the positive real axis.

Definition 2.1.8. The *exponential function* for imaginary arguments is defined by

$$re^{i\theta} := r\cos(\theta) + ir\sin(\theta) = x + iy. \tag{2.1}$$

Euler's famous formula, $e^{i\pi} = -1$, follows directly from this definition.

Example 2.1.9. Convert $z_1 = -1 + 2i$ and $z_2 = 4 - 3i$ to their polar representations and compute (a) their product and (b) their quotient. Compare your answer to Example 2.1.3.

Solution. The polar representations of z_1 and z_2 are

$$r_1 = \sqrt{z_1 z_1^*} = \sqrt{5}, \quad \theta_1 = \tan^{-1}(10/-5) \approx 2.03, \quad z_1 = \sqrt{5}e^{2.03i};$$
$$r_2 = \sqrt{z_2 z_2^*} = 5, \quad \theta_2 = \tan^{-1}(-3/4) \approx -0.64, \quad z_2 = 5e^{-0.64i}.$$

Their product and quotient are

(a) $z_1 z_2 \approx \left(\sqrt{5}e^{2.03i}\right)\left(5e^{-0.64i}\right) = 5\sqrt{5}e^{1.39i} = 5\sqrt{5}\cos(1.39) + i5\sqrt{5}\sin(1.39) = 2 + 11i$,

(b) $z_1/z_2 \approx \sqrt{5}e^{2.03i}/5e^{-0.64i} = \frac{1}{\sqrt{5}}e^{2.67i} = \frac{1}{\sqrt{5}}\cos(2.67) + i\frac{1}{\sqrt{5}}\sin(2.67) = -0.4 + 0.2i$.

Sometimes, it is convenient to express a complex number using a mixture of Cartesian and polar representations.

Example 2.1.10. Find $\tilde{r}$ and $\tilde{\theta}$ such that the point $\omega = 1 + 5i$ can be written as $\omega = -1 + \tilde{r}e^{i\tilde{\theta}}$.

Solution. Given that $1+5i = -1+\tilde{r}e^{i\tilde{\theta}}$, solve for $\tilde{r}e^{i\tilde{\theta}}$ to get $2+5i = \tilde{r}e^{i\tilde{\theta}}$. Solve for $\tilde{r}$ and $\tilde{\theta}$:

$$\tilde{r} = \sqrt{(2+5i)(2-5i)} = \sqrt{29} \approx 5.39, \quad \tilde{\theta} = \tan^{-1}(5/2) \approx 1.19\,\text{rad}.$$

Thus, $\omega \approx -1 + 5.39e^{i1.19}$.

Example 2.1.11. Express $z = (2+2i)e^{-i\pi/6}$ in (a) Cartesian and (b) polar representations.

Solution.

(a) $z = (2+2i)\left(\cos(-\frac{\pi}{6}) + i\sin(-\frac{\pi}{6})\right) = \left(2\cos(-\frac{\pi}{6}) - 2\sin(-\frac{\pi}{6})\right) + i\left(2\cos(-\frac{\pi}{6}) + 2\sin(-\frac{\pi}{6})\right) = (1+\sqrt{3}) + i(\sqrt{3}-1)$.

(b) $z = (2+2i)e^{-i\pi/6} = 2\sqrt{2}e^{i\pi/4}e^{-i\pi/6} = 2\sqrt{2}e^{i\pi/12}$.

2.1.3 *Parameterization of curves in the complex plane*

Definition 2.1.12. A *curve* in the complex plane is a set of points $z(t)$, where $a \le t \le b$ for some $a \le b$. A curve is said to be *closed* if $z(a) = z(b)$ and *simple* if it does not self-intersect, except possibly at the endpoints. In other words, the curve is simple if $z(t) \neq z(t')$ for $t \neq t'$ and $a < t, t' < b$. A curve is called a *contour* if it is continuous and piecewise smooth. By convention, all simple, closed contours are parameterized to be traversed counterclockwise, unless stated otherwise. Examples of curves in the complex plain are shown in Fig. 2.3.

Example 2.1.13. Parameterize the following curves:

(a) The infinite "vertical" line passing through $\pi/2$.
(b) The semi-infinite ray extending from the point $z = -1$ and passing through $3i$.
(c) The circular arc of radius ε centered at 0.

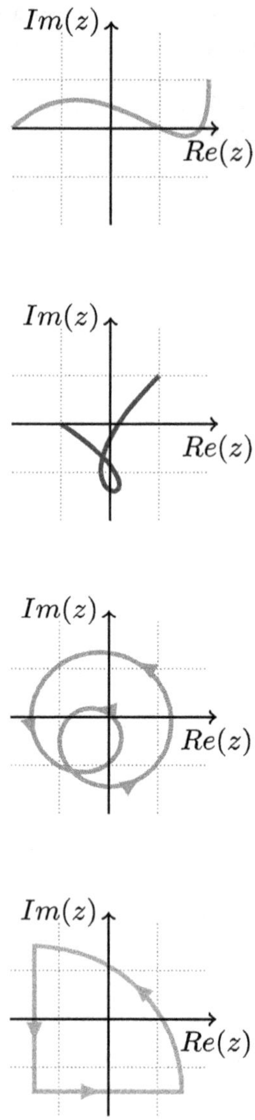

Figure 2.3. Examples of curves in the complex plane. The first two curves are open, while the last two curves are closed. The first and fourth curves are simple, whereas the second and third curves are not simple because they self-intersect at points other than the endpoints.

Solution.

(a) $z(s) = \pi/2 + is$ for $-\infty < s < \infty$.
(b) $z(\rho) = -1 + (1 + 3i)\rho$ for $0 \le \rho < \infty$.
(c) $z(\tau) = \varepsilon e^{i\tau}$ for $0 \le \tau \le 2\pi$.

The complex number system

Complex numbers provide a system that is closed under all possible algebraic operations. This means that any algebraic operation performed between two complex numbers will always result in another complex number. This property does not hold for other number systems. For example:

(i) The sum of two positive integers is always another positive integer, but the difference between two positive integers *is not necessarily* positive. Thus, positive integers are closed under addition but *not* under subtraction.
(ii) The set of integers is closed under subtraction and multiplication, but not under division, since the quotient of two integers is not always an integer.
(iii) Rational numbers are closed under division. However, taking limits of rational numbers may lead to non-rational numbers, which requires the introduction of real numbers to ensure closure under limits.
(iv) Taking non-integer powers of negative numbers does not yield a real number. To handle such operations, the system of complex numbers is introduced.

Additionally, the set of complex numbers is closed under other operations, such as solving algebraic equations, finding roots, and taking logarithms. In conclusion, the set of complex numbers is closed under all algebraic operations.

2.1.4 *Functions of a complex variable*

A function of a complex variable, denoted as $w = f(z)$, defines a mapping from a complex number z to another complex number w. Specifically, the function f takes a point in the z-plane (also known as the domain) and maps it to a corresponding point (or points) in the w-plane (the codomain). Since both z and w have Cartesian representations, with $z = x + iy$ and $w = u + iv$, the function $f(z)$ can be decomposed into two real-valued functions of two real variables. That is, the real and imaginary components of w are given by $f(z) = u(x, y) + iv(x, y)$, where $u(x, y)$ and $v(x, y)$ are real-valued functions representing the real and imaginary parts of $f(z)$, respectively.

2.1.5 *Complex exponentials*

In Eq. (2.1), we defined the exponential function $f(z) = e^z$ to preserve the property that $e^{z_1+z_2} = e^{z_1}e^{z_2}$ and, incidentally, that $e^1 = 2.718$. However, this is not the only property we could have used to define e^z. Several other properties could also motivate the definition, including:

- the function's Taylor series $\sum_{n=0}^{\infty} \frac{z^n}{n!}$;
- the limiting expression $\lim_{n\to\infty} \left(1 + \frac{z}{n}\right)^n$;
- the fact that $f(z) = e^z$ is the solution to the ordinary differential equation $f'(z) = f(z)$, subject to $f(0) = 1$.

We encourage the reader to verify that all these properties hold for the complex exponential and that any of them could have motivated our definition and led to the same results.

As a consequence of this, the natural definitions of the complex-valued trigonometric functions are

$$\cos(z) := \frac{e^{iz} + e^{-iz}}{2} \quad \text{and} \quad \sin(z) := \frac{e^{iz} - e^{-iz}}{2i}.$$

Example 2.1.14. Let $f(z) = \exp(iz)$, where $z = x + iy$. Express $f(z)$ as $u(x, y) + iv(x, y)$, where u and v are real-valued functions of x and y.

Solution.

$$\begin{aligned} f(z) &= \exp(i(x + iy)) = \exp(ix - y) = \exp(-y)\exp(ix) \\ &= \exp(-y)\cos(x) + i\exp(-y)\sin(x) \end{aligned} \qquad \square$$

Example 2.1.15. Evaluate the following functions along the specified curves: (a) $z \mapsto \sin z$ along the infinite vertical line passing through $\pi/2$, (b) $z \mapsto \exp(z + 1)$ along the semi-infinite ray extending from the point $z = -1$ and passing through $\sqrt{3}i$, and (c) $z \mapsto z^2$

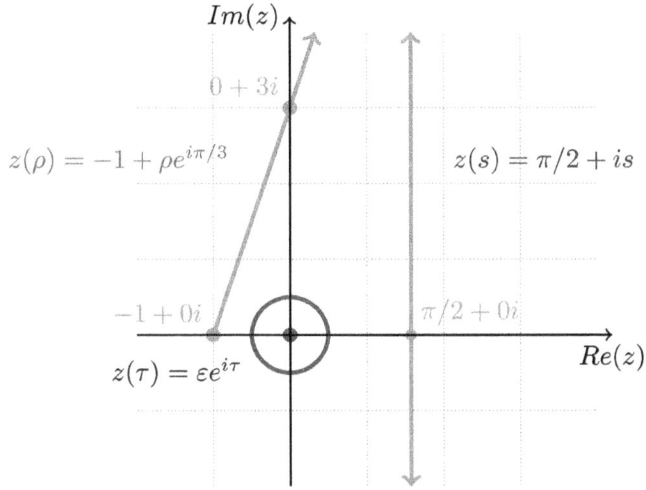

Figure 2.4. Parameterized curves for Example 2.1.13. Vertical line: The infinite "vertical" line passing through $\pi/2$ is parameterized by $z(s)=\pi/2+is$ for $-\infty < s < \infty$. Inclined line: The semi-infinite ray extending from the point $z=-1$ and passing through $3i$ is parameterized by $z(\rho)=-1+(1+3i)\rho$ for $0 \le \rho < \infty$. Circle: The circular arc of radius ε, centered at 0, is parameterized by $z(\tau)=\varepsilon e^{i\tau}$ for $0 \le \tau \le 2\pi$.

along a circular arc of radius ε centered at 0. See Example 2.1.13 and Fig. 2.4.

Solution.

(a) Parameterize the vertical line passing through $\pi/2$ by $\pi/2+is$ for $-\infty < s < \infty$:

$$f(s+i\pi/2)=\sin(\pi/2+is)=\frac{e^{i(\pi/2-s)}-e^{-i(\pi/2-s)}}{2i}=\frac{ie^{-s}+ie^{s}}{2i}$$
$$=\cosh(s)\quad \text{for } -\infty < s < \infty.$$

(b) Parameterize the semi-infinite ray extending from $z=-1$ and passing through $\sqrt{3}i$ by $-1+\rho e^{i\pi/3}$ for $0<\rho<\infty$:

$$f(-1+\rho e^{i\pi/3})=\exp(\rho e^{i\pi/3})=\exp\left(\rho\cos\left(\frac{\pi}{3}\right)+i\rho\sin\left(\frac{\pi}{3}\right)\right)$$
$$=e^{\rho/2}e^{i\sqrt{3}\rho/2}\quad \text{for } 0 \le \rho < \infty.$$

(c) Parameterize the circular arc of radius ε centered at 0 by $\varepsilon e^{i\tau}$ for $0 \le \tau \le 2\pi$:

$$f(\varepsilon e^{i\tau})=(\varepsilon e^{i\tau})^2=\varepsilon^2 e^{2i\tau}\quad \text{for } 0 \le \tau \le 2\pi.$$

Exercise 2.2. Investigate the asymptotic behavior of the following complex-valued functions as $|z| \to \infty$: (a) $f(z) = \exp(z)$, (b) $f(z) = \sin(z)$, and (c) $f(z) = \cos(z)$.

Hint: There are many ways in which $|z|$ can approach infinity. For example, you could write $z = x + iy$ and let $x \to \pm\infty$, with fixed or varying y, or let $y \to \pm\infty$, with fixed or varying x. Alternatively, you could write $z = re^{i\theta}$ and let $r \to \infty$ for fixed or varying θ. Consider each function from different perspectives to explore what happens as $|z| \to \infty$.

2.1.6 *Multi-valued functions and branch cuts*

Not all complex functions are single-valued. In many cases, we encounter multi-valued functions, meaning that for some z, there are two or more values w_i such that $f(z) = w_i$. For example, recall the parameterized curve in Example 2.1.15, specifically (c)(i), where we evaluated the function $f(z) = z^2$ along a circle of radius ε centered at the origin. In that case, the function returned to its original value: $f(\varepsilon e^{0i}) = f(\varepsilon e^{2\pi i}) = \varepsilon^2$. However, this is not always the case for all functions.

Example 2.1.16. Consider the function $\omega(z) = \sqrt{z}$. When z is written in polar form as $z = r\exp(i\theta)$, we know that θ is defined up to a shift by $2\pi n$ for any integer n. This gives us the following expression for $\omega(z)$:

$$\omega_n(z) = \sqrt{r}\exp\left(i\frac{\theta}{2} + i\pi n\right),$$

where different values of n yield two branches of $\sqrt{z}$: $\omega_1 = \sqrt{r}\exp(i\theta/2)$ and $\omega_2 = \sqrt{r}\exp(i\theta/2 + i\pi)$.

Now, if we choose one branch, say ω_1, and traverse a counterclockwise path around $z = 0$ (keeping $z = 0$ to the left), gradually increasing θ from 0 to 2π, ω_1 will transition to ω_2. After another full 2π turn, ω_2 will return to ω_1. Thus, the two branches switch after each full 2π rotation. The point $z = 0$ is called a second-order branch point of the two-valued function $\sqrt{z}$.

Definition 2.1.17. A multi-valued function $w(z)$ has a *branch point* at $z_0 \in \mathbb{C}$ if $w(z)$ changes continuously along a small circuit around z_0 but does not return to its original value after one complete circuit.

Definition 2.1.18. A *branch* of a multi-valued function $w(z)$ is a single-valued version of the function, obtained by restricting the image of $w(z)$ to only one set of values.

For multi-valued functions, traversing a small closed contour around a branch point results in a discontinuity. The location of this discontinuity depends on where the contour starts and ends. To illustrate this, consider the following two contours:

$$\alpha(\theta) = e^{i\theta}, \quad 0 \leq \theta \leq 2\pi,$$
$$\beta(\phi) = e^{i\phi}, \quad -\pi \leq \phi \leq \pi.$$

As illustrated in Fig. 2.5, if we apply the function $f(z) = \sqrt{z}$ to these two contours, we observe that the discontinuity occurs at $\theta = 0, 2\pi$ in the first case and at $\phi = -\pi, \pi$ in the second case. The location of the discontinuity depends entirely on the chosen contour. This brings us to the concept of a *branch cut*, which we introduce to manage the different branches of a multi-valued function. For most multi-valued functions, a branch cut prevents the function from being continuous across a specific contour.

Definition 2.1.19. A *branch cut* is a curve in the complex plane along which a multi-valued function is discontinuous.

Remark. Branch cuts are not unique; they are chosen based on how we wish to define the function. One branch is usually designated as the *principal branch*. Most mathematical software uses a predefined set of rules to select the principal branch of a multi-valued function.

Example 2.1.20. The function $\omega(z) = z^{1/n}$ generalizes Example 2.1.16. This function has n branches, $\omega_1(z), \ldots, \omega_n(z)$, making $z = 0$ an nth-order branch point of the n-valued function $z^{1/n}$.

Example 2.1.21. Another important example, illustrated in Fig. 2.6, is $\omega(z) = \log(z)$. When z is expressed in polar form as $z = re^{i(\theta+2\pi n)}$, we see that $\log(z)$ is a multi-valued function with infinitely many (though countable) values. These values are given by

$$\omega_n(z) = \log(r) + i(\theta + 2n\pi), \quad n = 0, \pm 1, \pm 2, \ldots.$$

In this case, $z = 0$ is an infinite-order branch point.

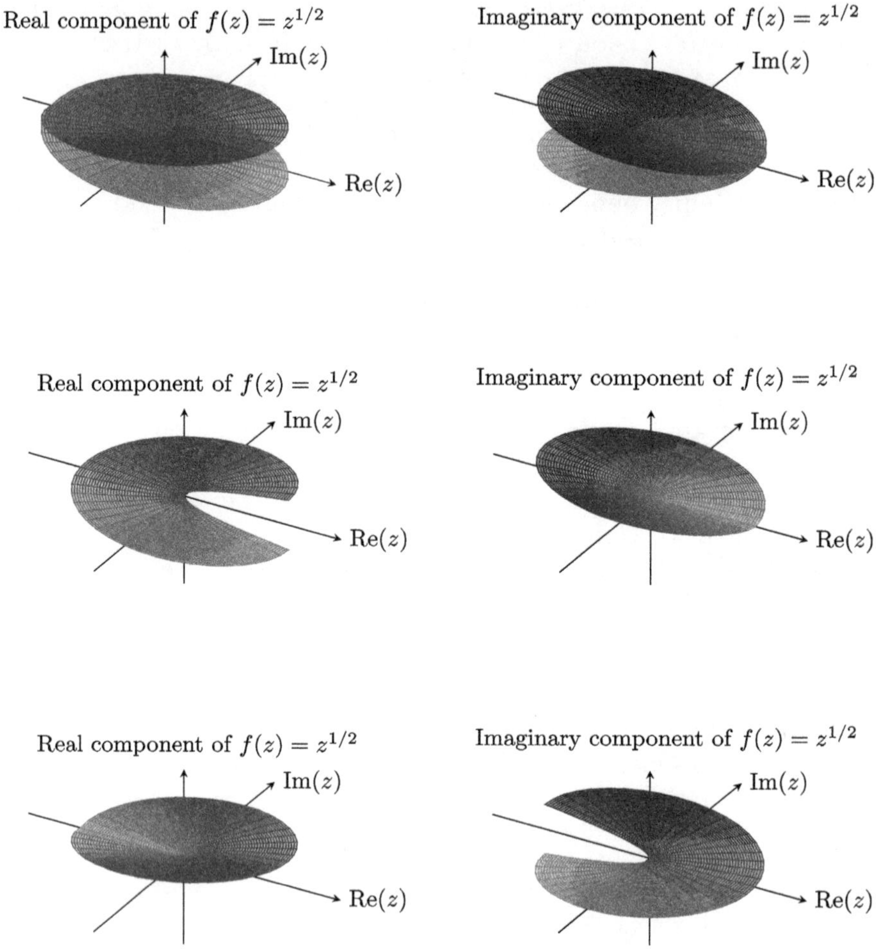

Figure 2.5. (a) Top row: The real (left) and imaginary (right) components of $z \mapsto z^{1/2}$. (b) Middle row: Representing $z = re^{-i\theta}$ with $0 \leq \theta < 2\pi$ gives a branch cut along the positive real axis, where a single branch is analytic everywhere except along the cut. (c) Bottom row: Representing $z = re^{-i\theta}$ with $-\pi \leq \theta < \pi$ gives a branch cut along the negative real axis, where another branch of $z \mapsto z^{1/2}$ is analytic everywhere except along the cut.

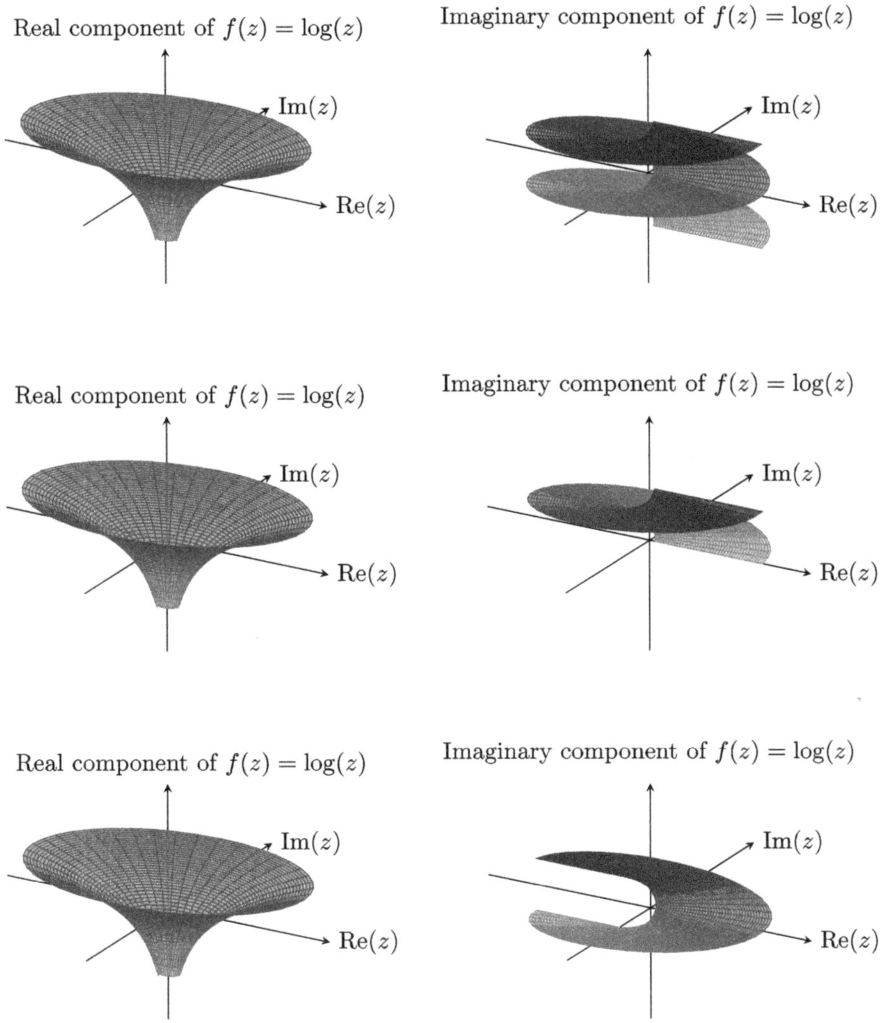

Figure 2.6. (a) Top row: The real (left) and imaginary (right) components of $z \mapsto \log(z)$. (b) Middle row: Representing $z = re^{-i\theta}$ with $0 \leq \theta < 2\pi$ gives a branch cut along the positive real axis, where a single branch of $z \mapsto \log(z)$ is analytic everywhere except along the cut. (c) Bottom row: Representing $z = re^{-i\theta}$ with $-\pi \leq \theta < \pi$ gives a branch cut along the negative real axis, where a single branch of $z \mapsto \log(z)$ is analytic everywhere except along the cut.

Definition 2.1.22. (Branch points at $z = \infty$.) Consider a multi-valued function $f(z)$. We say that f has a branch point at $z = \infty$ if the function $g(w) = f(1/w)$ has a branch point at $w = 0$.

Example 2.1.23. Find the branch points of $\log(z-1)$ and sketch possible branch cuts. Choose a branch cut and describe the resulting branches.

Solution. We can parameterize the function as follows: $\log(z-1) = \log\rho + i\phi$, where $z - 1 = \rho\exp(i\phi)$, with $\rho > 0$ and ϕ real. As ϕ increases by multiples of 2π when we traverse a closed path around $z = 1$, we conclude that $z = 1$ is a branch point of $\log(z-1)$.

Similarly, $z = \infty$ is also a branch point. This can be observed by substituting $z = 1/w$, where $w = 0$ becomes a branch point. Therefore, a valid branch cut should connect $z = 1$ and $z = \infty$, as illustrated in Fig. 2.7.

To describe the branches, let's choose a branch cut that starts at $z = 1$ and extends along the positive real axis to $z = +\infty$. The branches of z are given by

$$z_n = 1 + \rho\exp(i\phi + 2i\pi n), \quad n = 0, 1, \ldots.$$

For $f(z) = \log(z-1)$, the corresponding branches are

$$f_n(z) = \log\rho + i\phi + 2i\pi n.$$

Each branch is distinct, representing a single-valued, analytic function in $\mathbb{C}$ except along the branch cut.

Example 2.1.24. Consider the function $\log(z^2-1) = \log(z-1) + \log(z+1)$. As we travel around $z = 1$, both $\log(z-1)$ and $\log(z^2-1)$ change by 2π. Thus, $z = 1$ is a branch point of $\log(z^2-1)$. Similarly, $z = -1$ and $z = \infty$ are also branch points of $\log(z^2-1)$. Figure 2.8 shows two examples of branch cuts for $\log(z^2-1)$.

Two important remarks should be noted:

1. The function $\log(f(z))$ has branch points at the zeros of $f(z)$, at the points where $f(z)$ is infinite, and potentially at the branch points of $f(z)$ itself. However, be cautious about this third case – zeros and infinities refer to analytic zeros and poles, respectively. Other singular behaviors in $f(z)$ could lead to unexpected results. For example, examine what happens at $z = 0$ when $f(z) = \exp(1/z)$.

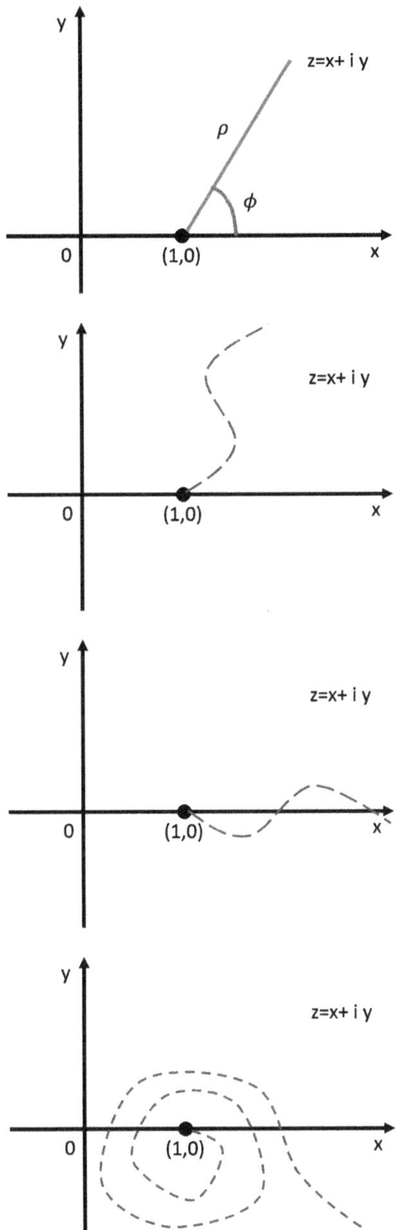

Figure 2.7. Polar parametrization of $\log(z-1)$ (left) and three examples of branch cuts for the function, connecting its two branch points at $z = 1$ and $z = \infty$.

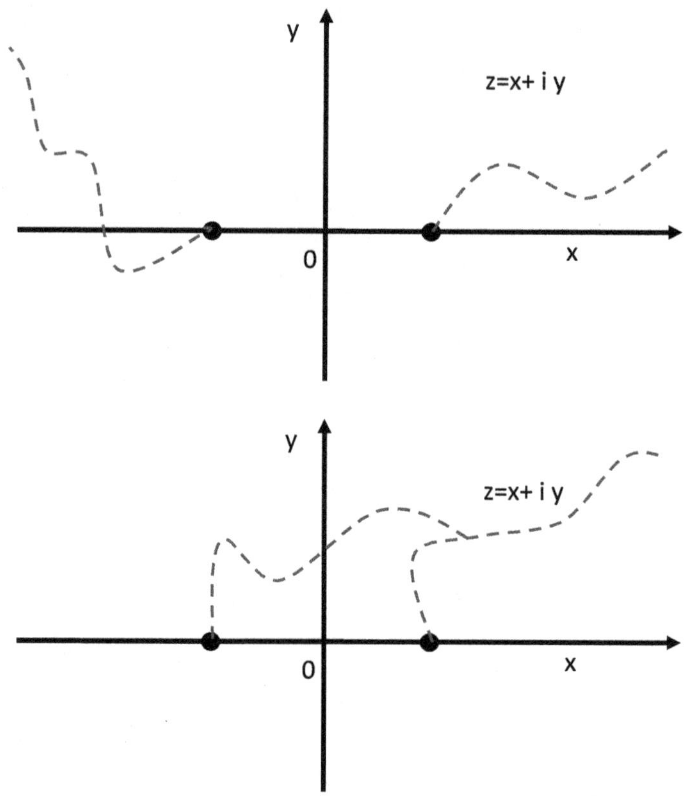

Figure 2.8. Two examples of branch cuts for $\log(z^2 - 1)$.

2. Whether or not a function $f(z)$ or its derivatives have a finite value at some point z_0 is irrelevant when determining if z_0 is a branch point of $f(z)$.

Exercise 2.3. Identify the branch points, introduce suitable branch cuts, and describe the resulting branches for the functions: (a) $f(z) = \sqrt{(z-a)(z-b)}$, and (b) $g(z) = \log((z-1)/(z-2))$.

The graphs of complex multi-valued functions form two-dimensional manifolds in the four-dimensional space $\mathbb{R}^4$. These manifolds are called Riemann surfaces. Riemann surfaces can be visualized in three-dimensional space using parallel projection, with the image rendered on the screen. For further details and visualizations using Mathematica, see http://matta.hut.fi/matta/mma/SKK_Mma Journal.pdf.

Example 2.1.25. Find all values of $z \in \mathbb{C}$ that satisfy the equation $\sin(z) = 3$.

Solution. Start by using the definition of the complex-valued sine function:

$$\frac{e^{iz} - e^{-iz}}{2i} = 3.$$

Multiply both sides by $2i\, e^{iz}$:

$$(e^{iz})^2 - 6ie^{iz} - 1 = 0.$$

This quadratic equation can be solved using the quadratic formula, which gives

$$e^{iz} = i(3 \pm 2\sqrt{2}).$$

Taking the natural logarithm of both sides, we get

$$iz = \ln(i) + \ln(3 \pm 2\sqrt{2}).$$

Using $\ln(z) = \ln(r) + i(\theta + 2n\pi)$ and noting that $\ln(i) = \ln(1) + i\left(\frac{\pi}{2} + 2n\pi\right)$, we find

$$z = \frac{\pi}{2} + 2n\pi \pm i\ln(3 + 2\sqrt{2}).$$

2.2 Analytic Functions and Integration along Contours

2.2.1 *Analytic functions*

The derivative of a real-valued function at a point x is defined using the limiting expression

$$f'(x) = \lim_{\Delta x \to 0} \frac{f(x + \Delta x) - f(x)}{\Delta x},$$

and we say that the function is differentiable at x if this limit exists and is independent of whether Δx approaches 0 from the positive or negative direction.

Definition 2.2.1. The derivative of a complex function is defined similarly via the limit

$$f'(z) = \lim_{\Delta z \to 0} \frac{f(z + \Delta z) - f(z)}{\Delta z}, \tag{2.2}$$

where z is a complex number. This limit only exists if $f'(z)$ is independent of the direction in the complex plane from which Δz approaches 0. (*Note*: There are infinitely many directions from which Δz can approach a point in $\mathbb{C}$.)

For example, if we set $\Delta z = \Delta x$, then from Eq. (2.2), we have

$$f'(z) = u_x + iv_x,$$

where $f = u + iv$. However, if we set $\Delta z = i\Delta y$, we obtain

$$f'(z) = -iu_y + v_y.$$

A consistent definition of the derivative requires that both expressions give the same result, which leads to the conditions

$$u_x = v_y, \quad u_y = -v_x, \tag{2.3}$$

which are known as the Cauchy–Riemann equations.

Theorem 2.2.2 (Cauchy–Riemann Theorem). *A function $f(z) = u(x, y) + iv(x, y)$ is differentiable at the point $z = x + iy$ if and only if the partial derivatives u_x, u_y, v_x, v_y are continuous, and the Cauchy–Riemann conditions* (2.3) *are satisfied in a neighborhood of z.*

In the derivation leading to the Cauchy–Riemann theorem, we only demonstrated the necessity of these conditions for differentiability. To complete the proof, we must also show that the Cauchy–Riemann equations are sufficient for differentiability. In other words, we need to show that any function $u(x, y) + iv(x, y)$ is **complex-differentiable** if the Cauchy–Riemann equations hold.

This sufficiency follows from the following transformations:

$$\begin{aligned}
\Delta f &= f(z + \Delta z) - f(z) \\
&= \frac{\partial f}{\partial x}\Delta x + \frac{\partial f}{\partial y}\Delta y + O\left((\Delta x)^2, (\Delta y)^2, \Delta x \Delta y\right) \\
&= \frac{1}{2}\left(\frac{\partial f}{\partial x} - i\frac{\partial f}{\partial y}\right)\Delta z + \frac{1}{2}\left(\frac{\partial f}{\partial x} + i\frac{\partial f}{\partial y}\right)\Delta z^* \\
&\quad + O\left((\Delta x)^2, (\Delta y)^2, \Delta x \Delta y\right) \\
&= \frac{\partial f}{\partial z}\Delta z + \frac{\partial f}{\partial z^*}\Delta z^* + O\left((\Delta x)^2, (\Delta y)^2, \Delta x \Delta y\right),
\end{aligned}$$

where $O\left((\Delta x)^2, (\Delta y)^2, \Delta x \Delta y\right)$ represents terms of second order or higher in Δx and Δy. In the last step, we changed variables from (x, y) to (z, z^*), using the relations

$$\frac{\partial}{\partial x} = \frac{\partial}{\partial z} + \frac{\partial}{\partial z^*},$$
$$\frac{\partial}{\partial y} = i\frac{\partial}{\partial z} - i\frac{\partial}{\partial z^*},$$

and their inverses (known as the Wirtinger derivatives):

$$\frac{\partial}{\partial z} = \frac{1}{2}\left(\frac{\partial}{\partial x} - i\frac{\partial}{\partial y}\right), \quad \frac{\partial}{\partial z^*} = \frac{1}{2}\left(\frac{\partial}{\partial x} + i\frac{\partial}{\partial y}\right).$$

To ensure that the derivative $f'(z)$ is well-defined, we require

$$\frac{\partial f}{\partial z^*} = 0,$$

which means that f does not depend on z^*. It is straightforward to check that this condition is equivalent to the Cauchy–Riemann equations (2.3). □

Definition 2.2.3 ((Complex) Analyticity). A function $f(z)$ is called analytic (or holomorphic) at a point z_0 if it is differentiable (as a complex function) in a neighborhood of z_0. It is analytic in a region of the complex plane if it is analytic at every point within that region.

Exercise 2.4. The *isolines* of a function $f(x, y) = u(x, y) + iv(x, y)$ are the curves where $u(x, y) = \text{const}$ and $v(x, y) = \text{const}'$. Show that for an analytic function, the isolines always intersect at right angles.

Example 2.2.4. Let $f(z) = u(x, y) + iv(x, y)$ be analytic. Given that $u(x, y) = x + x^2 - y^2$ and $f(0) = 0$, find $v(x, y)$.

Solution. We start by using the Cauchy–Riemann equations:

$$\frac{\partial u}{\partial x} = \frac{\partial v}{\partial y}, \quad \frac{\partial u}{\partial y} = -\frac{\partial v}{\partial x}.$$

This gives the system

$$\frac{\partial v}{\partial y} = 2x + 1, \qquad \frac{\partial v}{\partial x} = 2y.$$

Solving these, we obtain

$$v(x, y) = 2xy + y.$$

Thus, the function is $f(z) = (x + x^2 - y^2) + i(2xy + y)$.

Exercise 2.5. Let $f(z) = u(x, y) + iv(x, y)$ be analytic. Given that $v(x, y) = -2xy$ and $f(0) = 1$, find $u(x, y)$.

Examples of functions that are *not* analytic

Example 2.2.5. Determine whether (and where) $f(z) = z^*$ is analytic and compute its derivative where it exists.

Solution. Recall that if $z = x + iy$, then $z^* := x - iy$. We first compute the partial derivatives u_x, v_y, u_y and v_x and then assess their continuity:

$$u_x = 1, \quad v_y = -1,$$

$$u_y = 0, \quad v_x = 0.$$

The partial derivatives are continuous everywhere in $\mathbb{C}$. However, the Cauchy–Riemann conditions are *not* satisfied anywhere in $\mathbb{C}$ since $u_x = 1 \neq -1 = v_y$. Intuitively, the function z^* fails to be analytic because analytic functions can be locally approximated by smooth rotations and stretches in the complex plane, whereas z^* represents a reflection.

Example 2.2.6. Determine whether (and where) $f(z) = z^{1/2}$ is analytic and compute its derivative where it exists.

Solution. Applying the chain rule and using the trigonometric identity $\sin^2(\theta) + \cos^2(\theta) = 1$, we verify that the Cauchy–Riemann equations in polar coordinates are

$$\frac{\partial u}{\partial r} = \frac{1}{r}\frac{\partial v}{\partial \theta},$$

$$\frac{\partial v}{\partial r} = -\frac{1}{r}\frac{\partial u}{\partial \theta}.$$

We compute the relevant partial derivatives for $z^{1/2}$ and find that they are *not* defined at $z = 0$. Additionally, these derivatives cannot be continuous at a branch cut, as different branches of $z^{1/2}$ are discontinuous across the cut. However, the Cauchy–Riemann conditions are satisfied everywhere else. Therefore, a branch of $z^{1/2}$ is analytic in any region of $\mathbb{C} \setminus \{0\}$ that excludes the branch cut. The derivative in polar representation is given by

$$f'(z) = e^{-i\theta}\left(\frac{\partial u}{\partial r} + i\frac{\partial v}{\partial r}\right),$$

which simplifies to $f'(z) = \frac{1}{2}r^{-1/2}e^{-i\theta/2} = \frac{1}{2}z^{-1/2}$.

Example 2.2.7. Determine whether (and where) the function $f(z) = \frac{1}{z}$ is analytic.

Solution. The function $f(z)$ is not defined at $z = 0$, and $\lim_{z\to 0} f(z)$ does not exist. To analyze the function, we rationalize the denominator

$$f(z) = \frac{x}{x^2+y^2} - i\frac{y}{x^2+y^2}.$$

The relevant partial derivatives are

$$u_x = \frac{y^2-x^2}{(x^2+y^2)^2}, \quad v_y = \frac{y^2-x^2}{(x^2+y^2)^2},$$
$$u_y = \frac{-2xy}{(x^2+y^2)^2}, \quad v_x = \frac{2xy}{(x^2+y^2)^2}.$$

These partial derivatives exist and are continuous everywhere in $\mathbb{C} \setminus \{0\}$, and the Cauchy–Riemann conditions are satisfied for all $z \neq 0$. We compute the derivative $f'(z)$ and observe that $\lim_{z\to 0} f'(z)$ does not exist. We conclude that f has a *simple pole* at $z = 0$ since $zf(z)$ is analytic in a neighborhood of 0. This concept will be revisited in Section 2.2.5.

Example 2.2.8. Determine whether (and where) the functions (a) $\exp(z)$, (b) $z\exp(\bar{z})$ and (c) $\frac{\exp(z)-1}{z}$ are analytic. Compute their derivatives where they exist.

The Cauchy–Riemann theorem (Theorem 2.2.2) provides additional insights from both a geometrical perspective and through its connections to partial differential equations, as discussed in the sections that follow.

Geometry of complex: Conformal mapping

Let us now explore the concept of a conformal map, which describes a function of two variables, x, y, that locally preserves angles, though not necessarily distances. Remarkably, the rich family of conformal functions (maps) can be fully analyzed based on the Cauchy–Riemann conditions (2.3).

The Cauchy–Riemann conditions (2.3) can be succinctly expressed as

$$i\frac{\partial f}{\partial x} = \frac{\partial f}{\partial y},$$

where $f = u + iv$. The Jacobian matrix of the function $f : \mathbb{R}^2 \to \mathbb{R}^2$, i.e., the map from (x, y) to (u, v), is given by

$$J = \begin{pmatrix} \frac{\partial u}{\partial x} & \frac{\partial u}{\partial y} \\ \frac{\partial v}{\partial x} & \frac{\partial v}{\partial y} \end{pmatrix} = \begin{pmatrix} \frac{\partial u}{\partial x} & \frac{\partial u}{\partial y} \\ -\frac{\partial u}{\partial y} & \frac{\partial u}{\partial x} \end{pmatrix}.$$

Geometrically, the off-diagonal (skew-symmetric) part of the matrix represents rotation, while the diagonal part represents stretching or rescaling. The Jacobian of a function $f(z)$ transforms infinitesimal line segments at the intersection of two curves in the z-plane, rotating them to corresponding segments in the image plane $f(z)$. Therefore, a function that satisfies the Cauchy–Riemann equations with a nonzero derivative preserves the angle between curves in the plane. Such functions are called *conformal.* Thus, the Cauchy–Riemann conditions serve not only as criteria for the analyticity of a function but also for its conformality.

A fundamental result in complex analysis that builds the theoretical foundation for conformal maps is the following famous theorem, which was originally stated by Bernhard Riemann in his 1851 Ph.D. thesis and later proven by Constantin Carathéodory in 1912.

Theorem 2.2.9 (Riemann Mapping Theorem). *If A is a non-empty, simply connected open subset of $\mathbb{C}$ that is not the entire complex plane, then there exists a holomorphic (complex-analytic) function $f(z)$ that maps A onto the unit disk $D = \{z \in \mathbb{C} \mid |z| < 1\}$, such that $f : A \to D$. Moreover, $f^{-1} : D \to A$ is also holomorphic.*

This theorem provides a powerful tool for constructing conformal maps between simply connected domains by first mapping each

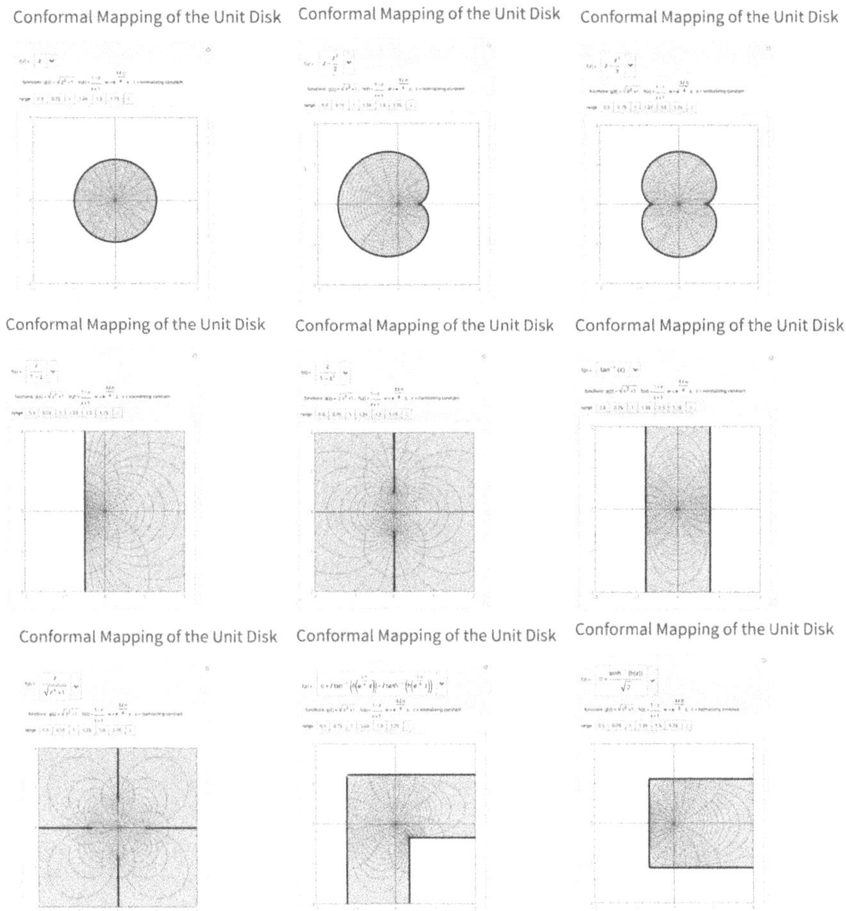

Figure 2.9. Examples of functions (maps) from the unit disk. Screenshots from https://demonstrations.wolfram.com/ConformalMappingOfTheUnitDisk/.

domain to the unit disk. Various conformal maps can be created by combining these mappings.

To develop a better geometrical intuition, it is helpful to study conformal maps associated with elementary functions. Figure 2.9 provides several illustrative examples. It is worth noting, however, that even relatively simple Riemann mappings, such as a map from the unit disk to the interior of a square, may not have explicit expressions in terms of elementary functions. In these cases, approximate numerical methods are often required to construct the maps. For instance, the `ConformalMaps.jl` package in Julia (https://github.com/sswatson/ConformalMaps.jl) offers tools to approximate the Riemann map from a simply connected planar domain to a disk.

The physics of complex functions: Harmonic functions

Let's take a brief detour into the realm of partial differential equations (PDEs), which we will explore in more detail later in the course. Consider the two-dimensional Laplace equation

$$(\partial_x^2 + \partial_y^2) f(x, y) = 0. \tag{2.4}$$

This equation defines what are known as *harmonic functions*. We introduce it here, during our study of complex calculus, because – quite remarkably – any analytic function is also a solution to Eq. (2.4). This result follows directly from the Cauchy–Riemann theorem (2.2.2).

To see this, recall that an analytic function can be written as $f = u + iv$, where $u(x, y)$ and $v(x, y)$ are real-valued functions. By applying the Cauchy–Riemann conditions (2.3) and differentiating with respect to x and y, we obtain

$$\begin{cases} u_x = \ \ v_y \\ u_y = -v_x \end{cases} \Rightarrow \begin{cases} u_{xx} = \ \ v_{xy} \\ u_{yy} = -v_{xy} \end{cases} \Rightarrow \quad u_{xx} + u_{yy} = 0,$$

$$\begin{cases} v_x = -u_y \\ v_y = \ \ u_x \end{cases} \Rightarrow \begin{cases} v_{xx} = -u_{xy} \\ v_{yy} = \ \ u_{xy} \end{cases} \Rightarrow \quad v_{xx} + v_{yy} = 0.$$

Thus, both $u(x, y)$ and $v(x, y)$ satisfy the Laplace equation (2.4), meaning that the real and imaginary components of any analytic function are harmonic functions.

The term "harmonic" in harmonic functions originates from the ancient Greeks, who used the word to describe the pleasant-sounding, periodic motion of a point on a vibrating string. These vibrations could be expressed in terms of sines and cosines – functions we now refer to as harmonics. Fourier analysis, which we will encounter later, expands periodic functions on the unit circle as a series of these harmonic functions. Since they satisfy the Laplace equation, the term "harmonic" has come to describe any function that satisfies this equation.

In mathematical physics, harmonic functions play a fundamental role, particularly in areas such as electromagnetism and fluid mechanics. For example, in electrostatics, the Laplace equation governs the distribution of the electrostatic potential in a region of space that is free of charge. In such a region, harmonic functions describe the behavior of the potential, while singular points correspond to "point charges" or continuously distributed charge densities.

Consider a point charge located at the origin, $z = x + iy = 0$. This introduces a singularity at the origin, meaning the solution to the Laplace equation (2.4) is not analytic there. Later in the course, we will see that the analytic function corresponding to a point charge at the origin is $f(z) = C \log z$, where C is a constant related to the charge. The real part of this function, $\mathrm{Re}(f(z)) = u = C \log r = (C/2) \log(x^2 + y^2)$, represents the electrostatic potential.

The family of harmonic (complex) functions is rich and diverse. Any harmonic function that satisfies Eq. (2.4) can be transformed into another harmonic function by multiplying by a constant, rotating or adding a constant. Inversion of a harmonic function also produces another harmonic function, with singularities mapped to their corresponding "mirror" images through a spherical inversion. Additionally, the sum of two harmonic functions is itself a harmonic function, preserving the harmonic property.

2.2.2 *Integration along contours*

In complex analysis, integration is performed along oriented contours in the complex plane, providing a powerful tool for evaluating functions in a broad range of applications.

Definition 2.2.10 (Complex Integration). Let $f(z)$ be analytic in the neighborhood of a contour C. The integral of $f(z)$ along C is defined as

$$\int_C f(z)\, dz := \lim_{n\to\infty} \sum_{k=0}^{n-1} f(\zeta_k)(\zeta_{k+1} - \zeta_k), \tag{2.5}$$

where for each n, $(\zeta_k | k = 0, \ldots, n)$ represents an ordered sequence of points along the contour, partitioning it into n intervals such that $\zeta_0 = a$, $\zeta_n = b$ and $\max_k |\zeta_{k+1} - \zeta_k| \to 0$ as $n \to \infty$.

Remark. We can use the parameterization of complex functions, as discussed earlier. Let $z(t)$ for $a \leq t \leq b$ be a parameterization of the contour C. Then, the integral in Definition 2.2.10 becomes equivalent to a Riemann integral of $f(z(t))z'(t)$ with respect to t:

$$\int_C f(z)\,dz = \int_a^b f(z(t))\,z'(t)\,dt.$$

Example 2.2.11. In Example 2.1.13, we parameterized curves and evaluated the functions $f_1(z) = \sin z$, $f_2(z) = \exp(z+1)$ and $f_3(z) = z^2$ along those curves. Now, compute: (a) $\int_{C_1} f_1(z)\,dz$, (b) $\int_{C_2} f_2(z)\,dz$ and (c) $\int_{C_3} f_3(z)\,dz$, where C_1 is the vertical line segment from $\pi/2 - iM$ to $\pi/2 + iM$, C_2 is the ray extending from $z = -1$ to $\sqrt{3}i$, and C_3 is the circular arc of radius ε centered at the origin.

Solution.

(a) Let $z = \pi/2 + is$. Then, $dz = i\,ds$ for $-M < s < M$:

$$\begin{aligned}\int_{C_1} \sin(z)dz &= \int_{-M}^{+M} \frac{e^{i\pi/2-s} - e^{-i\pi/2+s}}{2i} i\,ds \\ &= \int_{-M}^{+M} i\cosh(s)ds = 2i\cosh(M).\end{aligned}$$

(b) Let $z = -1 + \rho e^{i\pi/3}$ for $0 \leq \rho \leq 2$. Then, $dz = e^{i\pi/3}d\rho$:

$$\begin{aligned}\int_{C_2} e^{z+1}dz &= \int_0^2 e^{\rho e^{i\pi/3}} e^{i\pi/3} d\rho = \left[e^{\rho e^{i\pi/3}}\right]_0^2 \\ &= e^{2\cos(\pi/3)+i2\sin(\pi/3)} - 1 \\ &= e^1\cos(\sqrt{3}) - 1 + ie^1\sin(\sqrt{3}).\end{aligned}$$

(c) Let $z = \varepsilon e^{i\tau}$ for $0 \leq \tau < 2\pi$. Then, $dz = i\varepsilon e^{i\tau}d\tau$:

$$\begin{aligned}\int_{C_3} z^2\,dz &= \int_0^{2\pi} \left(\varepsilon e^{i\tau}\right)^2 i\varepsilon e^{i\tau}\,d\tau = \left[\tfrac{1}{3}\varepsilon^3 e^{3i\tau}\right]_0^{2\pi} \\ &= \tfrac{1}{3}\varepsilon^3 e^{6\pi i} - \tfrac{1}{3}\varepsilon^3 e^0 = 0.\end{aligned}$$

Exercise 2.6. Let C_+ and C_- represent the upper and lower unit semi-circles, centered at the origin and oriented from $z = -1$ to $z = 1$. Find the integrals of the functions: (a) z, (b) z^2, (c) $1/z$ and (d) $\sqrt{z}$ along C_+ and C_-. For $\sqrt{z}$, use the branch where z is represented by $re^{i\theta}$ with $0 \le \theta < 2\pi$.

Example 2.2.12. Let C be a circular closed contour of radius R centered at the origin. Show that

$$\oint_C \frac{dz}{z^m} = 0, \quad \text{for } m = 2, 3, \ldots$$

by parameterizing the contour in polar coordinates.

Solution. A possible parameterization of the contour is $z(\theta) = Re^{i\theta}$ for $0 \le \theta < 2\pi$. Then, $dz = iRe^{i\theta}d\theta$. Changing the integral to polar coordinates gives

$$\oint_C \frac{dz}{z^m} = \int_0^{2\pi} \frac{iRe^{i\theta}}{(Re^{i\theta})^m} d\theta.$$

Since $m = 2, 3, 4, \ldots$, we know $m - 1 > 0$. Therefore,

$$\int_0^{2\pi} \frac{iRe^{i\theta}}{(Re^{i\theta})^m} d\theta = \frac{i}{R^{m-1}} \int_0^{2\pi} \frac{1}{(e^{i\theta})^{m-1}} d\theta = 0.$$

A continuation of this example would be to confirm that the integral is nonzero when $m = 1$.

Example 2.2.13. Use numerical integration to approximate the integrals in the examples above and verify your results.

2.2.3 *Cauchy's theorem*

In general, the integral of a complex function along a path in the complex plane depends on the entire path, not only on the end points. This leads to a natural and fundamental question: Under what conditions does the integral depend only on the endpoints? Cauchy's theorem provides a profound answer to this question.

Theorem 2.2.14 (Cauchy's Theorem, 1825). *If $f(z)$ is analytic in a simply connected region $\mathcal{D}$ of the complex plane, then for any two paths C in this region with the same endpoints, the integral $\int_C f(z)\,dz$ will have the same value.*

In other words, the integrals of analytic functions are *path independent*, meaning the integral depends only on the endpoints, not on the specific path taken.

It's crucial to note that, for Cauchy's theorem to apply in the case of multi-valued functions, the integrand must be a single-valued function. The branch cuts discussed in the previous section are introduced precisely to ensure that the integration path stays within a single branch of a multi-valued function, guaranteeing the analyticity (and thus differentiability) of the function along the path.

This theorem can also be expressed in another important form:

Theorem 2.2.15 (Cauchy's Theorem (Closed Contour Version)). *If $f(z)$ is analytic in a simply connected region $\mathcal{D}$ and C is any closed contour that lies entirely within $\mathcal{D}$, then the integral of $f(z)$ around C is zero:*

$$\oint_C f(z)\,dz = 0.$$

To understand the connection between the two formulations of Cauchy's theorem, consider two paths between two points in the complex plane. From the definition of complex integrals (Eq. 2.5), we know that path integrals are oriented, and reversing the direction of a path reverses the sign of the integral. Thus, when we combine two such paths with opposite orientations, we create a closed contour, which leads to the closed contour version of Cauchy's theorem.

To sketch the proof of Cauchy's theorem for closed contours, consider subdividing the region enclosed by the contour C into small squares, each with its own contour C_k, oriented counterclockwise, as is the original contour C. Then,

$$\oint_C f(z)\,dz = \sum_k \oint_{C_k} f(z)\,dz,$$

where the integrals over adjacent inner sides cancel out, as they are traversed in opposite directions.

Next, for each small contour C_k, choose a point z_k and approximate $f(z)$ by its Taylor expansion around z_k:

$$f(z) = f(z_k) + f'(z_k)(z - z_k) + O\left((z - z_k)^2\right).$$

Since the length of each small contour C_k is proportional to Δ, the sum of these Taylor expansions gives

$$\oint_{C_k} f(z)\, dz = f(z_k) \oint_{C_k} dz + f'(z_k) \oint_{C_k} (z - z_k)\, dz + O(\Delta^3).$$

The first term vanishes because the integral of dz over a closed contour is zero. The second term also vanishes because $\oint_{C_k} (z - z_k)\, dz = 0$. Thus, the integral around each small contour C_k is of order $O(\Delta^3)$, and summing over all such small contours leads to the conclusion that

$$\oint_C f(z)\, dz = 0 \quad \text{as } \Delta \to 0. \quad \square$$

While the proof involves breaking the integration path into segments, in practical applications, we treat the integral in the limit, assuming it exists, without the need for this discretization. However, when questions arise about the details of the limiting process, returning to this discretized interpretation can clarify the limiting behavior.

One key consequence of Cauchy's theorem is that all familiar rules for real-valued integrals also apply to contour integrals in the complex plane. This is further facilitated by the following result:

Theorem 2.2.16 (Triangle Inequality).

(A) *(From Euclidean geometry) For any two complex numbers z_1 and z_2, we have*

$$|z_1 + z_2| \leq |z_1| + |z_2|,$$

with equality if and only if z_1 and z_2 lie on the same ray from the origin.

(B) *(Integral over interval) Suppose $g(t)$ is a complex-valued function of a real variable t, defined on $a \leq t \leq b$. Then,*

$$\left|\int_a^b g(t)\, dt\right| \leq \int_a^b |g(t)|\, dt,$$

with equality if and only if $g(t)$ takes values on the same ray from the origin for all t.

(C) *(Integral over curve/path) Let $f(z)$ be any complex function and γ be any curve. Then,*

$$\left|\int_\gamma f(z)\, dz\right| \leq \int_\gamma |f(z)||dz|,$$

where $dz = \gamma'(t)\, dt$ and $|dz| = |\gamma'(t)|\, dt$.

Proof. We take the Euclidean geometry version (A) of the theorem as given and briefly sketch the proofs for the integral formulations. For the interval version (B), we approximate the integral as a Riemann sum:

$$\left|\sum g(t_k)\Delta t\right| \leq \sum |g(t_k)|\Delta t \approx \int_a^b |g(t)|\, dt.$$

The contour version (C) follows immediately from the real-valued case by noting that

$$\int_\gamma f(z)\, dz = \int_a^b f(\gamma(t))\gamma'(t)\, dt.$$

Thus, by applying the result from part (B), we have

$$\left|\int_a^b f(\gamma(t))\gamma'(t)\, dt\right| \leq \int_a^b |f(\gamma(t))||\gamma'(t)|\, dt = \int_\gamma |f(z)||dz|.$$

□

2.2.4 *Cauchy's formula*

Recall from Definition 2.1.12 that a curve is called *simple* if it does not intersect itself, and it is referred to as a *contour* if it is piecewise smooth.

Theorem 2.2.17 (Cauchy's Formula, 1831). *Let $f(z)$ be analytic on and inside a simple closed contour C. Then, for any point z inside C, we have*

$$f(z) = \frac{1}{2\pi i} \int_C \frac{f(\zeta)\, d\zeta}{\zeta - z}.$$

To illustrate Cauchy's formula, consider one of the simplest and most important examples: the integral of $I = \oint \frac{dz}{z}$ around a closed contour in the complex plane. For the contour shown in Fig. 2.10(a), we can parameterize it explicitly in polar coordinates to compute the integral:

$$I = \oint \frac{dz}{z} = \int_0^{2\pi} \frac{rie^{i\theta} d\theta}{re^{i\theta}} = i \int_0^{2\pi} d\theta = 2\pi i. \tag{2.6}$$

Thus, the integral is not zero and equals $2\pi i$, a result we will return to in various contexts.

Next, recall that the indefinite integral $\int \frac{dz}{z} = \log z$. This is consistent with Eq. (2.6) and the fact that $\log(z)$ is a multi-valued function. To better understand this, consider the integral between two points in the complex plane, say $z = 1$ and $z = 2$. You can go directly from $z = 1$ to $z = 2$ in a straight line or by taking a detour around $z = 0$, perhaps circling it counterclockwise. The integral's value depends on how many times (and in which direction) you circle around $z = 0$, and each such detour adds an integer multiple of $2\pi i$ to the result.

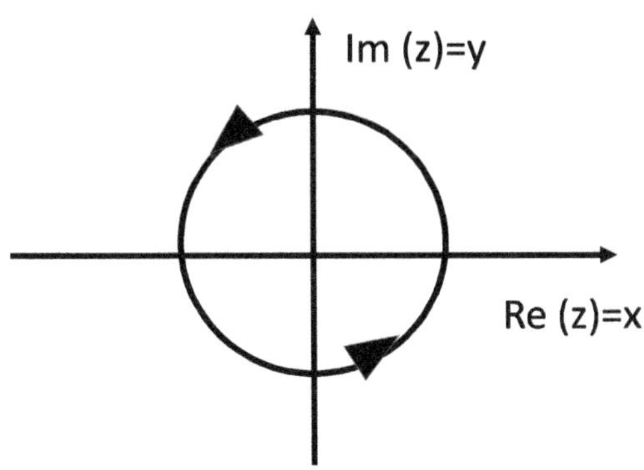

Figure 2.10. Illustration of contour integration and residues

Example 2.2.18. Compare the integrals $\oint \frac{dz}{z}$ along the two distinct paths shown in Fig. 2.11.

Solution. First, observe that $1/z$ is not analytic at $z = 0$, so the two contours cannot be continuously deformed into each other without passing through this singularity. Both curves are simple and closed, so Cauchy's formula applies. The left curve, C_1, encloses $z = 0$ and makes one full counterclockwise turn around it. By the calculation in Eq. (2.6), we know that

$$\oint_{C_1} \frac{dz}{z} = 2\pi i.$$

In contrast, the right curve, C_2, does not enclose $z = 0$, and since the function $1/z$ is analytic everywhere on and inside C_2, we conclude from Cauchy's theorem (Theorem 2.2.15) that

$$\oint_{C_2} \frac{dz}{z} = 0.$$

The "small square" discretization method, used earlier to prove the closed contour version of Cauchy's theorem (Theorem 2.2.15), is a powerful tool for handling integrals around singularities. However, it is important to note that not all integrals over closed contours are zero. Consider, for example,

$$\oint \frac{dz}{z^m}, \qquad m = 2, 3, \ldots.$$

The singularity at $z = 0$ makes this integral non-trivial. The indefinite integral is given by $\frac{z^{1-m}}{1-m} + C$, where C is a constant. This function is single-valued for $m \geq 2$, and thus, the integral over a closed

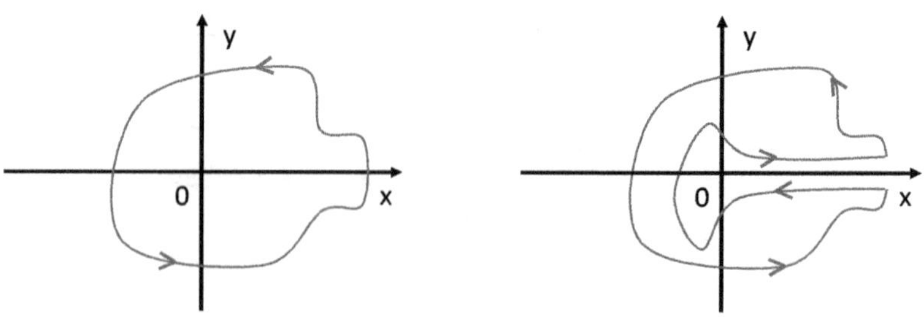

Figure 2.11. Two distinct paths for contour integration

contour is zero. For $m = 1$, however, the indefinite integral $\log(z)$ is multi-valued, leading to a nonzero result as seen in Eq. (2.6).

Cauchy's formula can be generalized to higher-order derivatives, yielding an important extension, as follows.

Theorem 2.2.19 (Cauchy's Formula for Derivatives, 1842). *Under the same conditions as in Theorem 2.2.17, the nth derivative of $f(z)$ is given by*

$$f^{(n)}(z) = \frac{n!}{2\pi i} \int_C \frac{f(\zeta)\, d\zeta}{(\zeta - z)^{n+1}}. \tag{2.7}$$

This result highlights the deep connection between complex integration and the behavior of analytic functions, showing that all higher derivatives of $f(z)$ can be computed directly from its values along a surrounding contour.

Theoretical implications of Cauchy's theorem and Cauchy's formulas

Cauchy's theorem and formulas have powerful and far-reaching consequences in complex analysis.

Theorem 2.2.20. *Suppose $f(z)$ is analytic in a region A. Then, $f(z)$ has derivatives of all orders in A.*

Proof. This follows directly from Cauchy's formula for derivatives (Theorem 2.2.19). Since we have an explicit formula for each derivative, it guarantees the existence of derivatives of all orders. □

Theorem 2.2.21 (Cauchy's Inequality). *Let C_R be the circle $|z - z_0| = R$. Assume that $f(z)$ is analytic on C_R and in its interior, i.e., on the disk $|z - z_0| \leq R$. Let $M_R = \max|f(z)|$ for z on C_R. Then, for all $n = 1, 2, \ldots$,*

$$|f^{(n)}(z_0)| \leq \frac{n! M_R}{R^n}.$$

Exercise 2.7. Prove Cauchy's inequality (Theorem 2.2.21) using Theorem 2.2.19. Additionally, provide an alternative argument for the theorem's validity using examples of $\exp(z)$ and $\cos(z)$ on a circle centered at the origin. (You are expected to argue informally, without referring to Theorem 2.2.19, why the inequality holds.)

Theorem 2.2.22 (Liouville's Theorem). *If $f(z)$ is entire (i.e., analytic at all finite points of the complex plane $\mathbb{C}$) and bounded, then $f(z)$ is constant.*

Proof. For any circle of radius R centered at z_0, Cauchy's inequality (Theorem 2.2.21) gives $|f'(z_0)| \leq M/R$. Since R can be arbitrarily large, we conclude that $|f'(z_0)| = 0$ for all $z_0 \in \mathbb{C}$. Hence, $f(z)$ is constant. □

Note that functions such as $P(z) = \sum_{k=0}^{n} a_k z^k$, $\exp(z)$ and $\cos(z)$ are entire but not bounded.

Theorem 2.2.23 (Fundamental Theorem of Algebra). *Any polynomial $P(z)$ of degree $n \geq 1$, i.e.,*

$$P(z) = \sum_{k=0}^{n} a_k z^k,$$

has exactly n roots (solutions of $P(z) = 0$).

Proof. The proof consists of two parts. First, we show that $P(z)$ has at least one root (see example below). Second, assuming $P(z)$ has exactly n roots, let z_0 be one of the roots. We can factor $P(z) = (z - z_0)Q(z)$, where $Q(z)$ is a polynomial of degree $n-1$. If $n-1 > 0$, we can apply the same reasoning to $Q(z)$. This process continues until the degree of the polynomial is zero. □

Example 2.2.24. Prove that $P(z) = \sum_{k=0}^{n} a_k z^k$ has at least one root.

Solution. We provide a hint rather than the full solution: Use proof by contradiction and apply Liouville's theorem (Theorem 2.2.22).

Theorem 2.2.25 (Maximum modulus principle (over disk)). *Suppose $f(z)$ is analytic on the closed disk C_r of radius r centered at z_0, i.e., the set $|z - z_0| \leq r$. If $|f|$ attains a relative maximum at z_0, then $f(z)$ is constant on C_r.*

In order to prove the theorem, we first establish the following result.

Theorem 2.2.26 (Mean value property). *Suppose $f(z)$ is analytic on the closed disk of radius r centered at z_0, i.e., the set $|z - z_0| \leq r$. Then,*

$$f(z_0) = \frac{1}{2\pi} \int_0^{2\pi} d\theta f\left(z_0 + r\exp(i\theta)\right).$$

Proof. Let C_r denote the boundary of the set $|z - z_0| \leq r$, parameterized as $z_0 + re^{i\theta}$ for $0 \leq \theta \leq 2\pi$. Then, by Cauchy's integral formula,

$$f(z_0) = \frac{1}{2\pi i} \int_{C_r} \frac{f(z)dz}{z - z_0}.$$

Substituting $z = z_0 + re^{i\theta}$ into the integral, we get

$$f(z_0) = \frac{1}{2\pi i} \int_0^{2\pi} \frac{f(z_0 + re^{i\theta}) ire^{i\theta} d\theta}{re^{i\theta}} = \frac{1}{2\pi} \int_0^{2\pi} f(z_0 + re^{i\theta}) d\theta.$$

□

Returning to Theorem 2.2.25, we now sketch the proof. We use both the mean value property (Theorem 2.2.26) and the triangle inequality (Theorem 2.2.16). Since z_0 is a relative maximum of $|f|$ on C_r, we know that $|f(z)| \leq |f(z_0)|$ for all $z \in C_r$. Applying the mean value property and the triangle inequality, we obtain

$$\begin{aligned}
|f(z_0)| &= \left| \frac{1}{2\pi} \int_0^{2\pi} f(z_0 + re^{i\theta}) d\theta \right| \quad \text{(mean value property)} \\
&\leq \frac{1}{2\pi} \int_0^{2\pi} |f(z_0 + re^{i\theta})| d\theta \quad \text{(triangle inequality)} \\
&\leq \frac{1}{2\pi} \int_0^{2\pi} |f(z_0)| d\theta \quad \left(\text{since } |f(z_0 + re^{i\theta})| \leq |f(z_0)|\right) \\
&= |f(z_0)|.
\end{aligned}$$

Since we begin and end with $|f(z_0)|$, all inequalities in the chain must be equalities. The first inequality is an equality if and only if

$f(z_0 + re^{i\theta})$ lies on the same ray from the origin for all θ, i.e., it has the same argument or equals zero. The second inequality becomes an equality only if $|f(z_0 + re^{i\theta})| = |f(z_0)|$ for all θ. Combining these observations, we conclude that $f(z_0 + re^{i\theta})$ has the same magnitude and argument for all θ, i.e., $f(z)$ is constant along the circle. Since $f(z_0)$ is the average of $f(z)$ over the circle, $f(z)$ must be constant on C_r. □

Two remarks are in order. First, based on previous experience (e.g., Theorem 2.2.22), it is reasonable to expect that Theorem 2.2.25 generalizes from a disk C_r to any simply connected domain. Second, one also expects that the maximum modulus can be attained at the boundary of a domain without implying that $f(z)$ is constant inside the domain. Indeed, consider $\exp(z)$ on the unit square $0 \leq x, y \leq 1$. The maximum $|\exp(x + iy)| = \exp(x)$ is achieved at $x = 1$ for any y, i.e., at the boundary of the domain. These remarks suggest the following extension of Theorem 2.2.25.

Theorem 2.2.27 (Maximum modulus principle (general)). *Suppose $f(z)$ is analytic on A, a bounded, connected, open set and continuous on $\bar{A} = A \cup \partial A$, where ∂A is the boundary of $\bar{A}$. Then, either $f(z)$ is constant, or the maximum of $|f(z)|$ on $\bar{A}$ occurs on ∂A.*

Proof. Let us sketch the proof. Cover A by disks whose centers form a path from the point where $|f(z)|$ is maximized to any other point in A, while remaining within A. If the maximum value of $|f(z)|$ is attained within A, then, by Theorem 2.2.25, $f(z)$ must be constant throughout A. If the maximum is achieved on ∂A, constancy is not required. □

Example 2.2.28. Find the maximum modulus of $\sin(z)$ on the square $0 \leq x, y \leq 2\pi$.

Solution. We write

$$\sin(z) = \sin(x + iy) = \sin(x)\cosh(y) + i\cos(x)\sinh(y).$$

We wish to find the maximum modulus of $\sin(z)$:

$$|\sin(x + iy)| = \sqrt{\sin^2(x)\cosh^2(y) + \cos^2(x)\sinh^2(y)}.$$

Using the identities $\cos^2(x) = 1-\sin^2(x)$ and $\cosh^2(x)-\sinh^2(x) = 1$, we simplify

$$|\sin(z)| = \sqrt{\sinh^2(y) + \sin^2(x)}.$$

This expression is maximized when $\sin^2(x) = 1$, i.e., when $x = \frac{n\pi}{2}$ for $n = 1, 2, 3, \ldots$, and $y = 2\pi$, the maximum value of y in our square. Thus,

$$\max(|\sin(z)|) = \sinh^2(2\pi) + 1 \approx 71688.33.$$

The point $z = \frac{n\pi}{2} + i2\pi$ lies on the boundary of the square, satisfying the maximum modulus principle.

2.2.5 *Laurent series*

The *Laurent series* of a complex function $f(z)$ about a point a is a representation of that function as a power series that includes terms of both positive and negative degrees.

Theorem 2.2.29. *A function $f(z)$ that is analytic on an annular region, $R_1 \leq |z-a| \leq R_2$, and within its interior, can be represented by a power series called the* Laurent series, *which converges within the annulus:*

$$f(z) = \sum_{k=-\infty}^{+\infty} c_k (z-a)^k.$$

The coefficients of the Laurent series are given by

$$c_k = \frac{1}{2\pi i} \oint_C \frac{f(z)}{(z-a)^{k+1}} dz,$$

where C is any contour contained within the annulus that encloses a.

Suppose we need to compute the contour integral

$$\oint_C f(z) dz,$$

where the contour C encircles $z = a$ in the positive (counterclockwise) direction, and there are no singular points of $f(z)$ inside C, except possibly at $z = a$. By expressing $f(z)$ in terms of its Laurent series, we observe that the only nonzero contribution comes from the term where $k = -1$:

$$\oint_C f(z)dz = \oint_C \sum_{k=-\infty}^{\infty} c_k(z-a)^k dz = c_{-1} \oint_C \frac{dz}{z-a} = 2\pi i c_{-1}.$$

Definition 2.2.30. The coefficient corresponding to the $k = -1$ term is of such importance in contour integration that it is given a special name – the *residue* of f at $z = a$ – and is denoted by $\text{Res}(f;a)$.

Note that if $f(z)$ has a simple pole at $z = a$, then

$$c_{-1} = \text{Res}(f;a) = \lim_{z\to a} (f(z)(z-a)).$$

2.3 Residue Calculus

2.3.1 *Singularities and residues*

Definition 2.3.1 (Singularity). Let $f : \mathbb{C} \to \mathbb{C}$, and consider $a \in \mathbb{C}$. If f is not analytic at a (i.e., $f'(a)$ does not exist), then a is called a *singular point* of f. If f is not analytic at a but is analytic in the region $0 < |z-a| < R$, then a is called an *isolated singular point* of f.

Definition 2.3.2 (Removable Singularity). Let a be a singular point of a function f, and let c_k be the coefficients of the Laurent expansion of f about a. If $c_k = 0$ for all $k < 0$, then a is called a *removable singularity* of f. (Note that f could become analytic if it were redefined at the single point a.)

Definition 2.3.3 (Simple Pole). Let a be a singular point of a function f, and let c_k be the coefficients of the Laurent expansion of f about a. If $c_{-1} \neq 0$ but $c_k = 0$ for all $k < -1$, then a is called a *first-order pole* or *simple pole* of f. See Fig. 2.12(a) for an example of a simple pole.

If f has a simple pole at $z = a$, we can represent f in the form

$$f(z) = \frac{g(z)}{z-a},$$

where $g(z)$ is analytic in a neighborhood of $z = a$ and $g(a) = c_{-1} \neq 0$.

Definition 2.3.4 (Higher-Order Pole). Let a be a singular point of a function f, and let c_k be the coefficients of the Laurent expansion of f about a. If, for some positive integer N, $c_{-N} \neq 0$ but $c_k = 0$ for all $k < -N$, then a is called an *Nth order pole* of f. See Fig. 2.12(b) for an example of a double pole.

If f has an Nth order pole at $z = a$, we can represent f in the form

$$f(z) = \frac{g(z)}{(z-a)^N},$$

where $g(z)$ is analytic in a neighborhood of $z = a$ and $g(a) = c_{-N} \neq 0$.

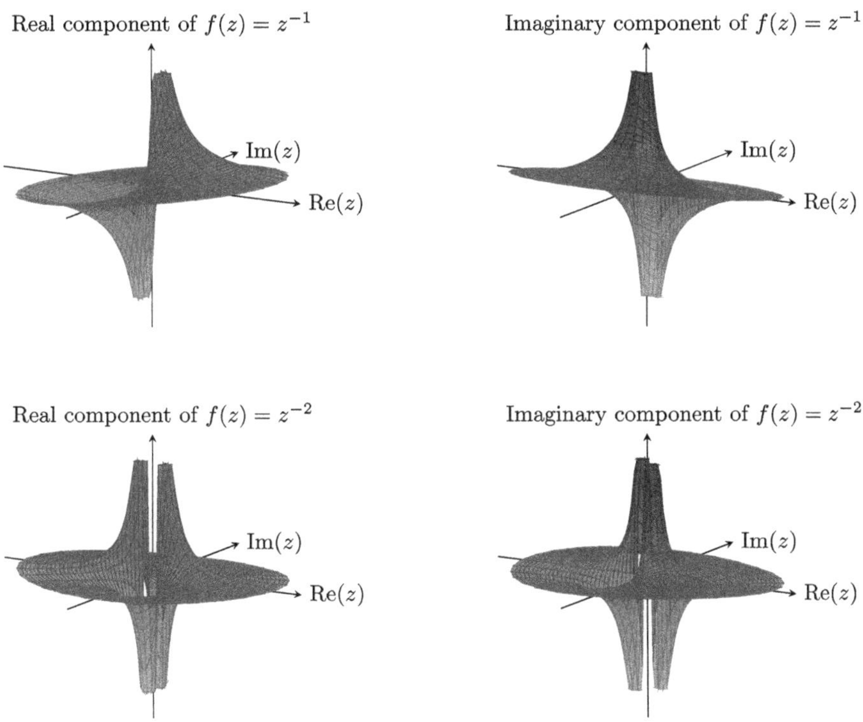

Figure 2.12. (a) Top row: The canonical example of a simple pole. The real (left) and imaginary (right) components of $z \mapsto z^{-1}$. (b) Bottom row: The canonical example of a double pole. The real (left) and imaginary (right) components of $z \mapsto z^{-2}$.

Example 2.3.5. Identify the removable singularity and determine the order of the poles in the function

$$f(z) = \frac{z-1}{(z^4-1)(z+1)}.$$

Solution. Observe that $z = 1$, $z = -1$, $z = i$ and $z = -i$ are singular points of f because f is not defined at these points.

- $z = 1$: For $z \neq 1$, $f(z) = g(z)$, where $g(z) = \frac{1}{(z+1)^2(z+i)(z-i)}$ is analytic in a neighborhood of $z = 1$. Therefore, $z = 1$ is a removable singularity of f.
- $z = i$: For $z \neq i$, $f(z) = \frac{g(z)}{z-i}$, where $g(z) = \frac{z-1}{(z-1)(z+1)^2(z+i)}$ is analytic in a neighborhood of $z = i$. Therefore, $z = i$ is a first-order pole (simple pole) of f.
- $z = -i$: This case is analogous to $z = i$.
- $z = -1$: For $z \neq -1$, $f(z) = \frac{g(z)}{(z+1)^2}$, where $g(z) = \frac{z-1}{(z-1)(z+i)(z-i)}$ is analytic in a neighborhood of $z = -1$. Therefore, $z = -1$ is a second-order pole (double pole) of f.

In summary, there is a removable singularity at $z = 1$, first-order poles at $z = \pm i$ and a second-order pole at $z = -1$.

Example 2.3.6. Use Cauchy's formula to compute

$$I = \oint \frac{\exp(z^2)\, dz}{z-1}$$

for the three contour examples shown in Fig. 2.13.

Solution.

(a) The integrand is analytic everywhere within the region enclosed by the contour, so by Cauchy's theorem, $I = 0$.

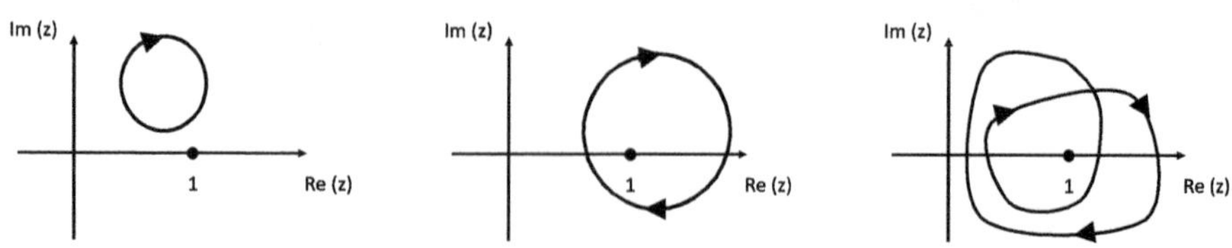

Figure 2.13. Three distinct contours for the integral in Example 2.3.6.

(b) The integrand has a single first-order pole at $z = 1$ within the region enclosed by the contour. Note that the contour is traversed in the negative (clockwise) direction. Therefore, by Cauchy's residue theorem,

$$I = -2\pi i \operatorname{Res}\left(\frac{\exp(z^2)}{z-1}, 1\right)$$
$$= -2\pi i \exp(1^2) = -2\pi i \exp(1) = -2\pi i e.$$

(c) The only singularity of the integrand is at $z = 1$, but the contour encloses the singularity twice. Thus, the integral is equivalent to traversing the contour in part (b) twice. Therefore, $I = 2(-2\pi i e) = -4\pi i e$.

Example 2.3.7. Use Cauchy's formula to compute

$$I = \oint \frac{dz}{\cosh z}$$

over the rectangular contour shown in Fig. 2.14, where a is a positive real number.

Solution. To find the singularities of the integrand, we solve $\cosh z_* = 0$, which gives $z_* = i\pi(n + 1/2)$ for $n = 0, \pm 1, \pm 2, \ldots$. These are all first-order poles. Only one pole, $z_* = i\pi/2$, lies inside

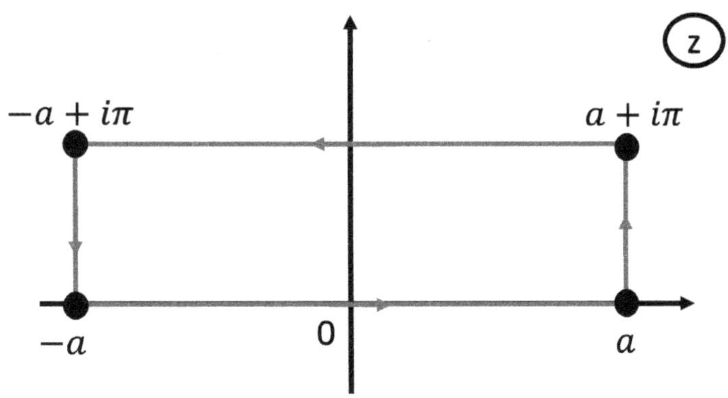

Figure 2.14. Contour for the integral in Example 2.3.7.

the contour. Therefore, by Cauchy's residue theorem and the residue formula for simple poles and using L'Hôpital's rule, we have

$$\begin{aligned} I &= 2\pi i \text{Res}\,(1/\cosh z, i\pi/2) = 2\pi i \lim_{z \to i\pi/2} \frac{z - i\pi/2}{\cosh z} \\ &= \frac{2\pi i}{\sinh(i\pi/2)} = \frac{2\pi}{\sin \pi/2} = 2\pi. \end{aligned}$$

Exercise 2.8. Compute the integral $\oint \frac{dz}{e^z - 1}$ over the circle of radius 4 centered at $3i$.

2.3.2 *Evaluation of real-valued integrals by contour integration*

Example 2.3.8. Evaluate the following real-valued integral using contour integration:

$$I = \int_{-\infty}^{\infty} \frac{e^{ikx}dx}{x^2 + 1}.$$

Solution. Let $f : \mathbb{C} \to \mathbb{C}$ be defined as $f(z) = \frac{e^{ikz}}{z^2+1}$. This function has simple poles at $z = \pm i$. Consider the contour $C = C_1 \cup C_R$ in the complex plane, as shown in Fig. 2.15. We now evaluate the integral

$$\int_C \frac{e^{ikz}dz}{z^2 + 1} = \int_{C_1} \frac{e^{ikz}dz}{z^2 + 1} + \int_{C_R} \frac{e^{ikz}dz}{z^2 + 1}.$$

Our objective is to apply Cauchy's theorem to show that the integral along C equals $2\pi i \,\text{Res}(f, i)$, where the residue at $z = i$ is $\text{Res}(f, i) = \lim_{z \to i} \frac{e^{ikz}(z-i)}{z^2+1} = \frac{e^{-k}}{2i}$. By showing that the integral along C_1 converges to I as $R \to \infty$ and that the integral along C_R tends to zero as $R \to \infty$, we can compute the original real-valued integral.

Parameterize C_1 by $z = x + 0i$ for $-R < x < R$. This gives $dz = dx$:

$$\int_{C_1} \frac{e^{ikz}dz}{z^2 + 1} = \int_{-R}^{R} \frac{e^{ikx}dx}{x^2 + 1}.$$

As $R \to \infty$, this integral converges to the desired real-valued integral I.

Next, parameterize C_R by $z = Re^{i\theta} = R\cos(\theta) + iR\sin(\theta)$ for $\theta \in [0, \pi]$. Thus, $dz = iRe^{i\theta}d\theta$:

$$\int_{C_R} \frac{e^{ikz}dz}{z^2+1} = \int_0^\pi \frac{e^{ikR(\cos(\theta)+i\sin(\theta))}iRe^{i\theta}d\theta}{(Re^{i\theta})^2+1}.$$

We now estimate the magnitude of this integral as $R \to \infty$:

$$\left| \int_0^\pi \frac{e^{ikR(\cos(\theta)+i\sin(\theta))}iRe^{i\theta}d\theta}{R^2e^{2i\theta}+1} \right|$$
$$\leq R\int_0^\pi \frac{e^{-kR\sin(\theta)}}{R^2+1}d\theta \quad (\text{since } |e^{ikR\cos(\theta)}| \leq 1).$$

For large R, $\left|\frac{1}{R^2e^{2i\theta}+1}\right| \leq \frac{1}{R^2-1}$. Thus, we continue:

$$\leq \frac{R}{R^2-1}\int_0^\pi e^{-kR\sin(\theta)}d\theta.$$

Since $\sin(\theta) \geq \frac{2\theta}{\pi}$ for all $\theta \in [0, \pi]$, we further bound the integral:

$$\leq \frac{2R}{R^2-1}\int_0^{\pi/2} e^{-kR\frac{2\theta}{\pi}}d\theta = \frac{2R}{R^2-1} \cdot \frac{\pi(1-e^{-kR})}{2kR} \to 0 \quad \text{as } R \to \infty.$$

Thus, we have $\int_{C_R} \frac{e^{ikz}dz}{z^2+1} \to 0$ as $R \to \infty$.

By Cauchy's residue theorem, we find

$$\int_C \frac{e^{ikz}dz}{z^2+1} = 2\pi i \operatorname{Res}(f, i) = 2\pi i \times \frac{e^{-k}}{2i} = \pi e^{-k}.$$

Thus, the desired integral is

$$I = \int_{-\infty}^\infty \frac{e^{ikx}dx}{x^2+1} = \pi e^{-k}.$$

Observe that Example 2.3.8 is an application of Jordan's lemma.

Lemma 2.3.9 (Jordan's Lemma). *Let C_R be a contour of an infinite semi-circle in the upper half-plane. Let $f(z) = e^{iaz}g(z)$, and assume that* $\lim_{R\to\infty} |g(Re^{i\theta})| = 0$. *Then,*

$$\left|\int_{C_R} f(z)dz\right| \leq \frac{\pi}{a} M_R,$$

where $M_R = \max_{\theta\in[0,\pi]} |g(Re^{i\theta})|$.

Example 2.3.10. Evaluate the following integral:

$$I_1 = \int_{-\infty}^{+\infty} \frac{\cos(\omega x)\, dx}{1+x^2}, \qquad \omega > 0.$$

Note: The indefinite form of this integral cannot be expressed using elementary functions, so we need an alternative method, specifically contour integration, to evaluate the definite integral.

Solution. We start by observing that

$$\int_{-\infty}^{+\infty} \frac{\sin(\omega x)\, dx}{1+x^2} = 0,$$

since the integrand is odd (skew-symmetric) over x. Adding the sine and cosine integrals together gives

$$I_1 = \int_{-\infty}^{+\infty} \frac{\cos(\omega x)\, dx}{1+x^2} + \int_{-\infty}^{+\infty} \frac{\sin(\omega x)\, dx}{1+x^2} = \int_{-\infty}^{+\infty} \frac{\exp(i\omega x)\, dx}{1+x^2}.$$

Now, consider the auxiliary contour integral

$$I_R = \oint \frac{\exp(i\omega z)\, dz}{1+z^2}, \qquad \omega > 0,$$

where the contour consists of a large semi-circle of radius R and a real line segment from $-R$ to R, as shown in Fig. 2.15. Since the

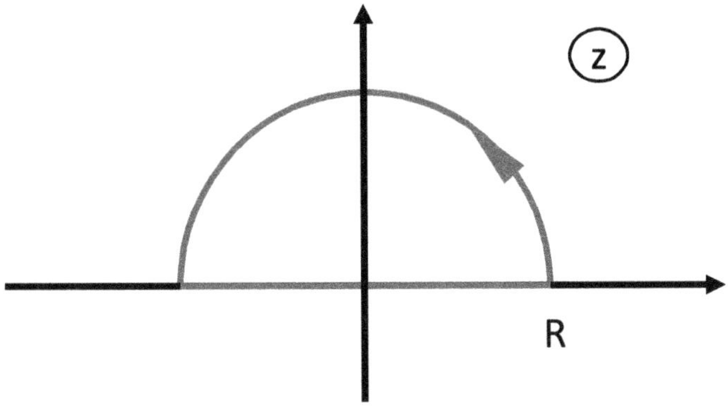

Figure 2.15. See Example 2.3.10.

integrand has two simple poles at $z = \pm i$, and only $z = i$ lies within the contour, we apply the residue theorem:

$$I_R = 2\pi i \, \mathrm{Res}\left[\frac{\exp(i\omega z)}{1+z^2}, i\right] = 2\pi i \cdot \frac{\exp(-\omega)}{2i} = \pi \exp(-\omega).$$

On the other hand, I_R can be split into two parts: one integral over $[-R, R]$ and the other over the semi-circular arc. As $R \to \infty$, the contribution from the semi-circle vanishes, leaving us with the desired result:

$$I_1 = \pi \exp(-\omega).$$

Example 2.3.11. Evaluate the integral

$$I = \int_{-\infty}^{+\infty} \frac{dx}{\cosh x}$$

by reducing it to a contour integral.

Solution. Consider the contour shown in Fig. 2.14 as $a \to \infty$. The integral along the real axis corresponds to the desired integral. The integrals over the left (up) and right (down) vertical portions of the contour tend to zero as $a \to \infty$ because the respective integrands decay exponentially. Note that

$$\cosh(x + i\pi) = \frac{\exp(x + i\pi) + \exp(-x - i\pi)}{2}$$
$$= -\frac{\exp(x) + \exp(-x)}{2} = -\cosh(x).$$

Thus, the fourth part of the contour becomes $\int_{i\pi+\infty}^{i\pi-\infty} \frac{dx}{\cosh(x+i\pi)} = I$. Summing all four parts and using the result of Example 2.3.7, we obtain

$$I + 0 + 0 + I = 2\pi, \quad \text{so} \quad I = \pi.$$

This integral can also be evaluated directly via the anti-derivative and definite integration:

$$I = 2\arctan\left(\tanh\left(\frac{x}{2}\right)\right)\Big|_{-\infty}^{+\infty} = \frac{\pi}{2} - \left(-\frac{\pi}{2}\right) = \pi.$$

Exercise 2.9. Evaluate the following integrals by reducing them to contour integrals:

(a) $\displaystyle\int_0^\infty \frac{dx}{1+x^4}$,

(b) $\displaystyle\int_0^\infty \frac{dx}{1+x^3}$,

(c) $\displaystyle\int_0^\infty \exp\left(ix^2\right) dx$,

(d) $\displaystyle\int_{-\infty}^\infty \frac{\exp(ikx)\, dx}{\cosh(x)}$,

Cauchy principal value

Consider the integral

$$\int_0^\infty \frac{\sin(ax)}{x}\, dx,$$

where $a > 0$. We evaluate this integral by constructing and evaluating a contour integral. Since $\frac{\sin(az)}{z}$ is analytic near $z = 0$ (recall L'Hôpital's rule or consult a reference), we can build the contour around the origin, as shown in Figure 2.16.

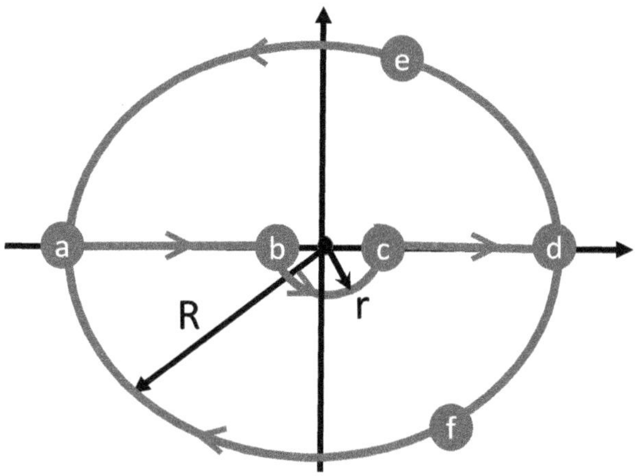

Figure 2.16. Contour for evaluating the Cauchy principal value.

Following the steps of the contour evaluation, we obtain

$$\begin{aligned}\int_0^\infty \frac{\sin(ax)}{x}\,dx &= \frac{1}{2}\int_{\text{contour}} \frac{\sin(az)}{z}\,dz\\ &= \frac{1}{4i}\int_{\text{contour}}\left(\frac{e^{iaz}}{z} - \frac{e^{-iaz}}{z}\right)dz\\ &= \frac{1}{4i}\int_{\text{upper semi-circle}} \frac{e^{iaz}}{z}dz - \frac{1}{4i}\int_{\text{lower semi-circle}} \frac{e^{-iaz}}{z}dz\\ &= \frac{1}{4i}\left(2\pi i - 0\right) = \frac{\pi}{2}.\end{aligned}$$

Here, several details in this chain of transformations are omitted. The reader is encouraged to fill in the missing steps. In particular, check that the integrals over the semi-circles vanish as $r \to 0$ and $R \to \infty$. For the latter, one can either compute the asymptotics or apply Jordan's lemma (see Lemma 2.3.9).

This limiting process defines what is known as the (Cauchy) *principal value* of the integral:

$$\text{PV}\int_{-\infty}^{\infty} \frac{e^{ix}}{x}\,dx = \lim_{R\to\infty}\int_{-R}^{R} \frac{e^{ix}}{x}\,dx = i\pi.$$

More generally, if the integrand $f(x)$ has a singularity at $x = c$, the principal value of the integral is defined as

$$\int_{-R}^{R} f(x)\,dx = \lim_{\varepsilon \to 0}\left(\int_{-R}^{c-\varepsilon} f(x)\,dx + \int_{c+\varepsilon}^{R} f(x)\,dx\right).$$

Let's consider a specific example:

$$\int_a^b \frac{dx}{x} = \log\frac{b}{a},$$

which diverges when $a < 0$ and $b > 0$. However, we can still define its principal value:

$$\text{PV}\int_a^b \frac{dx}{x} = \lim_{\varepsilon \to 0}\left(\int_a^{-\varepsilon} \frac{dx}{x} + \int_{\varepsilon}^{b} \frac{dx}{x}\right) = \log\frac{b}{|a|},$$

excluding the ε vicinity of zero. This example emphasizes the unambiguous nature of the principal value: The integration around $x = 0$ must maintain symmetry, ensuring that the limits $-\varepsilon$ and ε are taken with equal absolute values.

Finally, using complex variables, the path can be closed by a semicircle either above or below the real axis. Depending on the choice, the integral would receive a contribution of $\pm i\pi$. Thus, the full result is $\log(b/|a|) \pm i\pi$, and the *principal value* is defined as the mean of these two alternatives.

2.3.3 *Contour integration with multi-valued functions*

Contour integrals can be a powerful tool to evaluate certain definite integrals, especially when multi-valued functions and branch cuts are involved.

Integrals involving branch cuts

We now discuss a few examples of definite integrals that can be reduced to contour integrals, carefully avoiding branch cuts.

Example 2.3.12. Evaluate the integral

$$\int_0^{\infty} \frac{dx}{\sqrt{x}(x^2+1)}$$

by reducing it to a contour integral.

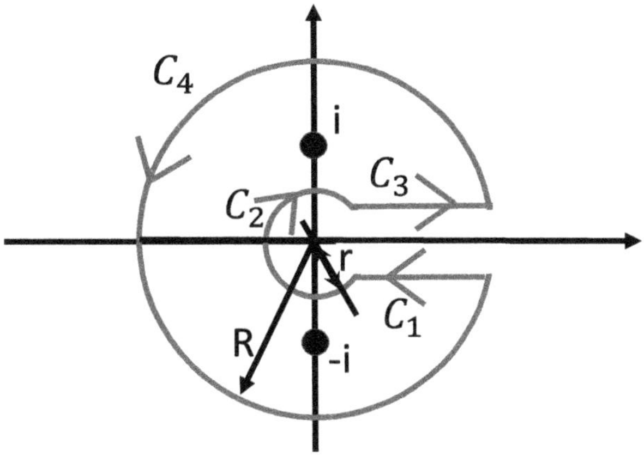

Figure 2.17. Keyhole contour for the evaluation in Example 2.3.12.

Solution. The square root function $\sqrt{z} = \exp\left(\frac{1}{2}\log z\right)$ is multi-valued, so the contour must avoid crossing branch cuts. We consider the contour integral:

$$\oint \frac{dz}{\sqrt{z}(z^2+1)},$$

where the contour is shown in Fig. 2.17. The contour is chosen such that

$$r \to 0: \quad \int_{C_2} \frac{dz}{\sqrt{z}(z^2+1)} \to 0,$$
$$R \to \infty: \quad \int_{C_4} \frac{dz}{\sqrt{z}(z^2+1)} \to 0,$$

where the total integral is split into four parts: $\int_{C_1} + \int_{C_2} + \int_{C_3} + \int_{C_4}$.

Letting $r \to 0$ and $R \to \infty$, the integral simplifies to

$$\oint \frac{dz}{\sqrt{z}(z^2+1)} = \int_{C_1} \frac{dz}{\sqrt{z}(z^2+1)} + \int_{C_3} \frac{dz}{\sqrt{z}(z^2+1)}$$
$$= 2\int_0^\infty \frac{dx}{\sqrt{x}(x^2+1)}.$$

The closed contour encloses two poles of the integrand at $z = \pm i$. Thus, by the residue theorem, we have

$$\oint \frac{dz}{\sqrt{z}(z^2+1)} = 2\pi i\,(\text{Res}\,(z=i) + \text{Res}\,(z=-i)).$$

We now calculate the residues:

$$\text{Res}\,(z=i) = \lim_{z\to i}\left(\frac{1}{\sqrt{z}(z+i)}\right) = \frac{\exp(3\pi i/4)}{2},$$

$$\text{Res}\,(z=-i) = \lim_{z\to -i}\left(\frac{1}{\sqrt{z}(z-i)}\right) = \frac{\exp(-3\pi i/4)}{2}.$$

Summing the residues gives

$$\oint \frac{dz}{\sqrt{z}(z^2+1)} = 2\pi i\left(\frac{\exp(3\pi i/4)}{2} - \frac{\exp(-3\pi i/4)}{2}\right)$$
$$= 2\pi i \cdot \frac{i}{\sqrt{2}} = \frac{2\pi}{\sqrt{2}}.$$

Thus, the original integral is

$$\int_0^\infty \frac{dx}{\sqrt{x}(x^2+1)} = \frac{\pi}{\sqrt{2}}.$$

Exercise 2.10. Evaluate the following integral:

$$\int_1^\infty \frac{dx}{x\sqrt{x-1}}.$$

Example 2.3.13. Evaluate the following integral by reducing it to a contour integral:

$$I = \int_0^1 \frac{dx}{x^{2/3}(1-x)^{1/3}}.$$

Solution. We analyze the contour integral with a similar integrand:

$$\oint \frac{dz}{z^{2/3}(z-1)^{1/3}} = \oint \frac{dz}{f(z)},$$

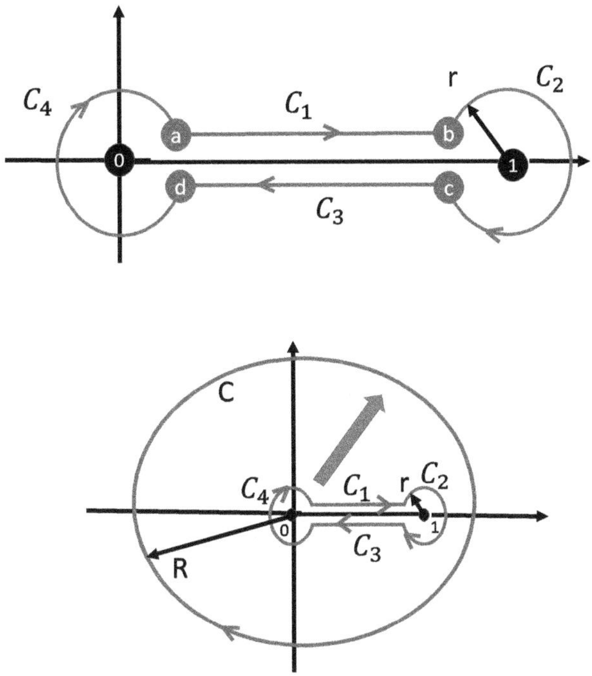

Figure 2.18. Contour for Example 2.3.13.

where the contour (shown in Fig. 2.18) encloses the cut connecting two branch points of $f(z)$, at $z = 0$ and $z = 1$ (both are branch points of order 3).

The branch cuts are introduced to make multi-valued functions, such as $f(z) = z^{2/3}(z-1)^{1/3}$, analytic within the complex plane, excluding the cut. The contour moves around the cut as follows, in the negative direction:

Sub-contour	Parametrization of z	Evaluation of $f(z)$
$C_1 := [a \to b]$	$x_1,\ x_1 \in [r, 1-r]$	$x_1^{2/3}\lvert 1-x_1\rvert^{1/3}\exp(i\pi/3)$
$C_2 := [b \to c]$	$1 + r\exp(i\theta_2),\ \theta_2 \in [\pi, -\pi]$	$r^{1/3}\exp(i\theta_2/3)$
$C_3 := [c \to d]$	$x_3,\ x_3 \in [1-r, r]$	$x_3^{2/3}\lvert 1-x_3\rvert^{1/3}\exp(-i\pi/3)$
$C_4 := [d \to a]$	$r\exp(i\theta_4),\ \theta_4 \in [2\pi, 0]$	$r^{2/3}\exp(i2\theta_4/3 + i\pi/3)$

Next, we compute the integrals over the sub-contours C_1, C_2, C_3, C_4:

$$\int_{C_1} \frac{dz}{f(z)} = \int_0^1 \frac{dx_1}{x_1^{2/3}(1-x_1)^{1/3}\exp(i\pi/3)} = \exp(-i\pi/3)I,$$

$$\int_{C_2} \frac{dz}{f(z)} = \int_\pi^{-\pi} \frac{ir\exp(i\theta_2)d\theta_2}{(1+r\exp(i\theta_2))^{2/3}(r\exp(i\theta_2))^{1/3}} \to_{r\to 0} 0,$$

$$\int_{C_3} \frac{dz}{f(z)} = \int_1^0 \frac{dx_3}{x_3^{2/3}|1-x_3|^{1/3}\exp(-i\pi/3)} = -\exp(i\pi/3)I,$$

$$\int_{C_4} \frac{dz}{f(z)} = \int_{2\pi}^0 \frac{ir\exp(i\theta_4)d\theta_4}{(r\exp(i\theta_4))^{2/3}(r\exp(i\theta_4)-1)^{1/3}} \to_{r\to 0} 0.$$

By Cauchy's integral theorem and the fact that $f(z)$ is analytic outside the cut $[0,1]$, we can relate the sum of these integrals to the integral over the large contour C (as shown in Fig. 2.18):

$$\int_{C_1} \frac{dz}{f(z)} + \int_{C_2} \frac{dz}{f(z)} + \int_{C_3} \frac{dz}{f(z)} + \int_{C_4} \frac{dz}{f(z)} = \int_C \frac{dz}{f(z)}.$$

The contour integral over C can be computed in the limit as $R \to \infty$:

$$\int_C \frac{dz}{f(z)} = \int_{2\pi}^0 \frac{iR\exp(i\theta)d\theta}{R^{2/3}\exp(2i\theta/3)(R\exp(i\theta)-1)^{1/3}} \to_{R\to\infty} -i\int_0^{2\pi} d\theta$$
$$= -2\pi i.$$

Finally, combining all the preceding formulas, we arrive at

$$I = \frac{-2\pi i}{-\exp(i\pi/3)+\exp(-i\pi/3)} = \frac{\pi}{\sin(\pi/3)} = \frac{2\pi}{\sqrt{3}}.$$

It may be useful to compare this derivation with the alternative method discussed in Ref. [4].

Exercise 2.11. Evaluate the following integral by reducing it to a contour integral:

$$\int_{-1}^1 \frac{dx}{(1+x^2)\sqrt{1-x^2}}. \tag{2.8}$$

2.4 Extreme-, Stationary-, and Saddle-Point Methods (*)

In this *auxiliary*[b] section, we study a family of methods used to approximate integrals dominated by contributions from special points and their vicinity. These methods are known as the extreme-point method (also called Laplace's method), the stationary-point method and the saddle-point method (also called the steepest-descent method). We begin by discussing the extreme-point method, which is used for estimating real-valued integrals over a real domain. We then move to the stationary-point method for oscillatory (complex-valued) integrals over a real interval and finally generalize to the saddle-point method for complex-valued integrals over a complex path.

The extreme-point method applies to integrals of the form

$$I_1 = \int_a^b dx \exp\left(f(x)\right), \tag{2.9}$$

where the real-valued, continuous function $f(x)$ achieves its maximum at a point $x_0 \in (a,b)$. To approximate the integral, we expand $f(x)$ around x_0 using the first few terms of its Taylor series:

$$f(x) = f(x_0) + \frac{(x-x_0)^2}{2} f''(x_0) + O\left((x-x_0)^3\right), \tag{2.10}$$

assuming $f'(x_0) = 0$. Since x_0 is a maximum, we have $f'(x_0) = 0$ and $f''(x_0) \le 0$, with $f''(x_0) < 0$ in the generic case. Substituting Eq. (2.10) into Eq. (2.9) and dropping the higher-order terms, we extend the integration domain to $(-\infty, \infty)$. Evaluating the resulting Gaussian integral, we obtain the extreme-point approximation:

$$I_1 \approx \sqrt{\frac{2\pi}{-f''(x_0)}} \exp\left(f(x_0)\right).$$

This approximation is valid when $|f''(x_0)| \gg 1$.

Example 2.4.1. Estimate the following integral using the extreme-point method for large positive α:

$$I = \int_{-\infty}^{+\infty} dx \exp(f(x)), \quad f(x) = \alpha x^2 - \frac{x^4}{2}.$$

---

[b]Auxiliary sections are marked with an asterisk (*). These sections can be skipped on a first reading and will not contribute to midterm or final exams.

Solution. First, we find the stationary points of $f(x)$. Solving $f'(x_s) = 0$, we obtain $x_s = 0$ or $x_s = \pm\sqrt{\alpha}$. Evaluating $f(x)$ at these points, we find

$$f(0) = 0, \quad f(\pm\sqrt{\alpha}) = \frac{\alpha^2}{2}.$$

Since $f(\pm\sqrt{\alpha})$ dominates, we focus on these points. Given the symmetry, we can pick one of the extreme points and multiply the result by two:

$$\begin{aligned} I &\approx 2\exp\left(\frac{\alpha^2}{2}\right)\int_{-\infty}^{+\infty} dx \, \exp\left(\frac{f''(\sqrt{\alpha})x^2}{2}\right) \\ &= 2\exp\left(\frac{\alpha^2}{2}\right)\int_{-\infty}^{+\infty} dx \, \exp(-2\alpha x^2) \\ &= 2\exp\left(\frac{\alpha^2}{2}\right)\sqrt{\frac{\pi}{2\alpha}}. \end{aligned}$$

Thus, the final result is

$$I = \exp\left(\frac{\alpha^2}{2}\right)\sqrt{\frac{2\pi}{\alpha}}.$$

We also used the fact that $f''(\pm\sqrt{\alpha}) = -4\alpha$.

The same principle, known as the *stationary-point method*, applies to highly oscillatory integrals of the form

$$I_2 = \int_a^b dx \, \exp(if(x)),$$

where $f(x)$ is real-valued, continuous and has a stationary point x_0 such that $f'(x_0) = 0$. The integrand oscillates least at the stationary point, making the vicinity of x_0 the dominant contributor to the integral.

While this may seem counterintuitive since the integrand oscillates significantly over x, the result is more apparent when we shift the integration contour into the complex plane. The contour is adjusted to cross the real axis at x_0 along a direction where the term $if''(x_0)(x - x_0)^2$ reaches a maximum at x_0. This causes the integrand to decay rapidly along the contour, away from x_0. The resulting approximation is

$$I_2 \approx \exp\left(if(x_0)\right) \int dx \exp\left(i\frac{f''(x_0)}{2}(x - x_0)^2\right)$$
$$= \sqrt{\frac{2\pi}{|f''(x_0)|}} \exp\left(if(x_0) + i\,\mathrm{sign}(f''(x_0))\frac{\pi}{4}\right), \qquad (2.11)$$

where the dependence on the interval's endpoints disappears in the limit of sufficiently large $|f''(x_0)|$.

Example 2.4.2. Estimate the integral

$$I_2 = \int_{-\infty}^{+\infty} dx \exp(if(x)), \quad f(x) = \alpha x^2 - \frac{x^4}{2}$$

at large positive α using the stationary-point method.

Solution. We can reuse the results from Example 2.4.1. The stationary points of the integrand are $x_s = 0$ and $x_s = \pm\sqrt{\alpha}$. The values of $f(x)$ at these points are $f(0) = 0$ and $f(\pm\sqrt{\alpha}) = \alpha^2/2$, leading to contributions of 1 and $\exp(i\alpha^2/2)$ to the integral.

For large α, we retain all three contributions. The second derivatives are $f''(0) = 2\alpha$ and $f''(\pm\sqrt{\alpha}) = -4\alpha$. Applying Eq. (2.11) to estimate the contributions, we find

$$I_2 \approx 2\exp\left(i\alpha^2/2\right) \int dx \exp(-2i\alpha x^2) + \int dx \exp(i\alpha x^2)$$
$$= 2\exp\left(i\alpha^2/2 - i\pi/4\right) \sqrt{\frac{\pi}{2\alpha}} + \sqrt{\frac{2\pi}{\alpha}} \exp(i\pi/4).$$

In the most general case, the *saddle-point method* (also known as the *steepest-descent method*) is used for contour integrals of the form

$$I_3 = \int_C dz \exp\left(f(z)\right),$$

where $f(z)$ is analytic along the contour C and within a domain $\mathcal{D}$ of the complex plane that contains C. Assume that there exists a saddle point $z_0 \in \mathcal{D}$, where $f'(z_0) = 0$. A saddle point is characterized by the fact that the iso-lines of $f(z)$ around z_0 show a saddle – having a minimum and maximum along two orthogonal directions.

By deforming C such that it passes through z_0 along the "maximal" path, we arrive at the saddle-point approximation:

$$I_3 \approx \sqrt{\frac{2\pi}{-f''(z_0)}} \exp\left(f(z_0)\right),$$

where the square root is taken along its principal branch.

Two remarks are in order:

(1) Both $f(z_0)$ and $f''(z_0)$ may be complex.
(2) There can be multiple saddle points within the region where $f(z)$ is analytic. In this case, the saddle point with the largest $f(z_0)$ contributes the most. If multiple saddle points have the same value of $f(z_0)$, the contour must pass through all of them, and the integral becomes the sum of all saddle-point contributions.

Exercise 2.12. Estimate the following integrals using the saddle-point approximation for large positive α:

(a)

$$\int_{-\infty}^{+\infty} dx \cos\left(\alpha x^2 - \frac{x^3}{3}\right),$$

(b)

$$\int_{-\infty}^{+\infty} dx \exp\left(-\frac{x^4}{4}\right) \cos(\alpha x).$$

In summary, the key takeaway from the extreme-, stationary-, and saddle-point analysis is the ability to identify regions that dominate the integral using the analyticity of the integrand. By shifting the contour of integration, we can pass through points where the absolute value of the integrand is maximized and ascend/descend along the steepest directions to capture the integral's asymptotic behavior.

Chapter 3

Fourier Analysis

Fourier analysis explores how functions can be represented or approximated by their oscillatory components. This process, known as decomposing a function into its oscillatory components (or basis functions), involves computing the correct coefficients for each component via an integral. Conversely, reconstructing the function from its orthogonal basis functions requires summing or integrating these components. When the oscillatory components span a continuous range of wave numbers (or frequencies), the decomposition and reconstruction are referred to as the Fourier transform and inverse Fourier transform, respectively. If the components are restricted to a discrete set of wave numbers (or frequencies), the process is called a Fourier series.

Fourier analysis originated from the study of Fourier series, which is attributed to Joseph Fourier, who demonstrated that heat transfer problems could be greatly simplified by expressing a function as a sum of trigonometric basis functions. Over time, the original concept of Fourier analysis has evolved and broadened, now encompassing more general and abstract applications. The field is often referred to as harmonic analysis.

As with other chapters of this book, our coverage of Fourier analysis is tuned to its further use in the following chapters, especially those devoted to differential equations. For further introductory reading on Fourier/harmonic analysis, we suggest *The Fourier Transform and Its Applications* by Bracewell [5], which offers a thorough treatment of the Fourier transform, with a special focus on practical applications, and *Introduction to Fourier Analysis and Generalised Functions* by Lighthill [6], which provides deeper theoretical coverage, particularly regarding the role of generalized functions, including the delta function, in Fourier analysis.

3.1 The Fourier Transform and Inverse Fourier Transform

Certain functions $f(\boldsymbol{x})$ can be expressed using a representation known as the Fourier integral:

$$f(\boldsymbol{x}) = \frac{1}{(2\pi)^d} \int_{\mathbb{R}^d} d\boldsymbol{k} \exp\left(i\boldsymbol{k}^T\boldsymbol{x}\right) \hat{f}(\boldsymbol{k}), \tag{3.1}$$

where $\boldsymbol{k} = (k_1, \ldots, k_d)$ is the "wave vector," $d\boldsymbol{k} = dk_1 \cdots dk_d$, and $\hat{f}(\boldsymbol{k})$ is the Fourier transform of $f(\boldsymbol{x})$, defined as follows:

$$\hat{f}(\boldsymbol{k}) := \int_{\mathbb{R}^d} d\boldsymbol{x} \exp\left(-i\boldsymbol{k}^T\boldsymbol{x}\right) f(\boldsymbol{x}). \tag{3.2}$$

Equations (3.1) and (3.2) are inverses of one another (meaning, for example, that substituting Eq. (3.2) into Eq. (3.1) will recover $f(\boldsymbol{x})$). For this reason, the Fourier integral is also called the *inverse Fourier transform.*

Proofs that these are indeed inverses, along with other important properties of the Fourier transform, rely on Dirac's delta function (δ-function), which, in d dimensions, is defined as

$$\delta(\boldsymbol{x}) := \frac{1}{(2\pi)^d} \int_{\mathbb{R}^d} d\boldsymbol{k} \exp(i\boldsymbol{k}^T\boldsymbol{x}). \tag{3.3}$$

We will discuss Dirac's δ-function in Section 3.3, primarily for the case of $d = 1$.

At first glance, it might seem that the appropriate class of functions for which Eq. (3.1) is valid would require both $f(\boldsymbol{x})$ and $\hat{f}(\boldsymbol{k})$ to be integrable. However, we will demonstrate how the definition of the δ-function allows Eq. (3.1) to be extended to a broader class of functions in Section 3.4.

For the sake of compact notation and clarity, we will present important properties of the Fourier transform for the one-dimensional (1D) case (Section 3.2), although these properties also apply to the general d-dimensional Fourier transform. Only a few multi-variate functions have Fourier transforms that can be expressed in closed form; these are discussed in Section 3.4.

Remark. There are alternative definitions for the Fourier transform and its inverse. Some authors place the multiplicative constant $(2\pi)^{-d}$ within the definition of $\hat{f}(\boldsymbol{k})$, while others use a "symmetric" definition, where both $f(\boldsymbol{x})$ and $\hat{f}(\boldsymbol{k})$ are multiplied by $(2\pi)^{-d/2}$. Still others may place a 2π factor inside the complex exponential. Therefore, be aware that the specific results you encounter will depend on the definitions adopted by the author.

3.2 Properties of the 1D Fourier Transform

In the $d = 1$ case, the variable x may represent either a spatial coordinate or time. When x is a spatial coordinate, the spectral variable k is often referred to as the wave number, which is the 1D version of the wave vector. When x represents time, k is commonly referred to as *frequency* and denoted by ω. The spatial and temporal terminologies are interchangeable.

The $d = 1$ Fourier Transform shows the following properties (see Tables 3.1 and 3.2):

- **Linearity:** Let $h(x) = af(x) + bg(x)$, where $a, b \in \mathbb{C}$, then

$$\begin{aligned}\hat{h}(k) &= \int_{\mathbb{R}} dx\, h(x)e^{-ikx} = \int_{\mathbb{R}} dx\, (af(x) + bg(x))\, e^{-ikx} \\ &= a\int_{\mathbb{R}} dx\, f(x)e^{-ikx} + b\int_{\mathbb{R}} dx\, g(x)e^{-ikx} = a\hat{f}(k) + b\hat{g}(k).\end{aligned}$$

- **Spatial/temporal translation:** Let $h(x) = f(x - x_0)$, where $x_0 \in \mathbb{R}$, then

$$\begin{aligned}\hat{h}(k) &= \int_{\mathbb{R}} dx\, h(x)e^{-ikx} = \int_{\mathbb{R}} dx\, f(x - x_0)e^{-ikx} \\ &= \int_{\mathbb{R}} dx'\, f(x')e^{-ikx' - ikx_0} = e^{-ikx_0}\hat{f}(k).\end{aligned}$$

Table 3.1. Summary of 1D Fourier transform properties.

Definition		
$\hat{f}(k) = \int_{-\infty}^{+\infty} f(x)e^{-ikx}\,dx \quad \Leftrightarrow \quad f(x) = \frac{1}{2\pi}\int_{-\infty}^{+\infty} \hat{f}(k)e^{ikx}\,dk$		
Property	**Original function**	**Fourier transform**
Transformations		
Linearity	$a\,f(x) + b\,g(x)$	$a\,\hat{f}(k) + b\,\hat{g}(k)$
Space-shifting	$f(x - x_0)$	$e^{-ikx_0}\,\hat{f}(k)$
k-space-shifting	$e^{ik_0x}\,f(x)$	$\hat{f}(k - k_0)$
Space reversal	$f(-x)$	$\hat{f}(-k)$
Space scaling	$f(ax)$	$\lvert a\rvert^{-1}\hat{f}(k/a)$
Calculus		
Convolution	$g(x) * h(x)$	$\hat{g}(k)\,\hat{h}(k)$
Multiplication	$g(x)h(x)$	$\frac{1}{2\pi}\left(\hat{g}(k) * \hat{h}(k)\right)$
Differentiation	$\frac{d}{dx}f(x)$	$ik\hat{f}(k)$
Integration	$\int_0^x f(\xi)\,d\xi$	$\left(\frac{1}{ik}\right)\hat{f}(k)$
Real-valued vs. Complex-valued functions		
Conjugation	$\left(f(x)\right)^*$	$\left(\hat{f}(-k)\right)^*$
Real and even	$f(x)$ real and even	$\hat{f}(k)$ real and even
Real and odd	$f(x)$ real and odd	$\hat{f}(k)$ imaginary and odd
Parseval's theorem/unitarity		
$\int_{-\infty}^{+\infty} \lvert f(x)\rvert^2\,dx = \frac{1}{2\pi}\int_{-\infty}^{+\infty} \lvert\hat{f}(k)\rvert^2\,dk$		

- **Frequency modulation:** For any real number k_0, if $h(x) = \exp(ik_0x)f(x)$, then

$$\begin{aligned}\hat{h}(k) &= \int_{\mathbb{R}} dx\,h(x)e^{-ikx} = \int_{\mathbb{R}} dx\,f(x)e^{ik_0x}e^{-ikx}\\ &= \int_{\mathbb{R}} dx\,f(x)e^{-i(k-k_0)x} = \hat{f}(k-k_0).\end{aligned}$$

- **Spatial/temporal rescaling:** For a nonzero real number a, if $h(x) = f(ax)$, then

$$\begin{aligned}\hat{h}(k) &= \int_{\mathbb{R}} dx\,h(x)e^{-ikx} = \int_{\mathbb{R}} dx\,f(ax)e^{-ikx}\\ &= \lvert a\rvert^{-1}\int_{\mathbb{R}} dx'\,f(x')e^{-ikx'/a} = \lvert a\rvert^{-1}\hat{f}(k/a).\end{aligned}$$

Table 3.2. Summary of Fourier transform properties.

Definitions	
$\hat{f}(k) = \int_{-\infty}^{+\infty} f(x)e^{-ikx}\,dx \quad \Leftrightarrow$	$f(x) = \frac{1}{2\pi}\int_{-\infty}^{+\infty} \hat{f}(k)e^{ikx}\,dk$
Fourier transform pairs	
$f(x) = \delta(x)$	$\hat{f}(k) = 1$
$f(x) = (2a)^{-1}\, H(1 - \lvert x/a\rvert)$	$\hat{f}(k) = \mathrm{sinc}(ak)$
$f(x) = (2a)^{-1}\, \exp(-\lvert x/a\rvert)$	$\hat{f}(k) = (1 + (ak)^2)^{-1}$
$f(x) = (\sqrt{2\pi}a)^{-1}\, \exp(-(x/a)^2/2)$	$\hat{f}(k) = \exp(-(ak)^2/2)$
$f(x) = (\sqrt{2\pi}a)^{-1}\, \mathrm{sech}(\pi x/2a)$	$\hat{f}(k) = \mathrm{sech}(\pi ak/2)$
$f(x) = (\pi a)^{-1}\, (1 + (x/a)^2)^{-1}$	$\hat{f}(k) = \exp(-\lvert ak\rvert)$
$f(x) = (\pi a)^{-1}\, \mathrm{sinc}(x/a)$	$\hat{f}(k) = H(1 - \lvert ak\rvert)$
$f(x) = 1$	$\hat{f}(k) = 2\pi\,\delta(k)$

The case $a = -1$ leads to the time-reversal property: if $h(t) = f(-t)$, then $\hat{h}(\omega) = \hat{f}(-\omega)$.

- **Complex conjugation:** If $h(x)$ is the complex conjugate of $f(x)$, i.e., $h(x) = (f(x))^*$, then

$$\hat{h}(k) = \int_{\mathbb{R}} dx\, h(x)e^{-ikx} = \int_{\mathbb{R}} dx\, (f(x))^* e^{-ikx} \int_{\mathbb{R}} dx \left(f(x)e^{ikx}\right)^*$$
$$= \left(\hat{f}(-k)\right)^*.$$

Exercise 3.1. Verify the following consequences of complex conjugation:

(a) If f is real, then $\hat{f}(-k) = (\hat{f}(k))^*$ (which implies that $\hat{f}$ is a Hermitian function).
(b) If f is purely imaginary, then $\hat{f}(-k) = -(\hat{f}(k))^*$.
(c) If $h(x) = \mathrm{Re}(f(x))$, then $\hat{h}(k) = \frac{1}{2}\left(\hat{f}(k) + (\hat{f}(-k))^*\right)$.
(d) If $h(x) = \mathrm{Im}(f(x))$, then $\hat{h}(k) = \frac{1}{2i}\left(\hat{f}(k) - (\hat{f}(-k))^*\right)$.

Exercise 3.2. Show that the Fourier transform of a radially symmetric function in two variables, i.e., $f(x_1, x_2) = g(r)$, where $r^2 = x_1^2 + x_2^2$, is also radially symmetric, i.e., $\hat{f}(k_1, k_2) = \hat{f}(\rho)$, where $\rho^2 = k_1^2 + k_2^2$. (We recall that in polar coordinates (r, θ), a radially symmetric function does not depend on the angle θ.)

We continue listing the features of the $d = 1$ Fourier transform:

- **Differentiation:** If $h(x) = f'(x)$ and assuming that $|f(x)| \to 0$ as $x \to \pm\infty$,

$$\hat{h}(k) = \int_{\mathbb{R}} dx\, h(x)e^{-ikx} = \int_{\mathbb{R}} dx\, f'(x)e^{-ikx}$$
$$= \Big[f(x)e^{-ikx}\Big]_{-\infty}^{\infty} - \int_{\mathbb{R}} dx\,(-ik)f(x)e^{-ikx} = (ik)\hat{f}(k).$$

- **Integration:** Substituting $k = 0$ in the definition, we obtain $\hat{f}(0) = \int_{-\infty}^{\infty} f(x)\,dx$. That is, evaluating the Fourier transform at the origin, $k = 0$, yields the integral of f over its entire domain.

Proofs for the following two additional properties of the $d = 1$ Fourier transform rely on the properties of the δ-function (which are to be discussed next in Section 3.3) and require careful consideration of integrability, which is beyond the scope of this introduction. These properties are included here for completeness:

- **Unitarity (Parseval/Plancherel theorem):** For any function f such that $\int |f|\,dx < \infty$ and $\int |f|^2 < \infty$,

$$\int_{-\infty}^{\infty} dx\,|f(x)|^2 = \int_{-\infty}^{\infty} dx\, f(x)(f(x))^*$$
$$= \int_{-\infty}^{\infty} dx \int_{-\infty}^{\infty} \frac{dk_1}{2\pi} e^{ik_1x}\hat{f}(k_1) \int_{-\infty}^{\infty} \frac{dk_2}{2\pi} e^{-ik_2x}\Big(\hat{f}(k_2)\Big)^*$$
$$= \frac{1}{2\pi}\int_{-\infty}^{\infty} dk\,|\hat{f}(k)|^2.$$

- **Convolution:** If $h(x) = (f * g)(x)$, then

$$\hat{h}(k) = \int_{\mathbb{R}} dx\, h(x)e^{-ikx} = \int_{\mathbb{R}} dx \int_{\mathbb{R}} dy\, f(x-y)g(y)e^{-ikx}$$
$$= \hat{f}(k)\hat{g}(k). \tag{3.4}$$

Here, convolution of a function f with a kernel g is defined as follows.

Definition 3.2.1. The *integral convolution* of the function f with the function g is defined as

$$(f * g)(x) := \int_{\mathbb{R}} dy\; g(x-y)f(y).$$

Let us now elaborate a bit more on convolution. Consider whether there exists a convolution kernel g such that $(g * f) = f$ for arbitrary functions f. If such a g exists, what properties would it have? Heuristically, we might argue that such a function must be both localized and unbounded: localized because for the convolution $\int dy\, g(x-y)f(y)$ to "pick out" $f(x)$, $g(x-y)$ must be zero for all $x \neq y$ and unbounded because $g(x-y)$ must be sufficiently large at $x = y$ to ensure that the integral in Eq. (3.4) is nonzero.

Such a degree of unboundedness over a localized point is impossible in traditional function theory. However, Paul Dirac introduced an unbounded function, denoted $\delta(x)$, into the context of quantum mechanics. Later, Laurent Schwartz developed a rigorous theory for such "functions," known as the theory of distributions. This function is commonly referred to as the (Dirac) δ-function. For more details, see Chapter 4 in Ref. [4].

3.3 Dirac's Delta Function

3.3.1 *The δ-function as the limit of a δ-sequence*

We begin our study of Dirac's δ-function by considering the sequence of functions given by

$$\delta_\epsilon(x) = \begin{cases} \frac{1}{\epsilon}, & |x| \leq \epsilon/2, \\ 0, & |x| > \epsilon/2. \end{cases} \tag{3.5}$$

The point-wise limit of δ_ϵ is clearly zero for all $x \neq 0$; therefore, the integral of the limit of δ_ϵ must also be zero, i.e.,

$$\lim_{\epsilon \to 0} \delta_\epsilon(x) = 0 \quad \Rightarrow \quad \int_{-\infty}^{\infty} dx \lim_{\epsilon \to 0} f_\epsilon(x) = 0. \tag{3.6}$$

However, for any $\epsilon > 0$, the integral of δ_ϵ is clearly unity; therefore, the limit of the integral of δ_ϵ must also be unity:

$$\int_{-\infty}^{\infty} dx\, \delta_\epsilon(x) = 1 \quad \Rightarrow \quad \lim_{\epsilon \to 0} \int_{-\infty}^{\infty} dx\, \delta_\epsilon(x) = 1. \tag{3.7}$$

Although Eq. (3.6) suggests that $\delta_\epsilon(x)$ may not be very interesting as a *function*, the behavior demonstrated by Eq. (3.7) motivates the use of $\delta_\epsilon(x)$ as a *functional*.[a] For any sufficiently nice function $\phi(x)$, define the functionals $\delta_\epsilon[\phi]$ and $\delta[\phi]$ by

$$\delta[\phi] := \lim_{\epsilon\to 0} \delta_\epsilon[\phi] := \lim_{\epsilon\to 0} \int_{-\infty}^{\infty} dx\, \delta_\epsilon(x)\phi(x).$$

The behavior of $\delta[\phi]$ can be demonstrated by approximating the corresponding integrals $\delta_\epsilon[\phi]$ for each $\epsilon > 0$:

$$\delta_\epsilon[\phi] = \int_{-\infty}^{\infty} dx\, \delta_\epsilon(x)\phi(x) = \int_{-\epsilon/2}^{\epsilon/2} dx\, \frac{1}{\epsilon}\phi(x).$$

Letting m_ϵ and M_ϵ represent the minimum and maximum values of $\phi(x)$ on the interval $-\epsilon/2 < x < \epsilon/2$ gives the bounds

$$m_\epsilon \le \delta_\epsilon[\phi] \le M_\epsilon.$$

If ϕ is continuous at $x = 0$, the limit $\delta_\epsilon[\phi]$ as $\epsilon \to 0$ is given by

$$\delta[\phi] = \lim_{\epsilon\to 0} \delta_\epsilon[\phi] = \phi(0).$$

In summary, $\delta[\phi]$ evaluates its argument at the point $x = 0$.

Now, compare $\delta_\epsilon(x)$ with the sequence of functions given by

$$\tilde{\delta}_\epsilon(x) = \frac{1}{\pi}\frac{\epsilon}{x^2 + \epsilon^2}.$$

The point-wise limit of $\tilde{\delta}_\epsilon(x)$ is also zero for every $x \ne 0$; therefore, as before, the integral of the limit must be zero:

$$\lim_{\epsilon\to 0} \tilde{\delta}_\epsilon(x) = 0 \quad \Rightarrow \quad \int_{-\infty}^{\infty} dx\, \lim_{\epsilon\to 0} \tilde{\delta}_\epsilon(x) = 0.$$

A suitable trigonometric substitution shows that the integral of $\tilde{\delta}_\epsilon(x)$ is also unity for each $\epsilon > 0$, and as before, the limit of the integrals

[a] A function takes numbers as inputs and gives numbers as outputs, whereas a functional takes functions as inputs and gives numbers as outputs.

must be unity:

$$\int_{-\infty}^{\infty} dx\, \tilde{\delta}_\epsilon(x) = 1 \quad \Rightarrow \quad \lim_{\epsilon\to 0}\int_{-\infty}^{\infty} dx\, \tilde{\delta}_\epsilon(x) = 1.$$

As with $\delta_\epsilon(x)$, we can use $\tilde{\delta}_\epsilon(x)$ to define the functionals $\tilde{\delta}_\epsilon[\phi(x)]$ and $\tilde{\delta}[\phi]$ by

$$\tilde{\delta}[\phi] := \lim_{\epsilon\to 0} \tilde{\delta}_\epsilon[\phi] := \lim_{\epsilon\to 0}\int_{-\infty}^{\infty} \tilde{\delta}_\epsilon(x)\phi(x)dx.$$

This time, it takes a little more thought to find the appropriate bounds, but with some effort, it can be shown that

$$\tilde{\delta}[\phi] = \lim_{\epsilon\to 0} \tilde{\delta}_\epsilon[\phi] = \phi(0).$$

That is, $\tilde{\delta}[\phi]$ also evaluates its argument at the point $x = 0$.

The sequences $\delta_\epsilon(x)$ and $\tilde{\delta}_\epsilon(x)$ both have the same limiting behavior as functionals and are examples of what is known as a δ-sequence. Their limiting behavior leads us to the definition of a δ-function, defined as $\delta[\phi] = \int_{\mathbb{R}} dx\delta(x)\phi(x) = \phi(0)$.

Remark. The δ-function only makes sense in the context of an integral. Although it is common practice to write expressions like $\delta(x)f(x)$, such expressions should always be interpreted as $\int_{\mathbb{R}} dx\, \delta(x)f(x)$.

Alternative definitions of the δ-function

We have defined the δ-function in Eq. (3.3) as the limit of a particular δ-sequence, namely the "top-hat" function given in Eq. (3.5). One may wonder whether there are other δ-sequences that give the same limit. For example, consider

$$\delta(t) = \lim_{\epsilon\to 0} \frac{2t^2\epsilon}{\pi(t^2+\epsilon^2)^2}. \tag{3.8}$$

To validate the suitability of Eq. (3.8) as an alternative definition of the δ-function, one needs to check (1) that $\delta(t) \to 0$ as $\epsilon \to 0$ for all $t \neq 0$ and (2) that $\int dt\, \delta(t) = 1$. This integral is easy to evaluate using a complex contour, observing that the integrand has a second-order pole at $t = i\epsilon$. Expanding it into a Laurent series and applying the Cauchy formula, we confirm that the integral equals unity.

Exercise 3.3. Validate the following asymptotic representations for the δ-function:

(a) $\delta(t) = \lim_{\epsilon\to 0} \frac{1}{\sqrt{\pi\epsilon}} \exp\left(-\frac{t^2}{\epsilon}\right)$,

(b) $\delta(t) = \lim_{n\to\infty} \frac{1-\cos(nt)}{\pi n t^2}$.

In many applications, we deal with periodic functions. In this case, one needs to consider relations within an interval. Given the extreme locality of the δ-function (just explored), all the relations discussed above extend to this case.

Example 3.3.1. Validate the following asymptotic representation for the δ-function on the interval $(-\pi, \pi)$:

$$\delta(\theta) = \lim_{r\to 1^-} \frac{1-r^2}{2\pi(1-2r\cos(\theta)+r^2)},$$

where $r \to 1^-$ means $r = 1-\epsilon$, $\epsilon > 0$, $\epsilon \to 0$.

Solution. For $0 < r < 1$, define $\delta_r(\theta) = \frac{1-r^2}{2\pi(1-2r\cos(\theta)+r^2)}$. To show that $\delta_r(\theta)$ is a δ-sequence, we must prove:

(1) $\lim_{r\to 1^-} \delta_r(\theta) = 0$ for each $\theta \neq 0$,

(2) $\int_{-\pi}^{\pi} \delta_r(\theta)\, d\theta = 1$ for $r < 1$.

(1) To show the point-wise limit of $\delta_r(\theta)$ is zero for each $\theta \neq 0$, note that for any $\theta \neq 0$, $\lim_{r\to 1^-} 1 - 2r\cos(\theta) + r^2 > 0$ but $\lim_{r\to 1^-} 1 - r^2 = 0$.

(2) To show $\delta_r(\theta)$ integrates to unity for each r, we evaluate the integral using a complex contour integral over the unit circle $z(\theta) = e^{i\theta}$ for $-\pi < \theta < \pi$. The integral evaluates to unity using the residue theorem. □

3.3.2 *Properties of the δ-function*

Example 3.3.2. For $b, c \in \mathbb{R}$, show that $c\delta(x-b)f(x) = cf(b)\delta(x-b)$.

Solution. We need to show that (a) both expressions are zero for $x \neq b$ (which is trivial) and (b) their integrals are equal:

$$\int_{-\infty}^{\infty} dx\, c\delta(x-b)f(x) = c\int_{-\infty}^{\infty} dy\, \delta(y)f(y+b).$$

Example 3.3.3. For $a \in \mathbb{R}$, show that $\delta(ax)f(x) = \delta(x)f(0)/|a|$.

Solution. We need to show that (a) the two functions are zero at $x \neq 0$ (which is trivial) and (b) their integrals are equal:

$$\begin{aligned}\int_{-\infty}^{\infty} dx\, \delta(ax)f(x) &= \int_{-\infty}^{\infty} \frac{dy}{|a|}\, \delta(y)f(y/a)\\ &= \int_{-\infty}^{\infty} dx\delta(x)f(0)/|a| = f(0)/|a|.\end{aligned}$$

Example 3.3.4. Show that the Fourier transform of a δ-function is a constant.

Solution.

$$\hat{\delta}(k) = \int_{-\infty}^{\infty} dx\, \delta(x)e^{-ikx} = e^{-ik\cdot 0} = 1.$$

Example 3.3.5. Show that

(a) $\delta(x) = \int_{-\infty}^{\infty} \frac{dk}{2\pi} \exp(ikx)$.

(b) The Fourier transform of a constant is a δ-function.

Solution.

(a) The expression on the right-hand side is identified as the inverse Fourier transform of the function $\hat{f}(k) = 1$: $f(x) = \frac{1}{2\pi}\int_{-\infty}^{\infty} dk\, 1e^{ikx}$. Although the constant function is not integrable in the traditional sense, the theory of distributions provides meaning to this integral. Since the δ-function satisfies $\int dx\, \delta(x)\phi(x) = \phi(0)$ for any suitable function $\phi(x)$, we conclude that $f(x) = \delta(x)$.

(b)

$$f[\phi(x)] = \int_{-\infty}^{\infty} dx\, \phi(x) \int_{-\infty}^{\infty} \frac{dk}{2\pi}\, 1e^{-ikx} = \int_{-\infty}^{\infty} \frac{dk}{2\pi}\, \hat{\phi}(k) = \phi(0).$$

Thus, $f[\phi] = \phi(0)$ for every suitable test function ϕ, so we assert that $f(x) = \delta(x)$.

3.3.3 Using δ-functions to prove properties of Fourier transforms

We now return to proving (1) that the Fourier transform and the inverse Fourier transform are inverses of each other, (2) Plancherel's theorem and (3) the convolution property.

Proposition 3.3.6. *The Fourier transform of the convolution of the function f with the function g is the product $\hat{f}(k)\hat{g}(k)$:*

$$\begin{aligned}
\widehat{(f*g)}(k) &= \int_{-\infty}^{\infty} dx \int_{-\infty}^{\infty} dy\, g(x-y)f(y)e^{-ikx} \\
&= \int_{-\infty}^{\infty} dx \int_{-\infty}^{\infty} dy \int_{-\infty}^{\infty} \frac{dk_1}{2\pi} \int_{-\infty}^{\infty} \frac{dk_2}{2\pi}\, \hat{f}(k_1)\hat{g}(k_2) \\
&\quad \times e^{-ikx+ik_1(x-y)+ik_2y} \\
&= \int_{-\infty}^{\infty} dk_1 \int_{-\infty}^{\infty} dk_2\, \hat{f}(k_1)\hat{g}(k_2) \int_{-\infty}^{\infty} \frac{dx}{2\pi} e^{-ikx+ik_1x} \\
&\quad \times \int_{-\infty}^{\infty} \frac{dy}{2\pi}\, e^{-ik_1y+ik_2y} \\
&= \int_{-\infty}^{\infty} dk_1 \hat{f}(k_1)\delta(k-k_1) \int_{-\infty}^{\infty} dk_2\, \hat{g}(k_2)\delta(k_1-k_2) \\
&= \hat{f}(k)\hat{g}(k).
\end{aligned}$$

In moving from the first line to the second, we exchange the order of integration, assuming that all the integrals involved are well defined.

Proposition 3.3.7 (Unitarity (Parseval/Plancherel theorem)).

$$\begin{aligned}
\int_{-\infty}^{\infty} dx|f(x)|^2 &= \int_{-\infty}^{\infty} dx f(x)\,(f(x))^* = \int_{-\infty}^{\infty} dx \int_{-\infty}^{\infty} \frac{dk_1}{2\pi} \\
&\quad \times e^{ik_1x}\hat{f}(k_1) \int_{-\infty}^{\infty} \frac{dk_2}{2\pi} e^{-ik_2x} \left(\hat{f}(k_2)\right)^* \\
&= \frac{1}{2\pi} \int_{-\infty}^{\infty} dk_1 \int_{-\infty}^{\infty} dk_2\, \hat{f}(k_1) \left(\hat{f}(k_2)\right)^* \frac{1}{2\pi} \\
&\quad \times \int_{-\infty}^{\infty} dx \exp(ix(k_1-k_2))
\end{aligned}$$

$$= \frac{1}{2\pi} \int_{-\infty}^{\infty} dk_1 \int_{-\infty}^{\infty} dk_2\, \hat{f}(k_1) \left(\hat{f}(k_2)\right)^* \delta(k_1 - k_2)$$
$$= \frac{1}{2\pi} \int_{-\infty}^{\infty} dk |\hat{f}(k)|^2.$$

Remark. Using the δ-function as the convolution kernel yields the self-convolution property:

$$f(x) = \int dy \delta(x - y) f(y). \tag{3.9}$$

Consider the δ-function of a function, $\delta(f(x))$. It can be transformed into the following sum over the zeros of $f(x)$:

$$\delta(f(x)) = \sum_n \frac{1}{|f'(y_n)|} \delta(x - y_n).$$

To prove this, recall that the δ-function is zero everywhere except where its argument vanishes. This implies that the result is a sum of δ-functions, with each term's weight determined by the zeros of $f(x)$. To establish the weights, integrate the resulting expression over a small vicinity of each zero, change variables and make use of the properties of the δ-function. This yields

$$\int dx \delta(f(x)) = \int \frac{df}{f'(x)} \delta(f(x)).$$

Since $\delta(f(x))$ is nonzero only at the zeros of $f(x)$, we can replace $f'(x)$ with its value at each zero and move it outside the integral. The remaining integral depends on the sign of the derivative. □

3.3.4 *The δ-function in higher dimensions*

The d-dimensional δ-function, which was instrumental in introducing the d-dimensional Fourier transform in Section 3.1, is simply a product of one-dimensional δ-functions, i.e., $\delta(\boldsymbol{x}) = \delta(x_1) \cdots \delta(x_n)$.

Example 3.3.8. Compute the δ-function in polar spherical coordinates.

Solution. The δ-function in Cartesian and polar coordinates is given by

$$\begin{aligned} f(\boldsymbol{r}) &= \int \delta(\boldsymbol{r}-\tilde{\boldsymbol{r}}) f(\tilde{\boldsymbol{r}}) d\tilde{\boldsymbol{r}} \\ &= \int \delta(x-\tilde{x})\delta(y-\tilde{y})\delta(z-\tilde{z}) f(\tilde{x},\tilde{y},\tilde{z}) d\tilde{x} d\tilde{y} d\tilde{z}, \\ &= \int \delta(\theta-\tilde{\theta})\delta(\phi-\tilde{\phi})\delta(r-\tilde{r}) f(\tilde{r},\tilde{\theta},\tilde{\phi}) d\tilde{r} d\tilde{\theta} d\tilde{\phi}. \end{aligned}$$

The volume element transformation from Cartesian to polar coordinates is

$$d\tilde{\boldsymbol{r}} = d\tilde{x} d\tilde{y} d\tilde{z} = \tilde{r}^2 \sin\tilde{\theta} d\tilde{r} d\tilde{\theta} d\tilde{\phi}.$$

Thus, we find that

$$\delta(\boldsymbol{r}-\tilde{\boldsymbol{r}}) = \frac{\delta(\theta-\tilde{\theta})\delta(\phi-\tilde{\phi})\delta(r-\tilde{r})}{\tilde{r}^2 \sin\tilde{\theta}}.$$

3.3.5 *Formal differentiation: The Heaviside function and the derivatives of the δ-function*

The δ-function is not a well-defined function but rather exists in the context of being integrated against a well-defined function. Using formal integration techniques, however, we can define a "derivative" of the δ-function. This technique is often referred to as *formal differentiation.*

Substituting $f(x) = 1$ into Eq. (3.9), we obtain $\int_{-\infty}^{\infty} dx\, \delta(x) = 1$. This motivates the introduction of the Heaviside, or step, function:

$$\theta(y) := \int_{-\infty}^{y} dx\, \delta(x) = \begin{cases} 0, & y < 0, \\ 1, & y > 0. \end{cases} \tag{3.10}$$

Exercise 3.4. Prove the following relation:

$$\left(\frac{d^2}{dt^2} - \gamma^2\right) \exp(-\gamma|t|) = -2\gamma\delta(t).$$

Hint: The step function will be useful in the proof.

Differentiating Eq. (3.10) gives $\theta'(x) = \delta(x)$. We can also differentiate the δ-function itself. Indeed, integrating Eq. (3.9) by parts and assuming the anti-derivative is bounded, we get

$$\int dy\, \delta'(y - x) f(y) = -f'(x).$$

Substituting $f(x) = xg(x)$ into the δ-derivative equation yields

$$x\delta'(x) = -\delta(x).$$

Expanding $f(x)$ in a Taylor series around $x = y$ and neglecting terms of second order or higher, we derive

$$f(x)\delta'(x - y) = f(y)\delta'(x - y) - f'(y)\delta(x - y).$$

Note that $\delta'(x)$ is skew-symmetric, and $f(x)\delta'(x - y)$ is not equal to $f(y)\delta'(x - y)$.

Thus far, we have assumed that $\delta'(x)$ is convolved with a continuous function. To extend this to piecewise continuous functions with jumps, one must be more careful with integration by parts at points of discontinuity. An example is the Heaviside function. If $f(x)$ has a jump at $x = y$, its derivative has the form

$$f'(x) = (f(y + 0) - f(y - 0))\delta(x - y) + g(x),$$

where $f(y+0) - f(y-0)$ represents the jump value and $g(x)$ is finite at $x = y$. A similar representation involving $\delta'(x)$ can be built for a function with a jump in its derivative.

Exercise 3.5. Express $t\delta''(t)$ in terms of $\delta'(t)$.

3.4 Closed-Form Representations for Selected Fourier Transforms

There are a few functions for which the Fourier transforms can be written in closed form.

3.4.1 *Elementary examples of closed-form representations*

Example 3.4.1. Show that the Fourier transform of a δ-function is a constant.

Solution. See Example 3.3.4, where we showed that $\hat{\delta}(k) = 1$.

Example 3.4.2. Show that the Fourier transform of a constant is a δ-function.

Solution. In Example 3.3.5, we showed that the inverse Fourier transform of unity is $\delta(x)$. Similarly, a calculation shows that $\hat{1}(k) = 2\pi\delta(k)$.

Example 3.4.3. Show that the Fourier transform of a square pulse function is a sinc function:

$$f(x) = \begin{cases} b, & |x| < a \\ 0, & |x| > a. \end{cases} \quad \Rightarrow \quad \hat{f}(k) = \frac{2b}{k}\sin(ka)$$

Solution.

$$\hat{f}(k) = \int_{\mathbb{R}} dx f(x) e^{-ikx} = b\int_{-a}^{a} dx e^{-ikx} = \frac{b}{-ik} e^{-ikx}\Big|_{-a}^{a}$$
$$= \frac{b}{-ik}(e^{-ika} - e^{ika}) = \frac{2b}{k}\sin(ka).$$

Example 3.4.4. Show that the Fourier transform of a sinc function is a square pulse:

$$g(x) = \frac{\sin(ax)}{ax} \quad \Rightarrow \quad \hat{g}(k) = \begin{cases} \frac{\pi}{a}, & |k| < a, \\ 0, & |k| > a. \end{cases}$$

There are multiple methods to solve this problem, and it is instructive to consider two different approaches.

Solution (Method 1: Using contour integration). The Fourier transform of $g(x)$ is defined as

$$\hat{g}(k) = \int_{-\infty}^{\infty} \frac{\sin(ax)}{ax} e^{-ikx}\, dx.$$

Using the identity $\sin(ax) = \frac{e^{iax} - e^{-iax}}{2i}$, we rewrite the integrand:

$$\hat{g}(k) = \frac{1}{2ia} \left[\int_{-\infty}^{\infty} \frac{e^{i(a-k)x}}{x}\, dx - \int_{-\infty}^{\infty} \frac{e^{i(a+k)x}}{x}\, dx \right].$$

These integrals are recognized as standard forms, solvable via contour integration. Denote them as I and II, respectively:

$$I = \int_{-\infty}^{\infty} \frac{e^{i(a-k)x}}{x}\, dx, \quad II = \int_{-\infty}^{\infty} \frac{e^{i(a+k)x}}{x}\, dx.$$

Both integrals have a pole at $x = 0$, and their values depend on whether $k < a$ or $k > a$.

By applying the residue theorem and closing the contours in the upper or lower half-plane as appropriate:

$$\hat{g}(k) = \begin{cases} \frac{\pi}{a}, & |k| < a, \\ 0, & |k| > a, \end{cases}$$

where the contributions from the contours vanish for $|k| > a$, yielding the square pulse.

Solution (Method 2: Differentiation under the integral sign). We start with the original integral:

$$I(a) = \int_{-\infty}^{\infty} \frac{\sin(ax)}{ax} e^{-ikx}\, dx.$$

Differentiating with respect to a eliminates the troublesome x-denominator:

$$\frac{d}{da}(aI(a)) = \int_{-\infty}^{\infty} \frac{\partial}{\partial a} \left(\frac{\sin(ax)}{x} \right) e^{-ikx}\, dx = \int_{-\infty}^{\infty} \cos(ax) e^{-ikx}\, dx.$$

The resulting integral is straightforward:

$$\frac{d}{da}(aI(a)) = \pi(\delta(a-k) + \delta(a+k)).$$

Taking the anti-derivative with respect to a, we recover the square pulse function:

$$I(a) = \frac{1}{a}\int_0^a \pi(\delta(\tilde{a}-k)+\delta(\tilde{a}+k))\,d\tilde{a} = \begin{cases} \frac{\pi}{a}, & |k| < a, \\ 0, & |k| > a. \end{cases}$$

Example 3.4.5. Find the Fourier transform of a Gaussian function:

$$f(x) = a\exp(-bx^2), \quad a, b > 0.$$

Solution. We complete the square in the exponent to obtain

$$\begin{aligned} \hat{f}(k) &= a\exp\left(-\frac{k^2}{4b}\right)\int_{-\infty}^{\infty} dx e^{-b\left(x+\frac{ik}{2b}\right)^2} = \frac{a}{\sqrt{b}}\exp\left(-\frac{k^2}{4b}\right)\sqrt{\pi} \\ &= a\sqrt{\frac{\pi}{b}}\exp\left(-\frac{k^2}{4b}\right). \end{aligned}$$

Example 3.4.6. Let $a > 0$. Show that:

$$\text{(a)}\ f(x) = \frac{1}{x^2+a^2} \quad \Rightarrow \quad \hat{f}(k) = \frac{\pi}{a}e^{-a|k|};$$
$$\text{(b)}\ g(x) := e^{-a|x|} \quad \Rightarrow \quad \hat{g}(k) := \frac{2a}{k^2+a^2}.$$

Solution.

(a) To compute the Fourier transform of $f(x)$, we use contour integration. First, consider the case when $k < 0$. We compute the integral using a semi-circular contour in the upper-half complex plane. The integrand has simple poles at $z = \pm ia$.

For the upper-half plane contour, we enclose the pole at $z = ia$, which has a residue $\text{Res}(z = ia) = \frac{e^{ka}}{2ia}$. Applying the residue theorem,

$$\hat{f}(k) = \int_{-\infty}^{\infty} \frac{e^{-ikx}}{x^2+a^2}\,dx = 2\pi i \cdot \frac{1}{2ia}e^{ka} = \frac{\pi}{a}e^{ka}.$$

Similarly, for $k > 0$, we use a semi-circular contour in the lower-half plane, enclosing the pole at $z = -ia$. The residue is $\text{Res}(z = -ia) = -\frac{e^{-ka}}{2ia}$, and we reverse the orientation of the contour, introducing a negative sign:

$$\hat{f}(k) = 2\pi i \cdot \frac{1}{2ia} e^{-ka} = \frac{\pi}{a} e^{-ka}.$$

Combining the two cases for $k < 0$ and $k > 0$, we obtain

$$\hat{f}(k) = \frac{\pi}{a} e^{-a|k|}.$$

(b) Now, consider the Fourier transform of $g(x) = e^{-a|x|}$. We split the integral into two parts, over $x > 0$ and $x < 0$:

$$\hat{g}(k) = \int_{-\infty}^{\infty} e^{-a|x|} e^{-ikx}\, dx = \int_{-\infty}^{0} e^{(a-ik)x}\, dx + \int_{0}^{\infty} e^{-(a+ik)x}\, dx.$$

Evaluating these two integrals,

$$\int_{-\infty}^{0} e^{(a-ik)x}\, dx = \frac{1}{a - ik}, \quad \int_{0}^{\infty} e^{-(a+ik)x}\, dx = \frac{1}{a + ik}.$$

Adding the results together gives

$$\hat{g}(k) = \frac{1}{a - ik} + \frac{1}{a + ik} = \frac{2a}{a^2 + k^2}.$$

Exercise 3.6. Find the Fourier transform of:

(a) $f(x) = \dfrac{1}{x^4 + a^4}$;
(b) $f(x) = \text{sech}(ax)$.

3.4.2 *More advanced examples of closed-form representations*

We can find closed-form representations of additional functions by combining the examples above with the properties discussed in Section 3.2.

Example 3.4.7. Let $f(x) = e^{-x^2}$ and define $g(x) = (f * f * f * f)(x)$. Find (a) $\hat{g}(k)$ and (b) $g(x)$.

Solution. From Example 3.4.5, we know that $\hat{f}(k) = \sqrt{\pi}e^{-k^2/4}$. Therefore, we can compute the Fourier transform of $g(x)$ as follows:

$$\begin{aligned}\hat{g}(k) &= (\widehat{f * f * f * f})(k) = \hat{f}(k)\cdot\hat{f}(k)\cdot\hat{f}(k)\cdot\hat{f}(k) = (\hat{f}(k))^4\\ &= \left(\sqrt{\pi}e^{-k^2/4}\right)^4 = \pi^2 e^{-k^2}.\end{aligned}$$

We recognize that $\hat{g}(k)$ is a Gaussian function (see Example 3.4.5), and we know that if $\hat{g}(k)$ is a Gaussian of the form $a\sqrt{\pi/b}e^{-k^2/(4b)}$, then $g(x)$ will be a Gaussian of the form ae^{-bx^2}. A little algebra shows we can rewrite $\hat{g}(k)$ in this form by setting $a = \frac{1}{2}\pi^{3/2}$ and $b = \frac{1}{4}$:

$$g(x) = \frac{1}{2}\pi^{3/2}e^{-x^2/4}. \qquad \square$$

Example 3.4.8. Let $g(x) = \max(0, 1 - |x|)$ (sometimes referred to as a "tent" function). Compute $\hat{g}(k)$.

Solution. Let $f(x)$ be defined as:

$$f(x) = \begin{cases} 1, & |x| < 1/2,\\ 0, & \text{otherwise.}\end{cases}$$

Note that $(f * f)(x) = g(x)$. From Example 3.4.3, we know that the Fourier transform of a square pulse is a sinc function. Therefore,

$$\hat{g}(k) = \hat{f}(k)\hat{f}(k) = \left(\frac{2\sin(k/2)}{k}\right)^2 = \operatorname{sinc}^2(k/2). \qquad \square$$

Example 3.4.9. Let $f(t)$ be given by:

$$f(t) = \begin{cases} \cos(\omega_0 t), & |t| < A,\\ 0, & \text{otherwise,}\end{cases}$$

where $\omega_0, A \in \mathbb{R}$ with $A > 0$.

(a) Compute $\hat{f}(k)$, the Fourier transform of f, as a function of ω_0 and A.

(b) Identify the relationship between the continuity of f, ω_0 and A, and discuss how this affects the decay of the Fourier coefficients as $|k| \to \infty$.

Solution.

(a) By the convolution theorem, we can express $f(t)$ as

$$f(t) = \cos(\omega_0 t) \cdot \text{rect}_A(t) \quad \Rightarrow \quad \hat{f}(k) = \widehat{\cos_{\omega_0}}(k) * \widehat{\text{rect}_A}(k),$$

where

$$\widehat{\cos_{\omega_0}}(k) = \pi \left[\delta(k - \omega_0) + \delta(k + \omega_0)\right], \quad \widehat{\text{rect}_A}(k) = \frac{2\sin(Ak)}{k}.$$

Therefore, the Fourier transform is

$$\hat{f}(k) = \frac{2\pi \sin(A(k - \omega_0))}{k - \omega_0} + \frac{2\pi \sin(A(k + \omega_0))}{k + \omega_0}.$$

(b) With some basic algebra, we can rewrite $\hat{f}(k)$ as

$$\hat{f}(k) = 2\pi \left(\frac{2k \sin(Ak)\cos(A\omega_0) - 2\omega_0 \cos(Ak)\sin(A\omega_0)}{k^2 - \omega_0^2} \right).$$

In general, the decay rate of $\hat{f}(k)$ as $|k| \to \infty$ is $\hat{f}(k) \sim 1/k$, except in cases where ω_0 and A satisfy resonance conditions such that $A\omega_0 = n\pi + \frac{\pi}{2}$, in which case $\hat{f}(k) \sim 1/k^2$. □

Exercise 3.7. Let $a \in \mathbb{C}$ with $\text{Re}(a) > 0$, and define

$$f_a(x) := \frac{2a}{a^2 + (2\pi x)^2}.$$

If $b \in \mathbb{C}$ with $\text{Re}(b) > 0$, show that $(f_a * f_b)(x) = f_{a+b}(x)$.

Exercise 3.8. Show that:

(a) $g(x) = \exp(iax) f(bx) \quad \Rightarrow \quad \hat{g}(k) = \frac{1}{|b|} \hat{f}\left(\frac{k-a}{b}\right)$,

(b) $f(x) = \frac{\sin^2(x)}{x} \quad \Rightarrow \quad \hat{f}(k) = -\frac{i\pi}{2}\left[\Pi(k-1) - \Pi(k+1)\right]$,

where

$$\Pi(k) = \begin{cases} 1, & |k| \le 1, \\ 0, & |k| > 1. \end{cases}$$

3.4.3 *Closed-form representations in higher dimensions*

Example 3.4.10. Let $\boldsymbol{x} = (x_1, x_2, \ldots, x_d) \in \mathbb{R}^d$, and use the notation $|\boldsymbol{x}|$ to represent $\sqrt{x_1^2 + x_2^2 + \cdots + x_d^2}$. Find the Fourier transform of $g(\boldsymbol{x}) = \exp(-|\boldsymbol{x}|^2)$.

Solution. The Fourier transform of $g(\boldsymbol{x})$ is given by

$$\begin{aligned}
\hat{g}(\boldsymbol{k}) &= \int_{\mathbb{R}^d} g(\boldsymbol{x}) e^{-i\boldsymbol{k}^T\boldsymbol{x}}\, d\boldsymbol{x} = \int_{\mathbb{R}^d} e^{-|\boldsymbol{x}|^2} e^{-i\boldsymbol{k}^T\boldsymbol{x}}\, d\boldsymbol{x} \\
&= \int_{\mathbb{R}^d} e^{-x_1^2 - x_2^2 - \cdots - x_d^2} e^{-i(k_1x_1 + k_2x_2 + \cdots + k_dx_d)}\, d\boldsymbol{x} \\
&= \int_{\mathbb{R}} \int_{\mathbb{R}} \cdots \int_{\mathbb{R}} \left(e^{-x_1^2} e^{-x_2^2} \ldots e^{-x_d^2}\right) \left(e^{-ik_1x_1} e^{-ik_2x_2} \ldots e^{-ik_dx_d}\right) \\
&\quad \times dx_1 dx_2 \ldots dx_d \\
&= \int_{\mathbb{R}} e^{-x_1^2} e^{-ik_1x_1}\, dx_1 \int_{\mathbb{R}} e^{-x_2^2} e^{-ik_2x_2}\, dx_2 \ldots \int_{\mathbb{R}} e^{-x_d^2} e^{-ik_dx_d}\, dx_d \\
&= \sqrt{\pi} \exp(-k_1^2/4) \cdot \sqrt{\pi} \exp(-k_2^2/4) \cdot \cdots \cdot \sqrt{\pi} \exp(-k_d^2/4) \\
&= \pi^{d/2} \exp(-|\boldsymbol{k}|^2/4).
\end{aligned}$$

3.5 Fourier Series: An Introduction

A Fourier series is a version of the Fourier integral used when a function is periodic or has finite support (i.e., nonzero only within a finite interval). As in the case of the Fourier integral or transform, we will mainly focus on the one-dimensional case. The generalization of the Fourier series approach to the multi-dimensional case is straightforward.

Consider a periodic function with period L. It can be represented as a series over a standard set of periodic exponentials (harmonics)

$\exp(i2\pi nx/L)$:

$$f(x) = \sum_{n=-\infty}^{\infty} f_n \exp(2\pi inx/L). \tag{3.11}$$

This representation, known as the Fourier series, shows that the Fourier series is a particular case of the Fourier integral:

$$\sum_{n=-\infty}^{\infty} f_n \exp(2\pi inx/L) = \int_{-\infty}^{\infty} dk\, \exp(2\pi ikx/L) \sum_{n=-\infty}^{\infty} f_n \delta(k-n).$$

Just as with the Fourier transform and inverse Fourier transform, we would like to invert Eq. (3.11) to express the Fourier coefficients f_n in terms of the function $f(x)$. By analogy with the Fourier transform, we consider integrating both sides of Eq. (3.11) over the interval $x \in [0, L]$ with the factor $\exp(-2\pi inx/L)/L$, where n is an integer. Applying this to the right-hand side of Eq. (3.11), we evaluate the following integral for each term in the resulting sum:

$$\int_0^L \frac{dx}{L} \exp(ikx2\pi/L) \exp(-inx2\pi/L) = \frac{e^{i(k-n)2\pi} - 1}{i(k-n)2\pi}$$

$$= \delta_{k,n} := \begin{cases} 1, & k = n, \\ 0, & k \neq n, \end{cases}$$

where the middle expression is resolved via L'Hôpital's rule, and $\delta_{k,n}$ is the Kronecker delta. Since only one term in the sum is nonzero, we obtain the desired formula for the Fourier coefficients:

$$f_n = \int_0^L \frac{dx}{L} f(x) \exp(-2\pi inx/L).$$

Note that the Fourier transform can be viewed as a limit of the Fourier series. When the characteristic scale of change of $f(x)$ is much smaller than that of L, many harmonics contribute significantly, and the Fourier series becomes the Fourier integral:

$$\sum_{n=-\infty}^{\infty} \cdots \quad \to \quad \frac{L}{2\pi} \int_{-\infty}^{\infty} dk \cdots .$$

Let us illustrate the expansion of a function into a Fourier series using the example $f(x) = \exp(\alpha x)$, defined on the interval

$0 < x < 2\pi$. The Fourier coefficients in this case are

$$f_n = \int_0^{2\pi} \frac{dx}{2\pi} \exp(-inx + \alpha x) = \frac{1}{2\pi} \frac{e^{2\pi\alpha} - 1}{\alpha - in}.$$

Note that as $n \to \infty$, $f_n \sim 1/n$. As discussed further in the following section, the slow decay of the Fourier coefficients is associated with the fact that, when considered as a periodic function over the reals with period 2π, $f(x)$ has discontinuities (jumps) at $0, \pm 2\pi, \pm 4\pi, \ldots$.

Exercise 3.9. Let $f(x) = x$ and $g(x) = |x|$, both defined on the interval $-\pi < x < \pi$.

(a) Expand both functions as Fourier series.
(b) Compare the dependence of the nth Fourier coefficient on n for the two functions.

To conclude this section, we remind the reader that our construction of the Fourier series assumes that the set of harmonic functions forms a complete set of basis functions. Proving this assumption is beyond the scope of this book.

3.6 Properties of the Fourier Series

The properties of Fourier series closely mirror those of Fourier transforms, discussed in Section 3.2. In this section, we summarize the properties of Fourier series in Table 3.3 for convenient reference. The presentation is formal and provided without proofs.

3.7 Riemann–Lebesgue Lemma

In general, the Fourier series of a function is an infinite series, meaning that it contains an infinite number of terms. This makes it computationally difficult to represent an arbitrary function exactly. However, a function may be approximated by truncating its Fourier series. The Riemann–Lebesgue lemma helps justify such truncation by stating that the Fourier coefficients f_n must decay as $n \to \infty$, for any integrable function f.

Theorem 3.7.1 (Riemann–Lebesgue lemma). *If $f(x) \in L^1$ (i.e., the Lebesgue integral of $|f|$ is finite), then* $\lim_{n\to\infty} f_n = 0$.

Table 3.3. Summary of Fourier series properties.

Definition

Let $f(x)$ be periodic with period L.

$$f(x) = \sum_{n=-\infty}^{+\infty} f_n \exp\left(2\pi i \frac{nx}{L}\right) \quad \Leftrightarrow \quad f_n = \int_0^L \frac{dx}{L} f(x) \exp\left(-2\pi i \frac{nx}{L}\right)$$

Property	Periodic function	Fourier series coefficients
Scaling		
Linearity	$af(x) + bg(x)$	$af_n + bg_n$
Space shifting	$f(x - x_0)$	$e^{2\pi i n x_0/L} f_n$
k-space shifting	$e^{2\pi i k/L} f(x)$	f_{n-k}
Space reversal	$f(-x)$	f_{-n}
Space scaling	$f(ax)$ (with period L/a)	f_n
Calculus		
Periodic convolution	$\int_0^L f(\xi)g(x-\xi)\, d\xi$	$Lf_n g_n$
Multiplication	$f(x)g(x)$	$\sum_{m=-\infty}^{\infty} f_m g_{n-m}$
Differentiation	$\frac{d}{dx} f(x)$	$\frac{2\pi i n}{L} f_n$
Integration	$\int_0^x d\xi f(\xi)$	$\left(\frac{L}{2\pi i n}\right) f_n$
Real-valued vs. Complex-valued functions		
Conjugation	$f^*(x)$	f^*_{-n}
Real and even	$f(x)$ real and even	f_n real; $f_n = f_{-n}$
Real and odd	$f(x)$ real and odd	f_n imaginary; $f_n = -f_{-n}$
Parseval's theorem/unitarity		

$$\frac{1}{L}\int_0^L dx |f(x)|^2 = \sum_{n=-\infty}^{\infty} |f_n|^2$$

We will not prove the Riemann–Lebesgue lemma here, but only provide useful hints. A standard proof involves the following steps:

1. Show that the lemma holds for the characteristic function of a finite open interval in $\mathbb{R}^1$, where $f(x)$ is constant within (a, b) and zero elsewhere.
2. Extend the proof to simple functions over $\mathbb{R}^1$, which are piecewise constant functions.
3. Build a sequence of simple functions, which are dense in L^1, that approximate $f(x)$ more accurately as the sequence progresses.

A useful corollary of the Riemann–Lebesgue lemma states: For any periodic function $f(x)$ with continuous derivatives up to order m, integration by parts can be applied m times to show that the nth Fourier coefficient is bounded for sufficiently large n, according to

$$|f_n| \leq \frac{C}{|n|^{m+2}},$$

where C is a constant.

In particular, and consistent with the earlier example, we observe that in the case of a "jump" (i.e., a continuous anti-derivative with $m = -1$), $|f_n| \sim 1/n$ asymptotically as $n \to \infty$. In the case of a "ramp" (i.e., $m = 0$ with a continuous function but a discontinuous derivative), $|f_n|$ becomes $O(1/n^2)$ as $n \to \infty$. For an analytic function with all derivatives continuous, $|f_n|$ decays faster than any polynomial as n increases.

3.8 Gibbs Phenomenon

When truncating a Fourier series, one must be cautious due to the so-called Gibbs phenomenon, first described by Willard Gibbs in 1889, though it was actually discovered earlier in 1848 by Henry Wilbraham. The phenomenon occurs when a truncated Fourier series is used to approximate a piecewise continuous periodic function, and it involves two key effects: an overshoot near a jump discontinuity, and the persistence of this overshoot as more terms are added to the series.

Consider the classic example of a square wave:

$$f(x) = \begin{cases} \pi/4, & 2n\pi \leq x \leq (2n+1)\pi, \quad n = 0, 1, 2, \ldots, \\ -\pi/4, & (2n+1)\pi \leq x \leq (2n+2)\pi, \quad n = 0, 1, 2, \ldots. \end{cases} \tag{3.12}$$

This 2π-periodic function can be written as a Fourier series:

$$f(x) = \sum_{n=0}^{\infty} \frac{\sin((2n+1)x)}{2n+1}.$$

Note that the function jumps by $\pi/2$ at $2n\pi$. Now, let us truncate the Fourier series in Eq. (3.12) by considering the Nth partial sum:

$$S_N(x) = \sum_{n=0}^{N} \frac{\sin((2n+1)x)}{2n+1}.$$

The Gibbs phenomenon refers to the fact that, as $N \to \infty$, the error near the jump discontinuities reduces in width and overall energy, but the maximum overshoot remains constant.

To illustrate this behavior, you can view an animated demonstration of the Gibbs phenomenon at https://en.wikipedia.org/wiki/Gibbs_phenomenon#/media/File:SquareWave.gif (from Wikipedia). This behavior can also be explored with a Julia snippet available in the course repository on D2L.

Let us now back up this simulation with an analytical estimate. We calculate the value of the partial Fourier series near a jump. By differentiating the truncated series, we obtain

$$\frac{d}{d\epsilon} S_N(\epsilon) = \sum_{n=0}^{N} \cos((2n+1)\epsilon) = \frac{\sin(2(N+1)\epsilon)}{2\sin(\epsilon)},$$

where we have used the formula for the sum of a geometric progression. Note that as $\epsilon \to 0$, the derivative tends to $N+1$, which implies that the slope of $S_N(\epsilon)$ becomes large as N increases. The first maximum of $S_N(x)$ near $\epsilon = 0$ occurs at $\epsilon_* = \pi/(2(N+1))$.

Now, we estimate the value of $S_N(\epsilon_*)$:

$$S_N(\epsilon_*) \approx \sum_{n=0}^{N} \frac{\sin\left(\frac{(2n+1)\pi}{2(N+1)}\right)}{2n+1} = \frac{1}{2}\int_0^{\pi} \frac{\sin t}{t} dt \approx \frac{\pi}{4} + 0.14,$$

which shows that the partial sum overshoots the value of the function by a fixed amount (≈ 0.14) near the jump at $x = 0$.

Exercise 3.10. Generalize the two functions from Exercise 3.9 to be 2π-periodic on the interval $[-5\pi, 5\pi)$. Compute the partial Fourier series $S_N(x)$ for selected N, and study how the amplitude and width of the oscillations near the jump points $x = m\pi$, $m \in \{-5, -4, \ldots, 4\}$ behave as $N \to \infty$.

The Fourier series has many important applications, including in solving differential equations. While most differential equations of interest can only be solved numerically, mathematicians often represent solutions as Fourier series, truncating the series to achieve a desired level of accuracy.

3.9 Laplace Transform

The Laplace transform is a powerful tool, particularly useful when dealing with functions that do not decay at infinity, such as $\exp(ix)$ or $\exp(-x)$. To evaluate Fourier transforms of such functions, we must consider the frequency k with a nonzero imaginary part to ensure the convergence of the resulting integrals.

While the Fourier series provides a way to handle periodic functions or functions of finite support, the Laplace transform offers an elegant method for functions that decay as $x \to +\infty$, such as $\exp(-x)$. Instead of working with functions defined over all real x (as in the Fourier transform), the Laplace transform operates on functions defined over the semi-infinite interval $x \in \mathbb{R}_{\geq 0} = [0, +\infty)$.

A Laplace transform can be thought of as a Fourier transform applied to functions that are nonzero only for $x \geq 0$:

$$\tilde{f}(k) = \int_0^\infty dx \, \exp(-kx) f(x).$$

Here, k is complex, and we require that the integral converges for sufficiently large $\mathrm{Re}(k)$. Thus, $\tilde{f}(k)$ is analytic for $\mathrm{Re}(k) > C$, where C is a positive constant.

The inverse Laplace transform is defined as a complex integral:

$$f(x) = \frac{1}{2\pi i} \int_C dk \, \exp(kx) \tilde{f}(k), \tag{3.13}$$

where the contour C, known as the Bromwich contour, is shown in Fig. 3.1. This contour can be deformed arbitrarily within the domain $\mathrm{Re}(k) > 0$, where $\tilde{f}(k)$ is analytic. The integral vanishes for $x < 0$ by construction, consistent with the fact that the Laplace transform is only defined for $x \geq 0$.

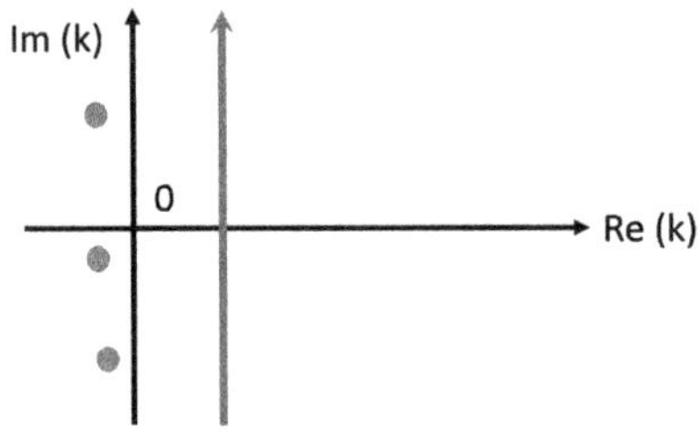

Figure 3.1. The Bromwich contour C in Eq. (3.13) is shown as a vertical straight line extending upward in the complex plane from $\varepsilon - i\infty$ to $\varepsilon + i\infty$, where $\varepsilon > 0$ is an infinitesimally small positive number. Possible singularities of the Laplace transform $\tilde{f}(k)$ are shown schematically as dots.

Table 3.4. Summary of Laplace transform properties.

Property	Function	Laplace transform
Scaling		
Linearity	$af(x) + bg(x)$	$a\tilde{f}(k) + b\tilde{g}(k)$
Space shifting	$f(x - x_0)\theta(x - x_0)$, $x_0 > 0$	$e^{-kx_0}\tilde{f}(k)$
k-space shifting	$h(x) = e^{k_0 x}f(x)$	$\tilde{h}(k) = \tilde{f}(k - k_0)$
Space scaling	$h(x) = f(ax)$, $a > 0$	$\tilde{h}(k) = \frac{1}{a}\tilde{f}(k/a)$
Differentiation	$h(x) = f'(x)$	$\tilde{h}(k) = k\tilde{f}(k) - f(0^-)$
Integration	$h(x) = \int_0^x f(y)\,dy$	$\tilde{h}(k) = \frac{1}{k}\tilde{f}(k)$
Convolution	$h(x) = (f * g)(x) = \int_0^x f(y)g(x-y)\,dy$	$\tilde{h}(k) = \tilde{f}(k)\tilde{g}(k)$

3.9.1 *Properties of the Laplace transform*

The Laplace transform shares many properties with the Fourier transform (see Section 3.2). These properties are summarized in Table 3.4.

3.9.2 *Laplace transform examples*

We illustrate the Laplace transform and compare it with the Fourier transform through several examples.

Consider the one-sided exponential function:

$$f(x) = \theta(x)\exp(-\alpha x), \quad \alpha > 0;\ \hat{f}(k) = \frac{\alpha - ik}{\alpha^2 + k^2};$$

$$\tilde{f}(k) = \frac{1}{k + \alpha},\ \mathrm{Re}(k + \alpha) > 0.$$

As $\alpha \to 0^+$, these relations reduce to the following for the Heaviside step function:

$$f(x) = \theta(x); \quad \hat{f}(k) = \pi\delta(k) - \frac{i}{k}; \quad \tilde{f}(k) = \frac{1}{k}.$$

By shifting and rescaling the step function, we obtain the Laplace transform of the signum function:

$$f(x) = 2\theta(x) - 1 = \mathrm{sign}(x); \quad \hat{f}(k) = -\frac{2i}{k}; \quad \tilde{f}(k) = \frac{1}{k}.$$

Exercise 3.11. Find the Laplace transform of the following functions:

(a) $f(x) = \exp(-\lambda x)$, where $\mathrm{Re}(\lambda) > 0$.
(b) $f(x) = x^n$, where $n \in \mathbb{Z}$.
(c) $f(x) = \cos(\nu x)$, where $\nu \in \mathbb{R}$.
(d) $f(x) = \cosh(\lambda x)$, where $\mathrm{Re}(\lambda) > 0$.
(e) $f(x) = 1/\sqrt{x}$.

Exercise 3.12. Find the inverse Laplace transform of $\frac{1}{k^2+a^2}$. Show details.

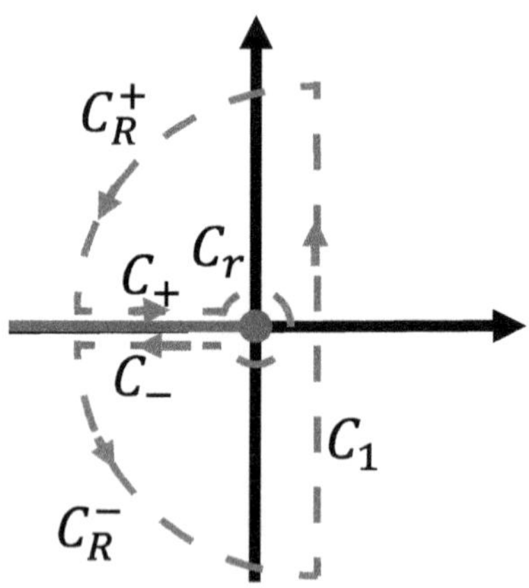

Figure 3.2. Contour transformation for the integral representation of the Γ function via the inverse Laplace transform. The integrand in Eq. (3.14) is analytic in the region enclosed by the (dashed) contour, resulting in a zero integral by Cauchy's theorem.

3.9.3 *Integral representations and asymptotics of special functions*

The Laplace transform is often used to describe and manipulate integral representations of special functions. Consider $f(x) = x^\nu$. Up to a basic multiplicative factor, its Laplace transform is related to the so-called Gamma function, Γ, with parameter ν:

$$\tilde{f}(k) = \int_0^\infty dx\, x^\nu e^{-kx} = k^{-\nu-1} \int_0^\infty dy\, y^\nu e^{-y} = \frac{\Gamma(\nu+1)}{k^{\nu+1}},$$

where the integrals converge for $k > 0$. Then, the inverse Laplace transform provides an integral representation of the Γ function in terms of a contour integral:

$$\frac{x^\nu}{\Gamma(\nu+1)} = \frac{1}{2\pi i} \int_{\varepsilon-i\infty}^{\varepsilon+i\infty} dk\, \frac{e^{kx}}{k^{\nu+1}}, \tag{3.14}$$

where $\varepsilon > 0$ ensures that the contour lies to the right of the singularity at $k = 0$, which is a pole if ν is a positive integer and a branch point if ν is non-integer.

For a branch point at $k = 0$, the integrand in Eq. (3.14) also has a branch point at $k = \infty$, necessitating a branch cut to ensure analyticity. We place the branch cut along the negative real axis. Then, the contour integral $\int_C dk\, \frac{e^{kx}}{k^{\nu+1}}$, taken over the contour C (shown dashed in Fig. 3.2), encloses no singularities and is therefore zero by Cauchy's theorem. Here, the contour C is composed of segments $C_1 + C_R^+ + C_+ + C_r + C_- + C_R^-$, where $R \to \infty$, $r \to 0$ and $C_1 = [\varepsilon - i\infty, \varepsilon + i\infty]$. By Jordan's lemma, the integrals along $C_R^\pm$ and C_r vanish as $R \to \infty$ and $r \to 0$. Consequently, the Bromwich contour in Eq. (3.14) can be replaced with the contour $-C_+ - C_-$, which runs along the negative real axis from $-\infty$ to 0, slightly below the cut, and returns from 0 to $-\infty$ slightly above the cut. We then arrive at

$$\frac{1}{\Gamma(\nu+1)} = \frac{1}{2\pi i} \int_{-C_- - C_+} dk\, \frac{e^k}{k^{\nu+1}}. \tag{3.15}$$

The result in Eq. (3.15) is useful for evaluating the asymptotic behavior of $f(x)$ as $x \to \infty$ using $\tilde{f}(k)$. This approach introduces a branch cut to the left of the Bromwich contour. Specifically, by considering the inverse Laplace transform formula (3.13), we see that as

$x \to +\infty$, the integral is dominated by the rightmost singularity in the complex k-plane. (This approach applies to both the branch points and poles of the integrand in Eq. (3.13).) We can then expand around this rightmost singularity k_0 of $\tilde{f}(k)$ as follows:

$$\tilde{f}(k) \approx \sum_{\nu} c_\nu (k - k_0)^{\lambda_\nu}, \tag{3.16}$$

where k_0 denotes the position of the rightmost singularity of $\tilde{f}(k)$ and λ_ν may be a non-integer. A loop integral around the branch point at k_0 leads to an asymptotic series, which can be obtained by integrating Eq. (3.16) term by term:

$$\begin{aligned} f(x) &= \frac{1}{2\pi i}\int_{C_{k_0}} dk\, \tilde{f}(k) e^{kx} \approx \frac{1}{2\pi i}\int_{C_{k_0}} dk \left(\sum_{\nu} c_\nu (k - k_0)^{\lambda_\nu}\right) e^{kx} \\ &= e^{k_0 x} \sum_{\nu} \frac{c_\nu}{\Gamma(-\lambda_\nu) x^{\lambda_\nu + 1}}, \end{aligned}$$

where C_{k_0} is a contour encircling the singularity at k_0 counterclockwise around the cut, and we have used Eq. (3.15) for the Γ function.

3.10 From Differential to Algebraic Equations with Fourier Transform, Fourier Series and Laplace Transform

The Fourier transform, Fourier series and Laplace transform introduced in this chapter are invaluable tools for simplifying and solving differential equations, which we will explore extensively in the following chapters.

To illustrate their power, consider a simple first-order linear differential equation relating a scalar function $f(x)$ to its derivative:

$$\frac{d}{dx} f(x) + q f(x) = g(x), \tag{3.17}$$

where $q > 0$ is a constant and $g(x)$ is a known function.

Assume that $g(x)$ has a well-defined Fourier transform, denoted $\hat{g}(k)$. Applying the Fourier transform to both sides of Eq. (3.17) transforms the differential equation into a simpler algebraic equation:

$$ik\hat{f}(k) + q\hat{f}(k) = \hat{g}(k), \tag{3.18}$$

which is easily solved:

$$\hat{f}(k) = \frac{\hat{g}(k)}{q + ik}.$$

To retrieve $f(x)$, we apply the inverse Fourier transform:

$$\begin{aligned} f(x) &= \frac{1}{2\pi}\int_{-\infty}^{+\infty} dk \frac{\hat{g}(k)}{q+ik} e^{ikx} = \frac{1}{2\pi}\int_{-\infty}^{+\infty} \frac{dk}{q+ik} \int_{-\infty}^{+\infty} dy\, g(y) e^{ik(x-y)} \\ &= \int_{-\infty}^{x} dy\, g(y) e^{-q(x-y)}, \end{aligned} \tag{3.19}$$

where we exchanged the order of integration and evaluated the pole integral at $k = iq$, ensuring that the integral is nonzero only for $y < x$. This yields the solution in integral form (quadrature) over the known function $g(y)$.

Thus, for $g(x)$ defined over $x \in \mathbb{R}$, the steps to solve the differential equation are straightforward:

1. Apply the Fourier transform to convert the differential equation into an algebraic equation.
2. Solve the algebraic equation for $\hat{f}(k)$.
3. Apply the inverse Fourier transform to recover $f(x)$.

Now, assume that $f(x)$ and $g(x)$ are nonzero only for $x \geq 0$. In this case, we use the Laplace transform instead of the Fourier transform. Applying the Laplace transform to Eq. (3.17), we obtain

$$\begin{aligned} f(x) &= \frac{\theta(x)}{2\pi i}\int_{\varepsilon-i\infty}^{\varepsilon+i\infty} dk\, \frac{f(0^+) + \tilde{g}(k)}{q+k} e^{kx} \\ &= \frac{\theta(x)}{2\pi i}\int_{\varepsilon-i\infty}^{\varepsilon+i\infty} \frac{dk}{q+k}\left(f(0^+)e^{kx} + \int_0^{+\infty} dy\, g(y) e^{k(x-y)}\right) \\ &= \theta(x)\left(f(0^+)e^{-qx} + \int_0^{x} dy\, g(y) e^{-q(x-y)}\right), \end{aligned}$$

where, in transitioning to the second line, we account for the additional $f(0^+)$ factor in the Laplace transform version of Eq. (3.18) and then evaluate the pole integrals at $k = -q$. This result is analogous to Eq. (3.19), but now applies to functions defined for $x \geq 0$.

Finally, consider the case where $f(x)$ and $g(x)$ are 2π-periodic. In this scenario, we apply the Fourier series instead of the Fourier transform. Assuming that $q = im$ (where $m \in \mathbb{Z}$) and $g(x)$ is 2π-periodic, applying the Fourier series to Eq. (3.17) yields a discrete algebraic equation similar to Eq. (3.18), but involving the Fourier coefficients f_k and g_k instead of the continuous functions $\hat{f}(k)$ and $\hat{g}(k)$:

$$f(x) = \sum_{k=-\infty}^{+\infty} \frac{g_k}{i(k+m)} e^{ikx} = e^{-imx} \sum_{k'=-\infty}^{+\infty} \frac{g_{k'-m}}{ik'} e^{ik'x}$$

$$= f(0) + \int_0^x dy\, e^{im(y-x)} g(y).$$

In the Fourier series case, as in the Laplace transform case, the final solution requires an additional condition on $f(x)$, which we impose at $x = 0$.

In both the above cases, the solution requires an additional condition at $x = 0$ (or at another specific point), reflecting the general fact that solving an nth order ordinary differential equation requires fixing n conditions. The solution based on Fourier transform in Eq. (3.19) does not explicitly show this dependence, as the Fourier transform implicitly assumes that $f(-\infty) = 0$ for the Fourier integral to be well defined.

Part II
Differential Equations

Chapter 4

Ordinary Differential Equations

A *differential equation* (DE) is an equation that relates an unknown function and its derivatives to other known functions or quantities. *Solving* a DE involves determining the unknown function. To fully determine a DE, it is necessary to provide auxiliary information, typically in the form of initial or boundary conditions.

Often, multiple DEs may be coupled together in a system of DEs. This is equivalent to a DE for a vector-valued function. Therefore, we will use the term "differential equation" to refer to both single equations and systems of equations and "function" to refer to both scalar- and vector-valued functions. We will only distinguish between singular and plural when necessary.

The function to be determined may depend on a single independent variable (e.g., $f = f(t)$ or $f = f(x)$), in which case the DE is called an *ordinary* differential equation (ODE). If it depends on two or more independent variables (e.g., $f = f(x, y)$ or $f = f(t, x, y, z)$), the DE is called a *partial* differential equation.

The *order* of a DE is the highest integer n for which the nth derivative of the unknown function appears in the equation.

A general DE is equivalent to the condition that a *nonlinear* function of the unknown function and its derivatives equals zero. An ODE is called *linear* if the condition is linear in both the function and its derivatives. It is further classified as linear *homogeneous* if the condition is also homogeneous. For homogeneous linear ODEs, if $f(t)$ is a solution, then $cf(t)$, where c is a constant, is also a solution. A linear ODE that is not homogeneous is called *inhomogeneous*. For instance, an inhomogeneous nth-order ODE can be written as

$$\alpha_n(t)f^{(n)}(t) + \cdots + \alpha_1(t)f'(t) + \alpha_0(t)f(t) = g(t),$$

where $\alpha_i(t), i = 0, \ldots, n$, and $g(t)$ are known functions.

Typical methods for solving linear DEs often leverage the fact that the linear combination of solutions to the homogeneous DE forms another solution, allowing the construction of particular solutions from a basis of general solutions. This approach does not work for nonlinear DEs. For such equations, analytic solutions are often customized for each specific case, with no single method applying broadly to nonlinear DEs. Given the difficulty of finding analytic solutions, qualitative and/or approximate methods are frequently used, such as dimensional analysis, phase plane analysis, perturbation methods or linearization. Generally, linear DEs exhibit simpler dynamics compared to nonlinear ones.

In this chapter, we focus on *ordinary differential equations* (ODEs). An ODE is a DE involving one or more functions of a *single* independent variable, along with their derivatives. The term "ordinary" distinguishes ODEs from *partial differential equations* (PDEs), which involve functions of *multiple* independent variables. PDEs will be discussed in Chapter 5. Our discussion of ODEs in this chapter is concise, emphasizing methods and solutions that will be relevant in later discussions of PDEs and variational calculus in Chapter 6. For readers seeking a more comprehensive and theoretical treatment of ODEs, we recommend several excellent resources. *Ordinary Differential Equations* by Arnold [7] provides a geometric perspective on the subject, offering deep insights into the qualitative theory of DEs. The classic text *Theory of Ordinary Differential Equations* by Coddington and Levinson [8] is another indispensable resource, covering the rigorous theory behind ODEs. For practical reference, *Handbook of Differential Equations* by Zwillinger [9] offers numerous methods and exact solutions for various types of ODEs. Lastly, Hartman's *Ordinary Differential Equations* [10] provides a comprehensive exploration of the subject, with a focus on the existence and stability of solutions.

4.1 ODEs: Simple Cases

To start, let us review some simple cases of ODEs that can be integrated directly.

4.1.1 *Separable differential equations*

A separable DE is a first-order DE that can be written such that the derivative function appears on one side of the equation, while the other side contains the product or quotient of two functions, one depending on the independent variable and the other on the dependent variable:

$$\frac{dx}{dt} = \frac{f(t)}{g(x)} \Rightarrow g(x)\,dx = f(t)\,dt \Rightarrow \int g(x)\,dx = \int f(t)\,dt. \tag{4.1}$$

Example 4.1.1. Solve the DE $\dot{x}(t) = ax(t)t^2$.

Solution.

$$\frac{dx(t)}{dt} = ax(t)t^2 \Rightarrow \frac{dx}{x} = at^2\,dt \Rightarrow \int^x \frac{d\xi}{\xi} = \int^t a\tau^2\,d\tau$$
$$\Rightarrow \log(x) + c_1 = \frac{a}{3}t^3 + c_2 \Rightarrow x(t) = c\,e^{at^3/3}.$$

4.1.2 *Variation of parameters*

Consider the following linear, inhomogeneous ODE:

$$\frac{dy}{dt} - p(t)y(t) = g(t), \quad y(t_0) = y_0, \tag{4.2}$$

Let us substitute

$$y(t) = c(t)\exp\left(\int_{t_0}^t d\tau\, p(\tau)\right), \tag{4.3}$$

where the second term is based on the solution of the homogeneous version of Eq. (4.2), i.e., $\frac{dy}{dt} = p(t)y(t)$. In this method, we allow $c(t)$, which would be a constant in the homogeneous case, to vary as a function of t. This leads to the following equation for $c(t)$:

$$\frac{dc(t)}{dt}\exp\left(\int_{t_0}^{t} d\tau\, p(\tau)\right) = g(t).$$

Applying the method of separable DEs (see Eq. (4.1)) and recalling the substitution (4.3), we obtain

$$y(t) = \exp\left(\int_{t_0}^{t} d\tau\, p(\tau)\right)\left(y_0 + \int_{t_0}^{t} d\tau\, g(\tau)\exp\left(-\int_{t_0}^{\tau} d\tau\, p(\tau)\right)\right).$$

This technique is known as the *method of variation of parameters* because $c(t)$, which is treated as a parameter in the homogeneous case, is allowed to vary in the inhomogeneous case.

Let us extend this idea to the second-order inhomogeneous DE of general form,

$$\frac{d^2y}{dt^2} - p(t)\frac{dy}{dt} - q(t)\,y(t) = g(t), \tag{4.4}$$

and attempt to find its general solution. Recall that the general solution of a linear inhomogeneous equation is the sum of the general solution to the homogeneous equation and a particular solution of the inhomogeneous equation.

First, consider the solution to the homogeneous equation (i.e., when $g(t) = 0$). Since this is a second-order homogeneous equation, it will have two independent (linearly independent) solutions. Denote these solutions as $y_1(t)$ and $y_2(t)$, and form their Wronskian:

$$W(t) = y_1y_2' - y_2y_1'.$$

Next, compute the derivative of the Wronskian and use the fact that y_1 and y_2 satisfy the homogeneous version of Eq. (4.4). This leads to

$$\frac{d}{dt}W = y_1y_2'' + y_1'y_2' - y_2'y_1' - y_2y_1'' = py_1y_2' - py_2y_1' = pW.$$

Thus, the Wronskian becomes

$$W(t) = \exp\left(\int_{t_0}^{t} d\tau\, p(\tau)\right), \tag{4.5}$$

where t_0 is chosen arbitrarily. Knowing the Wronskian, one can express one solution in terms of the other.

Next, we seek a particular solution to the inhomogeneous equation. We propose the following ansatz, where $y(t)$ is a linear combination of $y_1(t)$ and $y_2(t)$, multiplied by unknown functions $A(t)$ and $B(t)$:

$$y(t) = A(t)y_1(t) + B(t)y_2(t). \tag{4.6}$$

Substituting Eq. (4.6) into Eq. (4.4) results in

$$A''y_1 + B''y_2 + 2A'y_1' + 2B'y_2' - p\left(A'y_1 + B'y_2\right) = g. \tag{4.7}$$

To simplify, we relate $A(t)$ and $B(t)$ so that the dependence on $p(t)$ in Eq. (4.7) disappears, choosing

$$A'y_1 + B'y_2 = 0. \tag{4.8}$$

Then, Eq. (4.7) simplifies to

$$A'y_1' + B'y_2' = g. \tag{4.9}$$

Now, the order of derivatives is reduced. By expressing B' in terms of A', y_1 and y_2 using Eq. (4.8) and substituting it into Eq. (4.9), we derive

$$WA' + y_2 g = 0 \Rightarrow A(t) = -\int_{t_0}^{t} d\tau \frac{y_2(\tau)g(\tau)}{W(\tau)}. \tag{4.10}$$

Similarly, expressing A' in terms of B', y_1 and y_2, we find the B-analog of Eq. (4.10):

$$B(t) = \int_{t_0}^{t} d\tau \frac{y_1(\tau)g(\tau)}{W(\tau)}. \tag{4.11}$$

In summary, to construct the solution to Eq. (4.4), we follow these steps:

- Compute the Wronskian $W(t)$ using Eq. (4.5).
- Find a homogeneous solution $y_1(t)$, and express the linearly independent solution $y_2(t)$ using $y_1(t)$ and the Wronskian.

- Calculate the time-dependent factors $A(t)$ and $B(t)$ according to Eqs. (4.10) and (4.11) to find a particular solution for the inhomogeneous equation.
- The general solution is a sum of y_1 and y_2 multiplied by time-independent coefficients plus the particular solution just found.

The general approach is illustrated by the following two examples.

Example 4.1.2. Find the general solution to $t^2x''(t)+tx'(t)-x(t) = t$ (where $t \neq 0$) given that $x(t) = t$ is a solution.

Solution. Divide by t to normalize the leading coefficient:

$$x'' + t^{-1}x' - t^{-2}x = t^{-1}, \quad \text{where } t \neq 0.$$

Thus, $p(t) = -t^{-1}$. We compute the Wronskian:

$$W(t) = \exp\left(\int_{t_0}^{t} -\tau^{-1}d\tau\right) = t^{-1}.$$

The second linearly independent solution is

$$W(t) = y_1y_2' - y_2y_1' \Rightarrow \frac{1}{t} = ty_2' - y_2 \Rightarrow y_2(t) = -\frac{1}{2t}.$$

Now, compute $A(t)$ and $B(t)$:

$$A(t) = -\int_{t_0}^{t} \frac{y_2(\tau)g(\tau)}{W(\tau)}d\tau = \frac{1}{2}\log(t),$$

$$B(t) = \int_{t_0}^{t} \frac{y_1(\tau)g(\tau)}{W(\tau)}d\tau = \frac{t^2}{2}.$$

The general solution to the DE is

$$x(t) = c_1t + c_2t^{-1} + \frac{1}{2}t\log(t) - \frac{t}{4}.$$

Example 4.1.3. Find the general solution to $r''(\theta) + r(\theta) = \tan(\theta)$ for $-\pi/2 < \theta < \pi/2$.

Solution. We compute the Wronskian:

$$W(\theta) = \exp\left(\int_{\theta_0}^{\theta} 0\, d\theta'\right) = 1.$$

Let $r_1(\theta) = \cos(\theta)$ be the first linearly independent solution. The second solution is found as

$$1 = \cos(\theta) r_2'(\theta) - \sin(\theta) r_2(\theta) \Rightarrow r_2(\theta) = \sin(\theta).$$

Now, compute $A(\theta)$ and $B(\theta)$:

$$A(\theta) = -\int_{\theta_0}^{\theta} \frac{r_2(\theta')g(\theta')}{W(\theta')} d\theta' = \sin(\theta) - \log(\sec(\theta) + \tan(\theta)),$$

$$B(\theta) = \int_{\theta_0}^{\theta} \frac{r_1(\theta')g(\theta')}{W(\theta')} d\theta' = -\cos(\theta).$$

Thus, the general solution to the DE is

$$x(\theta) = c_1 \cos(\theta) + c_2 \sin(\theta) + \cos(\theta) \log(\sec(\theta) + \tan(\theta)).$$

Exercise 4.1. (a) Find the general solution $x(t)$ to the following ODE:

$$\frac{dx}{dt} - \lambda(t) x = \frac{f(t)}{x^2},$$

where $\lambda(t)$ and $f(t)$ are known functions of t.

(b) Solve the following general second-order, constant-coefficient, linear ODE:

$$\tau_0^2 \frac{d^2}{dt^2} y + \tau_1 \frac{d}{dt} y + y = g(t),$$

with initial conditions $y(0) = y_0$ and $\frac{d}{dt} y\big|_{t=0} = v_0$.

4.2 Direct Methods for Solving Linear ODEs

We now explore linear ODEs in more detail, gradually increasing complexity and introducing more technical methods.

4.2.1 *Homogeneous ODEs with constant coefficients*

Consider the nth-order homogeneous ODE with constant coefficients:

$$\mathcal{L}x(t) = 0, \quad \text{where} \quad \mathcal{L} := \sum_{m=0}^{n} a_{n-m} \frac{d^{n-m}}{dt^{n-m}}. \tag{4.12}$$

Here and in the following, $\mathcal{L}$ denotes a differential operator. We seek the general solution of Eq. (4.12) as a linear combination of exponentials:

$$x(t) = \sum_{k=1}^{n} c_k \exp(\lambda_k t), \tag{4.13}$$

where c_k are constants. Substituting Eq. (4.13) into Eq. (4.12), we obtain the condition that λ_k are the roots of the characteristic polynomial:

$$\sum_{m=0}^{n} a_{n-m} \lambda_k^{n-m} = 0.$$

If the λ_k are not degenerate (i.e., if there are n distinct solutions), Eq. (4.13) holds. In the case of degeneracy, we generalize Eq. (4.13) to include a sum of exponentials for non-degenerate λ_k and polynomials in t multiplied by the respective exponentials for degenerate λ_k. The degree of the polynomials matches the degree of root degeneracy:

$$x(t) = \sum_{k=1}^{m} \left(\sum_{l=0}^{d_k} c_k^{(l)} t^l \right) \exp(\lambda_k t),$$

where d_k is the degree of degeneracy of the kth root.

4.2.2 *Inhomogeneous ODEs*

Now consider an inhomogeneous version of a general linear ODE:

$$\mathcal{L}x(t) = f(t). \tag{4.14}$$

Recall that if $x_p(t)$ is a particular solution and $x_0(t)$ is a general solution of the homogeneous equation, then the general solution of Eq. (4.14) is

$$x(t) = x_0(t) + x_p(t).$$

Let us illustrate the utility of this result with an example:

Example 4.2.1. For (a) $\omega_0 \neq 3$ and (b) $\omega_0 = 3$, solve

$$\mathcal{L}x := \ddot{x} + \omega_0^2 x = \cos(3t). \tag{4.15}$$

Solution. The general solution to the homogeneous equation, $\mathcal{L}x = 0$, is

$$x_0(t) = c_1 \cos(\omega_0 t) + c_2 \sin(\omega_0 t).$$

For $\omega_0 \neq 3$, a particular solution to Eq. (4.15) is

$$x_p(t) = \frac{\cos(3t)}{\omega_0^2 - 9},$$

which can be found using variation of parameters (Section 4.1.2). Thus, for $\omega_0 \neq 3$, the solution to the inhomogeneous equation is

$$x(t) = c_1 \cos(\omega_0 t) + c_2 \sin(\omega_0 t) + \frac{\cos(3t)}{\omega_0^2 - 9}.$$

When $\omega_0 = 3$, the system resonates due to the coincidence of natural and forcing frequencies. In this case, the particular solution must be modified because it is already part of the homogeneous solution. The particular solution for $\omega_0 = 3$ is found using variation of parameters:

$$x(t) = c_1 \cos(3t) + c_2 \sin(3t) + \frac{1}{6} t \sin(3t).$$

4.3 Linear Dynamics via the Green Function

So far, our analysis of ODEs has been primarily formal. Often, though not always, we can associate an ODE with the dynamics of a system, hence the term "dynamical system." In this context, the ODE describes the evolution of the system variable x over time t, i.e., it considers x as a function of t, or $x(t)$.

The study of dynamics is extensive, and in this course, we will only touch the surface of the fascinating phenomena it involves. For example, in Section 4.6, we discuss the so-called "conservative" dynamics. An ODE may also describe a "dissipative" system, which relaxes to an "equilibrium." If a dissipative system is in equilibrium, its state remains constant over time. If the system is perturbed from a stable equilibrium and the perturbation is small, the system will relax back to equilibrium. The relaxation may not be monotonic; the system can exhibit oscillations. When the relaxational (dissipative with possible oscillations) dynamics is close to equilibrium, we model it using a linear ODE. There are also many interesting situations where linear ODEs describe oscillations that do not decay.

The method of Green functions, or "response functions," will be a central tool in our analysis of such dynamics when the ODE is linear. The Green function method provides a powerful and intuitive approach, which we will extend (in the following chapter) to the case of PDEs.

Let us begin with the following general idea. Given a linear differential equation $\mathcal{L}x(t) = f(t)$, the goal is to find an operator $\mathcal{L}^{-1}$ such that "$x(t) = \mathcal{L}^{-1}f(t)$." Since $\mathcal{L}$ is a differential operator, it is reasonable to expect that $\mathcal{L}^{-1}$ is an integral operator, which can be expressed as

$$\mathcal{L}^{-1}f(t) = \int d\tau\, G(t,\tau)f(\tau),$$

where $G(t,\tau)$ is the *Green function*, which we need to determine. Formal manipulations show that

$$f(t) = \mathcal{L}\mathcal{L}^{-1}f(t) = \mathcal{L}\int d\tau\, G(t,\tau)f(t) = \int d\tau\, \mathcal{L}G(t,\tau)f(t).$$

We have encountered this equation before, as one of the properties of the δ-function. That is,

$$f(t) = \int d\tau\, \mathcal{L}G(t,\tau)f(t) \Leftrightarrow \mathcal{L}G(t,\tau) = \delta(t-\tau).$$

The Green function for a differential operator $\mathcal{L}$ is the function $G(t,\tau)$ that solves the differential equation $\mathcal{L}G(t,\tau) = \delta(t-\tau)$, subject to prescribed boundary conditions. The Green function describes the response of the system at time t to an impulse applied at time τ.

Notation. Technically, the Green function is a function of two variables, t and τ, where τ represents the time of an impulse and t represents the time at which we observe the system's response to that impulse. Note that if $\mathcal{L}$ is a differential operator with constant (time-independent) coefficients, then the system's response depends only on the difference $t - \tau$, not on t and τ independently. In such cases, $G(t, \tau)$ reduces to the "homogeneous in time" or "time-invariant" $G(t - \tau)$.

We explore this method by revisiting the simple case of a constant-coefficient linear scalar-valued first-order ODE (4.2).

4.3.1 *Evolution of a linear scalar*

Consider the simplest example of scalar relaxation:

$$\frac{d}{dt}x + \gamma x = f(t), \tag{4.16}$$

where γ is a constant and $f(t)$ is a known function of t. This model arises, for instance, in the overdamped motion of a polymer, where Eq. (4.16) describes the balance of forces: $f(t)$ is the driving force, γx is the elastic (restoring) force for a polymer whose one end is fixed at the origin, and $\dot{x}$ represents the frictional force on the polymer from the medium. The general solution of this equation (recall the discussion of the integral operator above, and note the time-homogeneous form of the Green function) is

$$x(t) = \int_{-\infty}^{t} d\tau\, G(t - \tau) f(\tau), \tag{4.17}$$

where we assume that the evolution begins at $t = -\infty$, with $\lim_{t \to -\infty} x(t) = 0$. The Green function $G(t, \tau)$ satisfies

$$\frac{d}{dt}G(t, \tau) + \gamma G(t, \tau) = \delta(t - \tau), \tag{4.18}$$

where $\delta(t)$ is the Dirac delta function.

This is an *initial value problem* (or Cauchy problem). If we did not specify the initial condition $x(-\infty) = 0$, the solution to Eq. (4.16) would be ambiguous. For instance, if $x_s(t)$ is a particular solution of Eq. (4.16), then $x_s(t) + C \exp(-\gamma t)$, where C is a constant, is a family of solutions. The constant C represents a "zero mode" of the differential operator $d/dt + \gamma$, which is fixed by the initial condition.

Causality is another key principle. According to the "causality principle," the system's response at time t depends only on external driving forces $f(\tau)$ for $\tau \leq t$ (i.e., the past). The response cannot depend on future driving forces for $\tau > t$. This is reflected in the Green function, which is zero for $t < \tau$.

To solve Eq. (4.18), we first note that the Green function must satisfy $G(t-\tau) = A\exp(-\gamma(t-\tau))$ for $t > \tau$, where A is a constant. Due to causality, $G(t-\tau) = 0$ for $t < \tau$. By integrating both sides of Eq. (4.18) across the jump at $t = \tau$, we find that $A = 1$. Thus, the solution to Eq. (4.17) is

$$x(t) = \int_{-\infty}^{t} d\tau \exp(-\gamma(t-\tau))f(\tau).$$

This shows that the system "forgets" past disturbances at a rate determined by γ.

We now outline several methods to solve Eq. (4.18).

Method 1: Multiply the equation by an appropriate integrating factor, in this case $e^{\gamma t}$:

$$\mathcal{L}G(t,\tau) = \delta(t-\tau), \quad \frac{d}{dt}\left(e^{\gamma t}G(t,\tau)\right) = e^{\gamma t}\delta(t-\tau),$$

$$e^{\gamma t}G(t,\tau) = \int_{-\infty}^{t} e^{\gamma t'}\delta(t'-\tau)dt' = \int_{-\infty}^{\infty} \theta(t'-\tau)e^{\gamma t'}\delta(t'-\tau)dt',$$

$$G(t,\tau) = \theta(t-\tau)e^{-\gamma(t-\tau)}.$$

Method 2: Take the Fourier transform of both sides, solve the resulting algebraic equation for $\hat{x}(k)$ and use contour integration to compute the inverse Fourier transform:

$$\mathcal{F}\left[\dot{G}(t,\tau) + \gamma G(t,\tau)\right] = \mathcal{F}\left[\delta(t-\tau)\right],$$

$$ik\hat{G}(k,\tau) + \gamma\hat{G}(k,\tau) = e^{-ik\tau},$$

$$\hat{G}(k,\tau) = \frac{e^{-ik\tau}}{\gamma + ik},$$

$$G(t,\tau) = \frac{1}{2\pi}\int_{-\infty}^{\infty} \frac{e^{-ik\tau}}{\gamma + ik}e^{ikt}dk = \theta(t-\tau)e^{-\gamma(t-\tau)}.$$

We compute the contour integral by closing the contour with a semicircular arc of radius R and taking the limit $R \to \infty$. The contour is closed in the upper half-plane for $t > \tau$ and in the lower half-plane for $t < \tau$. The integrand has a simple pole at $k = i\gamma$, yielding the result for the Green function.

Method 3: Construct the Green function based on its essential properties:

(i) $G(t,\tau)$ solves $\mathcal{L}G(t,\tau) = 0$ for $t \neq \tau$.
(ii) $G(t,\tau)$ satisfies the initial condition.
(iii) $G(t,\tau)$ exhibits a jump of size one at $t = \tau$, as required by integrating around the delta function.

Generalization: The Green function for an nth-order differential operator has a jump in its $(n-1)$st derivative at $t = \tau$, while higher derivatives are continuous. The size of the jump equals the leading coefficient of the differential operator.

Let us now construct the Green function step-by-step. The solution to $\left(\frac{d}{dt} + \gamma\right) G(t,\tau) = 0$ is $Ae^{-\gamma t}$. For $t < \tau$, $G(t,\tau) = 0$. Applying the jump condition fixes $A = 1$. Thus, we arrive at

$$G(t,\tau) = \theta(t-\tau)e^{-\gamma(t-\tau)}.$$

Exercise 4.2. Solve Eq. (4.16) for $t > 0$, where $x(0) = 0$ and $f(t) = A\exp(-\alpha t)$. Analyze the dependence on α and γ, particularly for $\alpha \to \gamma$.

Note that Eq. (4.18) assumes that the Green function depends only on the difference $t - \tau$. This assumption is valid for this particular case but does not hold when the decay coefficient $\gamma(t)$ depends on t. In such situations, we must use the more general expression for the Green function $G(t,\tau)$. When γ is constant, the Green function depends on $t-\tau$ due to the time-translation symmetry of Eq. (4.18).

4.3.2 *Evolution of a vector*

Let us now generalize the previous discussion by considering the vector-valued system

$$\frac{d}{dt}\boldsymbol{x} + \hat{\boldsymbol{\Gamma}}\boldsymbol{x} = \boldsymbol{f}(t), \tag{4.19}$$

where $\boldsymbol{x} = (x_1, \ldots, x_n)^\top$ and $\boldsymbol{f} = (f_1, \ldots, f_n)^\top$ are n-dimensional vector-valued functions of t and $\hat{\boldsymbol{\Gamma}}$ is an $n \times n$ time-independent matrix. We consider two scenarios: (1) when $\hat{\mathbf{\Gamma}}$ is either diagonal or diagonalizable, and (2) when $\hat{\mathbf{\Gamma}}$ is not diagonalizable.

If $\hat{\mathbf{\Gamma}}$ is a diagonal matrix, the system of vector-valued differential equations decouples into n scalar differential equations of the form $\dot{x}_i(t) + \gamma_i x_i(t) = f_i(t)$, where $\gamma_1, \ldots, \gamma_n$ are the diagonal elements of $\hat{\mathbf{\Gamma}}$. Each scalar differential equation can be solved independently, as described in Section 4.3.1.

If $\hat{\mathbf{\Gamma}}$ is diagonalizable (but not necessarily diagonal), we compute its eigenvectors and eigenvalues:

$$\hat{\mathbf{\Gamma}}\boldsymbol{a}_i = \lambda_i \boldsymbol{a}_i,$$

where λ_i and $\boldsymbol{a}_i$ are the eigenvalues and eigenvectors, respectively. We can expand both $\boldsymbol{x}$ and $\boldsymbol{f}$ in terms of the eigenbasis $\{\boldsymbol{a}_i\}$:

$$\boldsymbol{x} = \sum_i y_i \boldsymbol{a}_i, \quad \boldsymbol{f} = \sum_i \phi_i \boldsymbol{a}_i. \tag{4.20}$$

Substituting these expansions into Eq. (4.19) reduces the vector equation to n independent scalar equations:

$$\frac{dy_i}{dt} + \lambda_i y_i = \phi_i, \tag{4.21}$$

which can be solved using methods similar to those discussed in Section 4.3.1.

If $\hat{\mathbf{\Gamma}}$ is not diagonalizable, we can transform it into Jordan canonical form, which occurs when two or more eigenvalues share an eigenvector. In this case, we introduce the Green function $\hat{\boldsymbol{G}}(t, \tau)$, which satisfies

$$\left(\frac{d}{dt} + \hat{\mathbf{\Gamma}}\right) \hat{\boldsymbol{G}}(t, \tau) = \delta(t - \tau)\hat{\mathbf{1}}. \tag{4.22}$$

The explicit solution of Eq. (4.22) is

$$\hat{\boldsymbol{G}}(t, \tau) = \theta(t - \tau) \exp\left(-\hat{\mathbf{\Gamma}}(t - \tau)\right), \tag{4.23}$$

which allows us to express the solution to Eq. (4.19) in the following form:

$$\boldsymbol{x}(t) = \int_{-\infty}^{t} d\tau\, \hat{\boldsymbol{G}}(t-\tau)\boldsymbol{f}(\tau) = \int_{-\infty}^{t} d\tau\, \theta(t-\tau) \exp\left(-\hat{\mathbf{\Gamma}}(t - \tau)\right) \boldsymbol{f}(\tau). \tag{4.24}$$

Note that the matrix exponential $\exp(-(t-\tau)\hat{\mathbf{\Gamma}})$, used in Eqs. (4.23) and (4.24), can be formally expanded as a Taylor series:

$$\exp\left(-(t-\tau)\hat{\mathbf{\Gamma}}\right) = \sum_{n=0}^{\infty} \frac{(-(t-\tau))^n \hat{\mathbf{\Gamma}}^n}{n!},$$

which is always convergent for matrices with finite elements.

To relate Eq. (4.24) to the eigen-decomposition of Eqs. (4.20) and (4.21), we introduce the eigen-decomposition

$$\hat{\mathbf{\Gamma}} = \hat{\mathbf{A}}\hat{\mathbf{J}}\hat{\mathbf{A}}^{-1},$$

where $\hat{\mathbf{J}}$ is a matrix of Jordan blocks and the columns of $\hat{\mathbf{A}}$ are the corresponding eigenvectors of $\hat{\mathbf{\Gamma}}$. Note that $\hat{\mathbf{\Gamma}}^n = \hat{\mathbf{A}}\hat{\mathbf{J}}^n\hat{\mathbf{A}}^{-1}$.

To illustrate the behavior in the degenerate case, consider the following Jordan matrix:

$$\hat{\mathbf{\Gamma}} = \begin{bmatrix} \lambda & 1 \\ 0 & \lambda \end{bmatrix},$$

which can be rewritten as $\lambda\hat{\mathbf{I}} + \hat{\mathbf{N}}$, where

$$\hat{\mathbf{N}} := \begin{bmatrix} 0 & 1 \\ 0 & 0 \end{bmatrix},$$

is a (2×2) nilpotent matrix, with $\hat{\mathbf{N}}^2 = \hat{\mathbf{0}}$. Using the nilpotent property of $\hat{\mathbf{N}}$, we compute the matrix exponential:

$$\exp\left(-(t-\tau)\hat{\mathbf{\Gamma}}\right) = e^{-\lambda(t-\tau)}\left(\hat{\mathbf{I}} - (t-\tau)\hat{\mathbf{N}}\right), \tag{4.25}$$

Substituting Eq. (4.25) into Eq. (4.24), we obtain the solution:

$$\boldsymbol{x}(t) = \int_{-\infty}^{t} d\tau\, \theta(t-\tau) e^{-\lambda(t-\tau)} \left(\hat{\mathbf{I}} - (t-\tau)\hat{\mathbf{N}}\right) \boldsymbol{f}(\tau).$$

Alternatively, we could write Eq. (4.19) in component form,

$$\frac{dx_1}{dt} + \lambda x_1 + x_2 = f_1, \quad \frac{dx_2}{dt} + \lambda x_2 = f_2,$$

integrate the second equation, substitute the result into the first and then change variables from x_1 to $\tilde{x}_1 = x_1 + tx_2$. This yields

$$\frac{d\tilde{x}_1}{dt} + \lambda\tilde{x}_1 = f_1 + tf_2.$$

Note the emergence of a secular term (a polynomial in t) on the right-hand side, which is typical in degenerate cases, which can then be integrated straightforwardly.

Exercise 4.3. Find the Green function of Eq. (4.19) for

$$\hat{\mathbf{\Gamma}} = \begin{pmatrix} \lambda & 1 & 0 \\ 0 & \lambda & 1 \\ 0 & 0 & \lambda \end{pmatrix}.$$

Note that vector-valued ODEs often arise from the "vectorization" of an nth-order scalar ODE for $y(t)$. This vectorization is done by defining $x_1 = y$, $x_2 = \frac{dy}{dt}, \ldots, x_n = \frac{d^{n-1}y}{dt^{n-1}}$. Thus, $d\boldsymbol{x}/dt$ can be expressed in terms of the components of $\boldsymbol{x}$, resulting in an equation of the form (4.19).

4.3.3 *Higher-order linear dynamics*

The Green function method, as illustrated earlier, can be applied to any inhomogeneous linear differential equation. Let us now examine how this approach works in the case of a second-order differential equation for a scalar variable. Consider the equation

$$\frac{d^2}{dt^2}x + \omega^2 x = f(t). \tag{4.26}$$

To solve Eq. (4.26), we note that its general solution is the sum of a particular solution and the solution to the homogeneous version of Eq. (4.26) with zero on the right-hand side. We choose a particular solution of Eq. (4.26) in the form of the convolution (as in Eq. 4.17) of the source term, $f(t)$, with the Green function of Eq. (4.26):

$$\left(\frac{d^2}{dt^2} + \omega^2\right) G(t) = \delta(t). \tag{4.27}$$

As previously established, $G(t) = 0$ for $t < 0$. By integrating Eq. (4.27) from $-\epsilon$ to ϵ, for small ϵ, and analyzing the balance of terms, we observe that $\dot{G}(t)$ experiences a discontinuity (jump) at $t = 0$, with the magnitude of this jump being unity. Additional integration around the singularity reveals that $G(t)$ is smooth (and zero) at $t = 0$.

Thus, for the second-order differential equation under consideration, its Green function satisfies $G(0^+) = 0$ and $\dot{G}(0^+) = 1$. Given that $\delta(0^+) = 0$, these two values can be treated as initial conditions for the homogeneous version of Eq. (4.27), defining $G(t)$ for $t > 0$. This results in the following expression for the Green function:

$$G(t) = \theta(t)\frac{\sin(\omega t)}{\omega}, \tag{4.28}$$

where $\theta(t)$ is the Heaviside function.

Furthermore, Eq. (4.17) provides the solution to Eq. (4.26) over an infinite time horizon. However, the Green function can also be used to solve the corresponding Cauchy problem (initial value problem). Since Eq. (4.26) is a second-order ODE, we must fix two values at $t = 0$, such as $x(0)$ and $\dot{x}(0)$. Considering that $G(0^+) = 0$ and $\dot{G}(0^+) = 1$, the general solution to the Cauchy problem for Eq. (4.26) is

$$x(t) = \dot{x}(0)G(t) + x(0)\dot{G}(t) + \int_0^t dt_1\, G(t - t_1)f(t_1).$$

Next, we generalize this approach for higher-order linear dynamics. Consider the general form:

$$\mathcal{L}x = f(t), \quad \mathcal{L} := \sum_{m=0}^{n} a_{n-m}\frac{d^{n-m}}{dt^{n-m}}, \tag{4.29}$$

where a_i are constants and $\mathcal{L}$ is an nth-order linear differential operator with constant coefficients, as discussed earlier in Section 4.2. We construct a particular solution of Eq. (4.29) as a convolution (similar to Eq. 4.17) of the source term $f(t)$ with the Green function $G(t)$ of Eq. (4.29):

$$\mathcal{L}G = \delta(t), \tag{4.30}$$

where $G(t) = 0$ for $t < 0$.

The solution to the corresponding homogeneous equation, $\mathcal{L}x = 0$, can be generally expressed as

$$x(t) = \sum_i b_i \exp(z_i t), \tag{4.31}$$

where b_i are arbitrary constants.

We now use the general representation (4.31) to construct the Green function by solving Eq. (4.30). Similar to the first- and second-order cases discussed earlier, we transition from the inhomogeneous equation to the homogeneous equation supplemented with initial conditions. Extending the integration approach around $t = 0$ (performed n times), we derive the initial conditions required for the nth-order differential equation:

$$a_{n-1}\frac{d^{n-1}}{dt^{n-1}}G(0^+) = 1, \quad \forall 0 \le m < n-1: \quad \frac{d^m}{dt^m}G(0^+) = 0. \tag{4.32}$$

Consider $\mathcal{L}$ formally as a polynomial in z, where z is the differential operator $z = d/dt$, i.e., $\mathcal{L}(z)$. For $t > 0^+$, the Green function satisfies the homogeneous equation $\mathcal{L}(d/dt)G = 0$, and its solution can be generally written as

$$G(t) = \sum_i b_i \exp(z_i t), \tag{4.33}$$

where the b_i constants are determined from a system of algebraic equations derived by substituting Eq. (4.33) into Eq. (4.32).

Example 4.3.1. Let $\gamma, \nu \in \mathbb{R}$ with $\gamma > 0$. Find the Green function for the differential operator and use it to solve the ODE

$$\frac{d^2}{dt^2}x + 2\gamma\frac{d}{dt}x + \nu^2 x = f(t), \tag{4.34}$$

subject to the conditions $\lim_{t\to-\infty} x(t) = 0$ and $\lim_{t\to-\infty} \dot{x}(t) = 0$. Consider the cases $\nu < \gamma$ and $\nu > \gamma$.

Notation. As before, $G(t,\tau)$ is a function of both t and τ, where τ represents the time when an impulse is applied. For any fixed τ, $G(t,\tau)$ is the response of the system at time t. Its Fourier transform, $\hat{G}(\omega,\tau)$, provides the decomposition of $G(t,\tau)$ into its oscillatory modes.

Solution. *Longer method: Solve $\mathcal{L}G = \delta(t - \tau)$ using Fourier transforms:*

To find $G(t, \tau)$, we start by taking the Fourier transform of the equation $\mathcal{L}G(t, \tau) = \delta(t - \tau)$ and solving for $\hat{G}(\omega, \tau)$:

$$\widehat{\mathcal{L}G}(\omega; \tau) = \hat{\delta}(\omega; \tau) \Rightarrow \left(-\omega^2 - 2i\gamma\omega + \nu^2\right) \hat{G}(\omega; \tau) = e^{-i\omega\tau},$$

yielding

$$\hat{G}(\omega; \tau) = \frac{-e^{-i\omega\tau}}{(\omega - \omega_+)(\omega - \omega_-)},$$

where $\omega_\pm = -i\gamma \pm \sqrt{\nu^2 - \gamma^2}$.

The inverse Fourier transform of $\hat{G}(\omega; \tau)$ can be computed using a contour integral. The contour should be closed by a semi-circular arc of radius R as $R \to \infty$. To ensure that the arc's contribution vanishes, we close the contour in the upper-half plane for $t < \tau$ and in the lower-half plane for $t > \tau$. The integrand has poles with associated residues, as follows:

- If $\nu > \gamma$,
  the poles are located at $\omega = \omega_\pm = -i\gamma \pm \sqrt{\nu^2 - \gamma^2}$ (complex values with real and imaginary parts):

$$\mathrm{Res}(f, \omega_-) = \frac{e^{-i\omega_-\tau}}{\omega_- - \omega_+} = -\frac{e^{-\gamma\tau} e^{-i\sqrt{\nu^2-\gamma^2}\tau}}{\sqrt{\nu^2 - \gamma^2}},$$

$$\mathrm{Res}(f, \omega_+) = \frac{e^{-i\omega_+\tau}}{\omega_+ - \omega_-} = \frac{e^{-\gamma\tau} e^{i\sqrt{\nu^2-\gamma^2}\tau}}{\sqrt{\nu^2 - \gamma^2}}.$$

- If $\nu < \gamma$,
  the poles are located at $\omega = \omega_\pm = -i\gamma \pm i\sqrt{\gamma^2 - \nu^2}$ (purely imaginary):

$$\mathrm{Res}(f, \omega_-) = \frac{e^{-i\omega_-\tau}}{\omega_- - \omega_+} = -\frac{e^{-\gamma\tau} e^{-\sqrt{\gamma^2-\nu^2}\tau}}{\sqrt{\gamma^2 - \nu^2}},$$

$$\mathrm{Res}(f, \omega_+) = \frac{e^{-i\omega_+\tau}}{\omega_+ - \omega_-} = \frac{e^{-\gamma\tau} e^{\sqrt{\gamma^2-\nu^2}\tau}}{\sqrt{\gamma^2 - \nu^2}}.$$

The Green function is then given by

$$G(t,\tau) = \frac{1}{2\pi}\int_{-\infty}^{\infty} e^{i\omega t}\hat{G}(\omega;\tau)\,d\omega = \frac{1}{2\pi}(2\pi\text{Res}(f,\omega_-)+2\pi\text{Res}(f,\omega_+)).$$

Since there are no singularities in the upper half-plane for $t < \tau$, we have $G(t,\tau) = 0$ in this case. This reflects causality, meaning that the system does not respond to an impulse that occurs in the future.

After simplifying (expressing the complex exponentials in terms of sines or hyperbolic sines where appropriate), we obtain

$$G(t-\tau) = \frac{\theta(t-\tau)e^{-\gamma(t-\tau)}}{\sqrt{|\nu^2-\gamma^2|}}\begin{cases}\sin\left(\sqrt{\nu^2-\gamma^2}\,(t-\tau)\right), & \gamma < \nu,\\ \sinh\left(\sqrt{\gamma^2-\nu^2}\,(t-\tau)\right), & \gamma > \nu.\end{cases}$$

Finally, the solution to the ODE is given by

$$x(t) = \begin{cases}\displaystyle\int_{-\infty}^{t}\frac{e^{-\gamma(t-\tau)}}{\sqrt{|\nu^2-\gamma^2|}}\sin\left(\sqrt{\nu^2-\gamma^2}\,(t-\tau)\right)f(\tau)\,d\tau, & \gamma < \nu,\\ \displaystyle\int_{-\infty}^{t}\frac{e^{-\gamma(t-\tau)}}{\sqrt{|\nu^2-\gamma^2|}}\sinh\left(\sqrt{\gamma^2-\nu^2}\,(t-\tau)\right)f(\tau)\,d\tau, & \gamma > \nu.\end{cases}$$

□

Solution. *Shorter method: Construct the Green function based on the properties it must satisfy.*

We solve for the case $\nu > \gamma$, with $\nu < \gamma$ following analogously. Given $\mathcal{L}G(t,\tau) = \delta(t-\tau)$, we note that $G(t,\tau)$ must satisfy:

(i) $G(t,\tau)$ solves $\mathcal{L}G(t,\tau) = 0$ for $t \neq \tau$.
(ii) $G(t,\tau)$ satisfies the initial conditions.
(iii) $G(t,\tau)$ is continuous at $t=\tau$, and $\frac{dG}{dt}$ has a jump of magnitude one at $t=\tau$.

In general, for an nth-order differential operator, the Green function has a jump in the $(n-1)$th derivative at $t=\tau$, with all higher derivatives being continuous.

Step 1: Find candidate solutions by solving $\left(\frac{d^2}{dt^2} + 2\gamma\frac{d}{dt} + \nu^2\right)$ $G(t,\tau) = 0$.

The two linearly independent solutions to this equation are $e^{-(\gamma\pm\sqrt{\nu^2-\gamma^2})t}$. We express these as

$$c_1e^{-\gamma(t-\tau)}\sin\left(\sqrt{\nu^2-\gamma^2}(t-\tau)\right)+c_2e^{-\gamma(t-\tau)}\cos\left(\sqrt{\nu^2-\gamma^2}(t-\tau)\right).$$

Step 2: Apply the initial conditions. For $t<\tau$, we require $G(t,\tau)=0$ due to the initial conditions. Therefore, we write

$$G(t,\tau)=\begin{cases}0, & t<\tau,\\ c_3e^{-\gamma(t-\tau)}\sin\left(\sqrt{\nu^2-\gamma^2}(t-\tau)\right), & t>\tau.\end{cases}$$

Step 3: Apply continuity and jump conditions. The continuity condition implies $c_4=0$. The jump condition for $\frac{dG}{dt}$ at $t=\tau$ requires $c_3=\frac{1}{\sqrt{\nu^2-\gamma^2}}$.

The Green function is thus

$$G(t-\tau)=\frac{\theta(t-\tau)e^{-\gamma(t-\tau)}}{\sqrt{\nu^2-\gamma^2}}\sin\left(\sqrt{\nu^2-\gamma^2}(t-\tau)\right),$$

and the solution to the ODE is

$$x(t)=\int_{-\infty}^{t}\frac{e^{-\gamma(t-\tau)}}{\sqrt{\nu^2-\gamma^2}}\sin\left(\sqrt{\nu^2-\gamma^2}(t-\tau)\right)f(\tau)\,d\tau.$$

A similar process applies to find the Green function for $\nu<\gamma$.

Exercise 4.4. Follow the approach of Example 4.3.1 and suggest two methods for finding the Green function (a) using Fourier transform and (b) based on the properties of the Green function for solving

$$\left(\frac{d^2}{dt^2}+\nu^2\right)^2x(t)=f(t),$$

with boundary conditions $x(0^-)=\frac{d}{dt}x(0^-)=\frac{d^2}{dt^2}x(0^-)=\frac{d^3}{dt^3}x(0^-)=0$.

4.3.4 *Laplace transform and Laplace's method*

So far, we have solved linear ODEs by employing the Green function approach, constructing the Green function as a solution of the homogeneous equation, supplemented with prescribed initial conditions (one fewer than the order of the differential equation). In this section, we explore an alternative method of solving ODEs using the Laplace transform, introduced in Section 3.9, to handle linear ODEs with constant coefficients. We also discuss Laplace's method for solving linear ODEs, where the coefficients depend linearly on the time or space variable. The connection between these two approaches extends beyond their shared name – Laplace contributed to the development of both – and lies in the fact that Laplace's method can be viewed as a generalization of the Laplace transform.

The Laplace transform is a natural tool for solving dynamical problems with a causal structure. Let us see how it works for finding the Green function defined by Eqs. (4.29) and (4.30). We apply the Laplace transform to Eq. (4.30) by integrating it over time with the Laplace weight $\exp(-kt)$ from a small positive value, ϵ, to ∞. The integral of the right-hand side is zero. Each term on the left-hand side can be transformed into a product of a monomial in k with $\tilde{G}(k)$, the Laplace transform of $G(t)$, by performing a series of integrations by parts. We also account for the boundary terms at $t = \epsilon$ and $t = \infty$. Assuming $G(\infty) = 0$ (which is always the case for stable systems), the contributions at $t = \infty$ vanish. All boundary terms at $t = \epsilon$ vanish as well, except for one: $d^{n-1}G(\epsilon)/dt^{n-1} = 1$. Altogether, we obtain the equation

$$L(k)\tilde{G}(k) = 1, \quad L(k) := \sum_{m=0}^{n} a_{n-m}(-k)^{n-m}.$$

This shows that $\tilde{G}(k)$ has poles in the complex plane of k associated with the zeros of the polynomial $L(k)$. To recover $G(t)$, we apply the inverse Laplace transform:

$$G(t) = \int_{c-i\infty}^{c+i\infty} \frac{dk}{2\pi i} \exp(kt)\tilde{G}(k). \tag{4.35}$$

Laplace's method allows us to solve ODEs where the coefficients are linear in t.

Remark. It is important to note that Laplace's method for differential equations should not be confused with the Laplace transform or Laplace's method for approximating integrals. While related, they are distinct concepts.

Consider an ODE of the form

$$\sum_{m=0}^{N} (a_m + b_m t) \frac{d^m y}{dt^m} = 0. \tag{4.36}$$

We seek solutions in the form of an integral:

$$y(t) = \int_C dk\, Z(k) e^{kt}, \tag{4.37}$$

where $Z(k)$ is a function of the complex variable k and C is a contour in the complex plane of k that depends on $Z(k)$ but is independent of t.

Remark. Note that the substitution in Eq. (4.37) is similar to the inverse Laplace transform in Eq. (4.35), but the contour C does not necessarily coincide with the one used in the inverse Laplace transform, which is typically placed along the right side of the imaginary axis.

Taking derivatives of $y(t)$ from Eq. (4.37), we obtain

$$\frac{d^m y}{dt^m} = \int_C dk\, Z(k) k^m e^{kt}.$$

Substituting these derivatives into the left-hand side of Eq. (4.36) results in

$$\int_C dk\, Z(k) \\ \times \left(\underbrace{(a_0 + a_1 k + \cdots + a_n k^n)}_{P(k)} + \underbrace{(b_0 + b_1 k + \cdots + b_n k^n)t}_{Q(k)} \right) e^{kt} = 0.$$

We introduce the notation $P(k)$ and $Q(k)$ for convenience. We then integrate by parts:

$$\begin{aligned} 0 &= \int_C dk\, Z(k)\Big(P(k) + Q(k)t\Big)e^{kt} \\ &= Z(k)Q(k)e^{kt}\Big|_{k_1}^{k_2} + \int_C dk\, \left(Z(k)P(k) - \frac{d}{dk}\Big(Z(k)Q(k)\Big)e^{kt}\right), \end{aligned} \tag{4.38}$$

where k_1 and k_2 are the end points of the contour C. If we can choose $Z(k)$ and a contour C such that Eq. (4.38) holds, then we can use them to express the solution to Eq. (4.36) as in Eq. (4.37). In summary, we must find $Z(k)$ and a contour C such that

$$\frac{d}{dk}\Big(Q(k)Z(k)\Big) - P(k)Z(k) = 0 \quad \text{and} \quad Z(k)Q(k)e^{kt}\Big|_{k_1}^{k_2} = 0.$$

The equation for $Z(k)$ can be solved either by finding an integrating factor or by separation of variables:

$$d\,(QZ) = PZdk \Rightarrow \frac{d\,(QZ)}{QZ} = \frac{P}{Q}dk \Rightarrow \ln(QZ) = \int \frac{P}{Q}dk + \text{const}.$$

Thus, $Z(k)$ is given by

$$Z(k) = \frac{c}{Q(k)} \exp\left(\int dk\, \frac{P(k)}{Q(k)}\right).$$

Once $Z(k)$ is determined, we must find a contour with endpoints k_1 and k_2 such that $Q(k_1)Z(k_1)e^{k_1 t} = Q(k_2)Z(k_2)e^{k_2 t}$.

Example 4.3.2. Use Laplace's method to find the general solution to the boundary value problem

$$x\frac{d^3}{dx^3}u + 2u = 0, \quad u(0) = 1, \quad u(\infty) = 0.$$

Solution. For this problem, we compute

$$\begin{aligned} &P(k) = 2, \quad Q(k) = k^3, \quad Z(k) = -\frac{c\, e^{-1/k^2}}{k^3}, \\ &Q(k)Z(k)e^{kt} = e^{kx-1/k^2}. \end{aligned}$$

We observe that $e^{kx-1/k^2} \to 0$ as $k \to -\infty$ and $e^{kx-1/k^2} = 0$ as $k \to 0$. Therefore, we take the contour of integration along the negative real axis. The solution is

$$u(x) = -2c \int_{-\infty}^{0} \frac{e^{kx-1/k^2}}{k^3} dk,$$

which can be expressed as

$$u(x) = \int_{0}^{\infty} e^{-x/\sqrt{z}-z} dz$$

via the change of variables $z = t^{-2}$. The constant $c = 1/2$ is chosen to satisfy the boundary condition $u(0) = 1$.

Exercise 4.5. Consider the sum

$$S(x) = \sum_{n=0}^{\infty} \frac{x^n}{(n!)^2}.$$

Find a second-order linear differential equation that, when supplied with the proper initial conditions at $x = 0$, yields $S(x)$ as a solution. Solve the initial value problem using Laplace's method, representing $S(x)$ as an integral.

Example 4.3.3. (a) Use Laplace's method to find the general solution to the Hermite equation

$$\frac{d^2y}{dt^2} - 2t\frac{dy}{dt} + 2ny = 0. \tag{4.39}$$

(b) Simplify your result for the case where n is a non-negative integer.

Solution. (a) Using Laplace's method, we derive the following expressions:

$$P(k) = k^2 + 2n, \quad Q(k) = -2k, \quad Z(k) = -\frac{c\,e^{-k^2/4}}{2k^{n+1}},$$

$$Q(k)Z(k)e^{kt} = \frac{e^{kt-k^2/4}}{k^n}. \tag{4.40}$$

This leads to the explicit solution of Eq. (4.39), up to a multiplicative constant:

$$y(t) = \int_C dk\, \frac{e^{kt-k^2/4}}{k^{n+1}}.$$

Next, we make a change of variables $k \to z$ using $k = 2(t - z)$, which gives

$$y(t) = e^{t^2} \int_{C'} \frac{e^{-z^2} dz}{(z-t)^{n+1}}, \tag{4.41}$$

where C' is a suitable contour in the complex z-plane, which is not yet specified. We have the freedom to choose the contour (just as we did for contour C in the complex k-space).

(b) When n is a non-negative integer, the integrand in Eq. (4.41) develops a simple pole. Choosing the contour to enclose the pole satisfies the boundary conditions and allows us to evaluate the integral using residue calculus. Applying Cauchy's residue theorem, we arrive at the Hermite polynomials:

$$y(t) = H_n(t) = (-1)^n e^{t^2} \frac{d^n}{dt^n} e^{-t^2}, \tag{4.42}$$

where the scaling (a degree of freedom in linear differential equations) is chosen according to a normalization constraint, as introduced in the following exercise. □

Hermite polynomials will reappear in the context of the Sturm–Liouville problem in Section 4.5.5.

Example 4.3.4. Consider another case of Eq. (4.36) that can be solved using Laplace's method:

$$\frac{d^2}{dt^2} y - ty = 0.$$

Solution. Following the general Laplace method, we find

$$P(k) = k^2, \quad Q(k) = -1, \quad Z(k) = -\exp(-k^2/3). \tag{4.43}$$

According to Eq. (4.37), the general solution of Eq. (4.43) can be written as

$$y(t) = \text{const} \int_C dk \exp\left(kt - \frac{k^3}{3}\right), \tag{4.44}$$

where we choose an infinite integration path C, as shown in Fig. (4.1), such that the values of the integrand at the infinite end points coincide (and are zero). This ensures that the end points of the contour lie

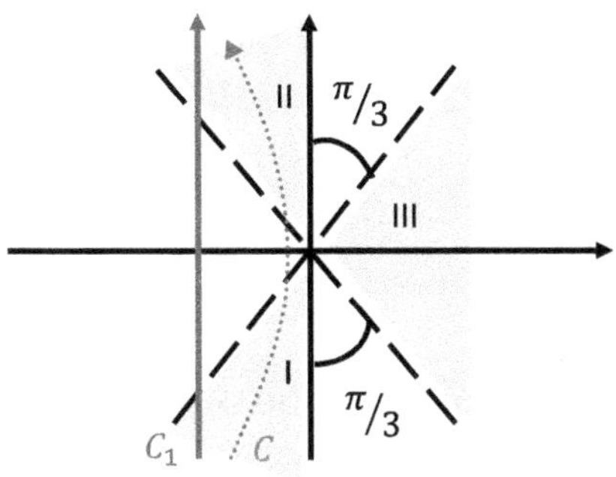

Figure 4.1. Layout of contours in the complex plane of k needed for saddle-point estimations of the Airy function described in Eq. (4.44).

in regions where $\text{Re}(k^2) > 0$ (shaded regions I, II and III in Fig. 4.1). By selecting a contour that starts in region I and ends in region II (shown as the dotted contour C in Fig. 4.1), we ensure that the solution remains finite as $t \to +\infty$.

Note that the contour can be shifted arbitrarily, provided that the end points remain in sectors I and II. In particular, we can shift the contour to coincide with the imaginary axis in the complex k plane (shown in Fig. 4.1). Then, Eq. (4.44) becomes (up to a constant) the Airy function:

$$Ai(t) = \frac{1}{\pi}\int_0^\infty dz \cos\left(\frac{z^3}{3} + zt\right) = \frac{1}{2\pi}\text{Re}\left(\int_{-\infty}^\infty dz \exp\left(i\frac{z^3}{3} + itz\right)\right). \tag{4.45}$$

The asymptotic expression for the Airy function for $t > 0$, $t \gg 1$, can be derived using the saddle-point method (Section 2.4). At $k = \pm\sqrt{t}$, the integrand in Eq. (4.44) has an extremum along the path of steepest descent, which follows the imaginary axis. Since the contour endpoints must stay in sectors I and II, we shift the contour slightly left from the imaginary axis while keeping it parallel to it. (See the solid contour C_1 in Fig. 4.1, which crosses the real axis at $k = -\sqrt{t}$.)

The integral is dominated by the saddle point at $k = -\sqrt{t}$. After substituting $k = \sqrt{t} + iz$, changing the integration variable to z, expanding the integrand near the saddle point, and keeping the quadratic term in z, we evaluate a Gaussian integral to obtain the asymptotic estimate for the Airy function:

$$t > 0,\ t \gg 1: \quad Ai(t) \approx \frac{1}{2\pi}\int_{-\infty}^{\infty} \exp\left(-\frac{2}{3}t^{3/2} - \sqrt{t}z^2\right) dz$$
$$= \frac{\exp\left(-\frac{2}{3}t^{3/2}\right)}{t^{1/4}\sqrt{4\pi}}.$$

Note that the contribution from the second saddle point at $k = \sqrt{t}$ can be disregarded by observing that the Gaussian integral along the steepest descent from this saddle point gives zero real contribution, as required by Eq. (4.45). □

4.4 Linear Static Problems

We now turn to static problems, which often arise when a dynamic system reaches equilibrium. Many natural and engineered systems exhibit spatial characteristics that are both non-trivial and informative, making them worthy of detailed analysis. In this section, we explore a few linear, one-dimensional spatial problems relevant to various applications.

4.4.1 *One-dimensional Poisson equation*

The Poisson equation is fundamental in fields such as electrostatics, where it describes the potential field generated by a given charge distribution.

Consider a function $u(x)$ that describes a system over a finite spatial interval, satisfying the following set of equations:

$$\frac{d^2}{dx^2}u(x) = f(x), \quad \forall x \in (a, b), \quad \text{with} \quad u(a) = u(b) = 0. \tag{4.46}$$

We introduce the Green function, $G(x; y)$, which satisfies

$$\forall a < x, y < b: \quad \frac{d^2}{dx^2}G(x; y) = \delta(x - y), \quad G(a; y) = G(b; y) = 0. \tag{4.47}$$

Note that the Green function now depends on both variables x and y, where y represents the location of the source and x is the observation point.

Since Eq. (4.47) holds for $x \neq y$, the Green function $G(x;y)$ must be a linear function of x when $x \neq y$. By applying the boundary conditions, we derive

$$\begin{aligned} x > y: &\quad G(x;y) = B(x-b), \\ y > x: &\quad G(x;y) = A(x-a). \end{aligned}$$

Furthermore, because the DE in Eq. (4.47) is second-order, $G(x;y)$ must be continuous at $x = y$, with a jump in its first derivative equal to unity at this point. Combining these conditions, we obtain

$$G(x;y) = \frac{1}{b-a} \begin{cases} (y-b)(x-a), & x < y, \\ (y-a)(x-b), & x > y. \end{cases}$$

The solution to Eq. (4.46) is then given by the convolution of the source term with the Green function:

$$u(x) = \int_a^b dy\, G(x;y) f(y). \tag{4.48}$$

Example 4.4.1. Find the Green function for the equation $\mathcal{L}u(x) = f(x)$, where the operator is $\mathcal{L} = -\frac{d}{dx}\left(x\frac{d}{dx}\right)$ and the boundary conditions are $u(1) = 0$ and $u'(2) = 0$.

Solution. To construct the Green function, we need to find a function that satisfies the necessary properties.

Property (i): The Green function satisfies $\mathcal{L}G(x;y) = 0$ for $x \neq y$. Two linearly independent solutions to the homogeneous equation are $u(x) = \text{const}$ and $u(x) = \log(x)$. Hence, for any y in $1 < y < 2$, we can write

$$G(x;y) = \begin{cases} c_1 + c_2 \log(x), & x < y, \\ c_3 + c_4 \log(x), & x > y. \end{cases}$$

Property (ii): The Green function satisfies the boundary conditions. Applying the boundary conditions $u(1) = 0$ and $u'(2) = 0$ gives $c_1 = c_4 = 0$, so that

$$G(x;y) = \begin{cases} c_2 \log(x), & x < y, \\ c_3, & x > y. \end{cases}$$

Property (iii): $G(x;y)$ is continuous at $x = y$, and its derivative jumps by $-1/y$ at $x = y$. To ensure continuity at $x = y$, we set $c_2 \log(y) = c_3$. For the derivative jump condition, we compute the derivative of $G(x;y)$ as $x \to y^-$ and $x \to y^+$. We find $\lim_{x\to y^-} G_x = c_2/y$ and $\lim_{x\to y^+} G_x = 0$, so we must set $c_2 = 1$ to ensure that the derivative jumps by $-1/y$ at $x = y$

$$: G(x;y) = \begin{cases} \log(x), & x < y, \\ \log(y), & x > y. \end{cases} \qquad \square$$

Remark. *Explanation of Property (iii):* To determine the magnitude of the derivative jump at $x = y$, integrate the differential equation over the interval $[y - \epsilon, y + \epsilon]$ and take the limit $\epsilon \to 0$:

$$-\frac{d}{dx}\left(x\frac{d}{dx}u(x)\right) = \delta(x-y),$$

$$\lim_{\epsilon\to 0}\int_{y-\epsilon}^{y+\epsilon} -\frac{d}{dx}\left(x\frac{d}{dx}u(x)\right)dx = \lim_{\epsilon\to 0}\int_{y-\epsilon}^{y+\epsilon}\delta(x-y)dx,$$

$$\lim_{\epsilon\to 0}\left[-x\frac{d}{dx}u(x)\right]_{y-\epsilon}^{y+\epsilon} = 1,$$

$$\left.\frac{d}{dx}u(x)\right|_{x=y} = -\frac{1}{y}.$$

Exercise 4.6. Find the Green function for the equation $\mathcal{L}u(x) = f(x)$, where the operator is $\mathcal{L} = -\frac{d^2}{dx^2} - \kappa^2$, with the following boundary conditions:

(a) $u(0) = u'(1) = 0$,
(b) $u(x)$ is periodic with period 2π.

4.5 Sturm–Liouville (Spectral) Theory

We now delve into the study of differential operators, which map functions to other functions. In doing so, it is essential to first introduce the Hilbert space, where these functions reside.

4.5.1 *Hilbert space and completeness*

Let us begin by reviewing some fundamental properties of a Hilbert space, especially the condition of completeness. A linear (vector) space is called a Hilbert space, denoted by $\mathcal{H}$, if the following conditions are satisfied:

1. For any two elements f and g, there exists a scalar product (f, g) that satisfies:
   (a) linearity in the second argument,
   $$(f, \alpha g_1 + \beta g_2) = \alpha(f, g_1) + \beta(f, g_2),$$
   for any $f, g_1, g_2 \in \mathcal{H}$ and $\alpha, \beta \in \mathbb{C}$;
   (b) conjugate symmetry (Hermitian symmetry),
   $$(f, g) = (g, f)^*;$$
   (c) non-negativity of the norm, $\|f\|^2 := (f, f) \geq 0$, with $(f, f) = 0$ if and only if $f = 0$.
2. $\mathcal{H}$ has a countable basis B, i.e., a set of elements $B := \{f_n, n = 1, 2, \dots\}$ such that any element $g \in \mathcal{H}$ can be written as a linear combination of the basis functions f_n. That is, for any $g \in \mathcal{H}$, there exist coefficients c_n such that $g = \sum c_n f_n$.

Remark. The Hilbert space defined here for complex-valued functions can also be used for real-valued functions. We will interchangeably use the two cases as required.

Any basis B can be transformed into an orthonormal basis with respect to a given scalar product. This leads to the representation

$$x = \sum_{n=1}^{\infty} (x, f_n) f_n, \quad \|x\|^2 = \sum_{n=1}^{\infty} |(x, f_n)|^2.$$

The Gram–Schmidt process is a standard method for orthonormalization.

One prominent example of a Hilbert space is $L^2(\Omega)$, the space of complex-valued functions $f(x)$ defined on $\Omega \subset \mathbb{R}^n$, for which

$$\int_\Omega dx\, |f(x)|^2 < \infty,$$

which loosely means that the square modulus of the function is integrable. In this space, the scalar product is defined as

$$(f, g) := \int_\Omega dx\, f^*(x) g(x).$$

The properties in points 1a–c are satisfied by construction, and the completeness condition in point 2 can be proved (this is a standard proof in mathematical analysis courses).

Given an orthonormal sequence of functions

$$\{f_n, n = 1, 2, \ldots, \infty\}, \quad (f_n, f_m) = \delta_{nm},$$

this sequence forms a basis in $L^2(\Omega)$ if the following completeness relation holds:

$$\sum_{n=1}^{\infty} f_n^*(x) f_n(y) = \delta(x - y). \tag{4.49}$$

As is typical for the Dirac delta function (and other generalized functions), Eq. (4.49) should be understood in the sense of distributions, meaning that integrals of both sides against a test function from $L^2(\Omega)$ are equal.

4.5.2 *Hermitian and non-Hermitian differential operators*

Consider a function from the Hilbert space $L^2(a, b)$ defined on the real interval $a \le x \le b$ with an integrable square modulus. A linear differential operator $\hat{L}$ acts on this function.

A differential operator is called Hermitian (self-adjoint) if for any two functions (from a suitable class, e.g., $L^2(a, b)$), the following relation holds:

$$(f, \hat{L}g) := \int_a^b dx\, f(x) \hat{L} g(x) = \int_a^b dx\, g(x) \hat{L} f(x) = (g, \hat{L} f). \tag{4.50}$$

It is clear that the Hermiticity condition in Eq. (4.50) depends on both the class of functions and the operator $\hat{L}$. For instance, for functions f and g with zero boundary conditions or periodic boundary conditions, where the derivative is also periodic, the operator

$$\hat{L} = \frac{d^2}{dx^2} + U(x), \tag{4.51}$$

where $U(x)$ is a real-valued potential, is Hermitian.

The natural generalization of the Schrödinger operator in Eq. (4.51) is the Sturm–Liouville operator:

$$\hat{L} = \frac{d^2}{dx^2} + Q(x)\frac{d}{dx} + U(x).$$

This operator is not Hermitian, i.e., Eq. (4.50) does not hold in general. However, under zero boundary conditions or periodic boundary conditions on $f(x)$ and $g(x)$ and their derivatives, the following skew-Hermitian generalization holds:

$$\int_a^b dx\, \rho(x) f(x) \hat{L} g(x) = \int_a^b dx\, \rho(x) g(x) \hat{L} f(x), \tag{4.52}$$

$$\text{where} \quad \frac{d}{dx}\rho(x) = Q(x)\rho(x), \quad \rho(x) = \exp\left(\int dx\, Q(x)\right). \tag{4.53}$$

4.5.3 *Eigenfunctions and orthogonality*

Now, consider the eigenfunctions f_n of the operator $\hat{L}$, satisfying

$$\hat{L} f_n = \lambda_n f_n,$$

where λ_n is the eigenvalue corresponding to the eigenfunction f_n of the Sturm–Liouville operator, indexed by n. We assume that $\lambda_n \neq \lambda_m$ for all $n \neq m$.

The conditions in Eqs. (4.52) and (4.53) imply that

$$\begin{aligned}\int dx\, \rho(x) f_n(x) \hat{L} f_m(x) &= \lambda_m \int dx\, \rho(x) f_n(x) f_m(x) \\ &= \lambda_n \int dx\, \rho(x) f_n(x) f_m(x),\end{aligned}$$

which leads to the orthogonality condition for the eigenfunctions:

$$\int dx\, \rho(x) f_n(x) f_m(x) = 0, \quad \forall n \neq m. \tag{4.54}$$

In the Hermitian case, distinct eigenfunctions are orthogonal with respect to the uniform weight $\rho(x) = 1$.

4.5.4 *Completeness of eigenfunctions*

A set of eigenfunctions $\{f_n(x)\}$ is called complete over a given function class if any function in that class can be expanded as a series over the eigenfunctions:

$$f(x) = \sum_n c_n f_n(x). \tag{4.55}$$

Multiplying both sides of Eq. (4.55) by $\rho(x) f_n(x)$, integrating over the domain and using the orthogonality condition in Eq. (4.54) gives

$$c_n = \frac{\int dx\, \rho(x) f_n(x) f(x)}{\int dx\, \rho(x) f_n^2(x)}. \tag{4.56}$$

For the operator $\hat{L}_0 = \frac{d^2}{dx^2}$, with $Q(x) = U(x) = 0$ and 2π-periodic boundary conditions, the eigenfunctions are $\cos(nx)$ and $\sin(nx)$, with eigenvalues $\lambda_n = -n^2$. In this case, the expansion in Eq. (4.55) corresponds to a Fourier series.

Substituting Eq. (4.56) back into (4.55) leads to

$$f(x) = \int dy\, \rho(y) \sum_n \frac{f_n(x) f_n(y)}{\int dx\, \rho(x) f_n^2(x)} f(y).$$

If the set $\{f_n(x)\}$ is complete, this relation must hold for any function in the class, implying

$$\sum_n \frac{f_n(x) f_n(y)}{\int dx\, \rho(x) f_n^2(x)} = \frac{1}{\rho(y)} \delta(x - y). \tag{4.57}$$

This completeness condition ensures that the eigenfunctions form a complete basis for the function space.

Exercise 4.7. Check the validity of Eq. (4.57) and thus the completeness of the respective set of eigenfunctions for our enabling example of $\hat{L}_0 = d^2/dx^2$ over the functions which are 2π-periodic.

4.5.5 *Hermite polynomials*

Let us now build on the previous example and consider the case where $Q(x) = -2x$ and $U(x) = 0$, i.e.,

$$\hat{L}_2 = \frac{d^2}{dx^2} - 2x\frac{d}{dx}, \quad \rho(x) = \exp\left(-x^2\right),$$

where $\rho(x)$ is the weighting function. We are working with functions mapping from $\mathbb{R}$ to $\mathbb{R}$, which also decay sufficiently fast as $x \to \pm\infty$. This leads us to study the equation

$$\hat{L}_2 f_n = \lambda_n f_n. \tag{4.58}$$

To simplify this, let's change the function $f_n(x)$ to $\Psi_n(x)$ via the transformation $\Psi_n(x) = f_n(x)\sqrt{\rho(x)}$. This yields the following equation for Ψ_n:

$$\begin{aligned} e^{-x^2/2}\hat{L}_2 f_n(x) &= e^{-x^2/2}\hat{L}_2\left(e^{x^2/2}\Psi_n(x)\right) \\ &= \frac{d^2}{dx^2}\Psi_n + (1-x^2)\Psi_n = \lambda_n\Psi_n. \end{aligned} \tag{4.59}$$

We observe that when $\lambda_n = -2n$, this equation coincides with the Hermite equation (4.39).

Now, let's seek a solution to Eq. (4.58) in the form of a Taylor series expansion around $x = 0$:

$$f_n(x) = \sum_{k=0}^{\infty} a_k x^k.$$

Substituting this series into the Hermite equation and matching terms with equal powers of x, we derive a recurrence relation for the expansion coefficients:

$$\forall k = 0, 1, \cdots: \quad a_{k+2} = \frac{2k + \lambda_n}{(k+2)(k+1)} a_k. \tag{4.60}$$

This recurrence relation leads to two linearly independent solutions of Eq. (4.58) – one even and one odd with respect to the transformation $x \to -x$. These solutions are expressed as series:

$$f_n^{(e)}(x) = a_0\left(1 + \frac{\lambda_n}{2!}x^2 + \frac{\lambda_n(4+\lambda_n)}{4!}x^4 + \cdots\right),$$

$$f_n^{(o)}(x) = a_1\left(x + \frac{(2+\lambda_n)}{3!}x^3 + \frac{(2+\lambda_n)(6+\lambda_n)}{5!}x^5 + \cdots\right),$$

where a_0 and a_1 are the first coefficients, which remain as free parameters.

Observe that these series terminate if $\lambda_n = -4n$ for the even solutions and $\lambda_n = -4n - 2$ for the odd solutions, where $n = 0, 1, \ldots$. In such cases, the functions $f_n^{(e)}$ and $f_n^{(o)}$ are polynomials – the *Hermite polynomials.* These can be combined into the standard notation for Hermite polynomials of order n, denoted by $H_n(x)$, which satisfy Eq. (4.59).

According to Exercise 4.8, Hermite polynomials are both normalized and orthogonal with respect to the weighting function $\rho(x) = \exp(-x^2)$.

Exercise 4.8. (a) Prove that

$$\int_{-\infty}^{+\infty} dt\, e^{-t^2} H_n(t) H_m(t) = 2^n n! \sqrt{\pi} \delta_{nm}, \tag{4.61}$$

where δ_{nm} is the Kronecker delta.

(b) Verify that the set of functions

$$\left\{ \Psi_n(x) = \frac{1}{\pi^{1/4}\sqrt{2^n n!}} \exp(-x^2/2) H_n(x) \mid n = 0, 1, \ldots \right\} \tag{4.62}$$

satisfies the completeness relation:

$$\sum_{n=0}^{\infty} \Psi_n(x)\Psi_n(y) = \delta(x - y).$$

Hint: The following identity may be useful:

$$\frac{d^n}{dx^n} \exp(-x^2) = \sqrt{\pi} \int_{-\infty}^{+\infty} \frac{dq}{2\pi} (iq)^n \exp\left(-q^2/4 + iqx\right).$$

A corollary of Exercise 4.8 is the statement of completeness: The set of functions in Eq. (4.62) forms an orthogonal basis in the Hilbert space of square-integrable functions $f(x) \in L^2(\mathbb{R})$, i.e., satisfying $\int_{-\infty}^{\infty} |f(x)|^2 dx < \infty$. More formally, an orthogonal basis for L^2 is a complete set, meaning that the only function orthogonal to all functions in the set is the zero function.

It is also worth noting that the final form of Eq. (4.59) is the spectral version of the imaginary-time Schrödinger equation in a quadratic potential:

$$\partial_x^2 \Psi(t;x) + (1 - x^2)\Psi(t;x) = -\partial_t \Psi(t;x),$$

which will be discussed in the following section.

4.5.6 *Case study: Schrödinger equation in 1D**

The Schrödinger equation* is given by

$$\frac{d^2\Psi(x)}{dx^2} + (E - U(x))\Psi(x) = 0, \tag{4.63}$$

where $\Psi(x)$ is the complex-valued wave function describing the spatial distribution of a quantum particle in one dimension ($x \in \mathbb{R}$), with energy E in a potential $U(x)$. We seek solutions where $|\Psi(x)| \to 0$ as $x \to \infty$. Our goal is to describe the spectrum (allowed values of E) and the respective eigenfunctions.

As a simple but instructive example, consider a quantum particle in a rectangular potential, where $U(x) = U_0$ for $x \notin [0, a]$ and zero elsewhere. The general solution of Eq. (4.63) becomes

$\underline{U_0 > E > 0:}$

$$\Psi_E(x) = \begin{cases} c_L \exp(x\sqrt{U_0 - E}), & x < 0, \\ a_+ \exp(ix\sqrt{E}) + a_- \exp(-ix\sqrt{E}), & x \in [0, a], \\ c_R \exp(-x\sqrt{U_0 - E}), & x > a, \end{cases} \tag{4.64}$$

$\underline{U_0 < E:}$

$$\Psi_E(x) = \begin{cases} c_{L+} \exp(ix\sqrt{E - U_0}) + c_{L-} \exp(-ix\sqrt{E - U_0}), & x < 0, \\ a_+ \exp(ix\sqrt{E}) + a_- \exp(-ix\sqrt{E}), & x \in [0, a], \\ c_{R+} \exp(ix\sqrt{E - U_0}) + c_{R-} \exp(-ix\sqrt{E - U_0}), & x > a. \end{cases} \tag{4.65}$$

We impose the condition that $E \geq 0$, as negative values of E are not allowed by the ODE. In the regime $U_0 > E > 0$, we choose the solution that decays as $x \to \pm\infty$.

*This auxiliary subsection is optional for a first reading and will not be included in midterm or final exams.

The solutions for each interval must be "glued" together, i.e., both $\Psi(x)$ and $d\Psi(x)/dx$ must be continuous across the boundaries. These continuity conditions, when applied to Eq. (4.64) or (4.65), yield algebraic consistency conditions for E. We expect a continuous spectrum for $E > U_0$ and a discrete spectrum for $U_0 > E > 0$.

Exercise 4.9. Complete the calculations for the case of $U_0 > E > 0$ and determine the allowed values of the discrete spectrum. What is the condition for the appearance of at least one discrete level?

Example 4.5.1. Find the eigenfunctions and energy levels of stationary states for the Schrödinger equation describing a harmonic oscillator:

$$\frac{d^2\Psi(x)}{dx^2} + (E - x^2)\Psi(x) = 0, \tag{4.66}$$

where $x \in \mathbb{R}$ and $\Psi : \mathbb{R} \to \mathbb{C}^2$.

Solution. As seen in the previous section, analyzing Eq. (4.66) reduces to solving the Hermite equation, as discussed in the context of Eq. (4.58). However, we follow a different approach here by introducing the so-called "creation" and "annihilation" operators,

$$\hat{a} = \frac{i}{\sqrt{2}}\left(\frac{d}{dx} + x\right), \quad \hat{a}^\dagger = \frac{i}{\sqrt{2}}\left(\frac{d}{dx} - x\right),$$

and rewriting the Schrödinger equation (4.66) as

$$\hat{H}\Psi(x) = \hat{a}^\dagger\hat{a}\Psi(x) = \left(2E - \frac{1}{2}\right)\Psi(x).$$

It is straightforward to verify that the operator $\hat{H}$ is positive definite for all functions in L^2:

$$\int dx\, \Psi^\dagger(x)\hat{H}\Psi(x) = \int dx\, \Psi^\dagger(x)\hat{a}\hat{a}^\dagger\Psi(x) = \int dx\, |\hat{a}\Psi(x)|^2 \geq 0,$$

with equality holding only if

$$\hat{a}\Psi_0(x) = \frac{i}{\sqrt{2}}\left(\frac{d}{dx} + x\right)\Psi_0(x) = 0,$$

leading to $\Psi_0(x) = A\exp(-x^2/2)$ and $E_0 = 1/4$. This gives us the ground state eigenfunction and its corresponding lowest energy level.

To find the excited states, we examine the commutation relations:

$$\begin{aligned}
\hat{a}\hat{a}^\dagger\Psi(x) &= \hat{a}^\dagger\hat{a}\Psi(x) + \Psi(x), \\
\hat{a}^\dagger\hat{a}\left(\hat{a}^\dagger\right)^n\Psi(x) &= \left(\hat{a}^\dagger\right)^2\hat{a}\left(\hat{a}^\dagger\right)^{n-1}\Psi(x) + \left(\hat{a}^\dagger\right)^n\Psi(x) \\
&= n\left(\hat{a}^\dagger\right)^n\Psi(x) + \left(\hat{a}^\dagger\right)^{n+1}\hat{a}\Psi(x). \qquad (4.67)
\end{aligned}$$

Define $\Psi_n(x) := (\hat{a}^\dagger)^n\Psi_0(x)$. Since $\hat{a}\Psi_0(x) = 0$, the commutation relations in Eq. (4.67) show that

$$\left(2E_n - \frac{1}{2}\right)\Psi_n(x) = \hat{H}\Psi_n(x) = \hat{a}^\dagger\hat{a}\Psi_n(x) = n\Psi_n(x).$$

We find that the eigenfunctions $\Psi_n(x)$ correspond to energy levels $2E_n = n + 1/2$, and these eigenfunctions are expressed in terms of the Hermite polynomials $H_n(x)$, as introduced in Eq. (4.42):

$$\begin{aligned}
\Psi_n(x) &= A_n\left(\frac{i}{\sqrt{2}}\left(\frac{d}{dx} - x\right)\right)^n \exp\left(-\frac{x^2}{2}\right) \\
&= A_n\frac{i^n}{2^{n/2}}\exp\left(\frac{x^2}{2}\right)\frac{d^n}{dx^n}\exp(-x^2),
\end{aligned}$$

where we have used the identity $(\frac{d}{dx} - x)\exp(x^2/2) = \exp(x^2/2)\frac{d}{dx}$.

From the orthogonality condition of Hermite polynomials in Eq. (4.61), we derive $A_n = (n!\sqrt{\pi})^{-1/2}$.

4.6 Phase Space Dynamics for Conservative and Perturbed Systems

4.6.1 *Integrals of motion*

Consider a system that describes the conservative dynamics of a particle of unit mass in a potential. A conservative system implies no dissipation of energy, and its dynamics can be expressed as

$$\dot{x} = v, \quad \dot{v} = -\frac{\partial U(x)}{\partial x}, \qquad (4.68)$$

where x is the position and v is the velocity of the particle. The energy of this particle is given by

$$E = \frac{v^2}{2} + U(x), \tag{4.69}$$

where the first term represents kinetic energy and the second term represents potential energy. It is straightforward to verify that the energy is conserved, i.e., $dE/dt = 0$. Therefore, we can express the velocity as

$$\dot{x} = \pm\sqrt{2(E - U(x))},$$

where the sign depends on the initial condition for $\dot{x}(0)$. This equation is separable and can be integrated to yield the particle's position as a function of time:

$$\int_{x_0}^{x} \frac{dx'}{\sqrt{2(E - U(x'))}} = \pm t,$$

which depends on the initial position x_0 and the conserved energy E.

In this example, the total energy E is an *integral of motion*, or a *first integral*, which is defined as a quantity that remains constant along solutions to the system of differential equations. Here, E is conserved along the trajectory $x(t)$ of the particle.

The concept of integrals of motion extends to more general conservative systems described by ODEs. (In this section and the next, we follow Refs. [4, 11].) For instance, consider a Hamiltonian system, where the Hamiltonian H is a smooth function of $2n$ variables – momenta $p_1, \ldots, p_n$ and coordinates $q_1, \ldots, q_n$. The corresponding system of equations, known as Hamilton's canonical equations, is given by

$$\dot{p}_i = -\frac{\partial H}{\partial q_i}, \quad \dot{q}_i = \frac{\partial H}{\partial p_i} \quad (\forall i = 1, \ldots, n).$$

The time derivative of the Hamiltonian is

$$\frac{dH}{dt} = \sum_{i=1}^{n} \left(\frac{\partial H}{\partial p_i}\dot{p}_i + \frac{\partial H}{\partial q_i}\dot{q}_i \right) = \sum_{i=1}^{n} (-\dot{q}_i\dot{p}_i + \dot{p}_i\dot{q}_i) = 0,$$

showing that H is conserved and thus serves as an integral of motion.

The system in Eq. (4.68), which describes a particle of unit mass in a potential, is an example of a Hamiltonian system with a single degree of freedom. In this case, the energy (4.69), as a function of x and v, is the Hamiltonian, with x and v corresponding to q and p, respectively. We further explore this system in the following section.

For readers interested in a comprehensive mathematical introduction to Hamiltonian dynamics, we refer to Ref. [11].

4.6.2 *Phase portrait*

We now discuss phase portraits for conservative Hamiltonian systems, focusing on the example from Eq. (4.68). As we have established, the energy (or Hamiltonian) is conserved, making it insightful to study the energy level curves (isolines) in the two-dimensional (x, v) space, defined by the equation

$$\frac{v^2}{2} + U(x) = E.$$

To draw these level curves, we fix E and track how (x, v) evolves over time.

To build intuition, consider an analogy with a bead sliding frictionlessly along a rigid wire shaped like the potential curve. Starting the bead at any position with some initial velocity, its trajectory in (x, v) space traces a level curve of the energy.

Consider the potential $U(x) = \frac{k}{2}x^2$. For $k > 0$, the phase portrait is illustrated in Fig. 4.2 (top). The curves, except at the equilibrium point $(x, v) = (0, 0)$, are smooth ellipses. For $k < 0$, as shown in Fig. 4.2 (bottom), the phase space splits into quadrants, and the level curves are hyperbolas centered at $(0, 0)$.

In the $k > 0$ case, the energy is non-negative, $E \geq 0$. The particle has two turning points, $x_\pm = \pm\sqrt{2E/k}$, where its velocity changes direction. The motion is periodic, and the period T can be calculated as

$$T = \int_{x_-}^{x_+} \frac{dx}{\sqrt{2(E - \frac{kx^2}{2})}} = 2\pi.$$

Notably, the period is independent of the energy.

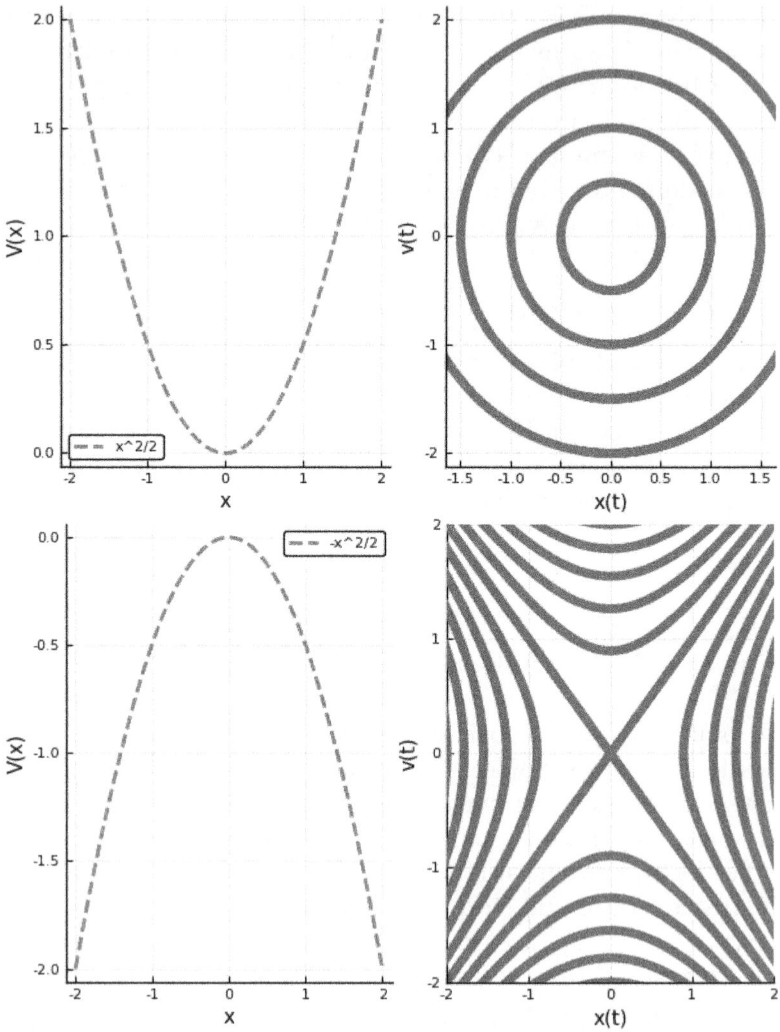

Figure 4.2. Phase portraits: (x, v) level curves for the conservative system (4.68) with potential $U(x) = \frac{k}{2}x^2$ for $k > 0$ (top) and $k < 0$ (bottom).

For $k < 0$, all values of E are accessible, and the critical point remains at $(x, v) = (0, 0)$. For positive E, the particle's velocity increases without bound as $x(t)$ grows unbounded over time.

By sketching the level curves for various potentials $U(x)$, we can qualitatively study the dynamics of more complex systems.

Figure 4.3. Illustration of energy level curves (phase portraits) for various potentials.

Exercise 4.10. Sketch level curves of the energy for the Kepler potential, $U(x) := -\frac{1}{x}+\frac{C}{x^2}$, and for the potentials shown in Fig. (4.3).

4.6.3 *Small perturbation of a conservative system*

Following Section 12.10 in Ref. [11], let us analyze a simple but illustrative example of a system that deviates slightly from a quadratic potential with $k = 1$:

$$\dot{x} = v + \varepsilon f(x, v), \quad \dot{v} = -x + \varepsilon g(x, v), \tag{4.70}$$

where $\varepsilon \ll 1$ and $x^2 + v^2 \leq R^2$.

For $\varepsilon = 0$ and with the initial condition $x(0) = x_0$, we find

$$x(t) = x_0 \cos(t), \quad v(t) = -x_0 \sin(t).$$

The energy $E = (x^2+v^2)/2$ is conserved in this unperturbed system, resulting in periodic motion with period $T = 2\pi$.

In the general case where $0 < \varepsilon \ll 1$, the system is no longer conservative. To examine how the energy changes over time, we compute

$$\frac{dE}{dt} = x\dot{x} + v\dot{v} = \varepsilon\,(xf + vg) = \varepsilon\left(x^{(0)}f + v^{(0)}g\right) + O(\varepsilon^2),$$

where $x^{(0)}$ and $v^{(0)}$ are the solutions of the unperturbed system ($\varepsilon = 0$). Integrating this over one period, we obtain the energy gain (or loss) over a cycle:

$$\Delta E = \varepsilon \int_0^{2\pi} dt\,\left(x^{(0)}f + v^{(0)}g\right) + O(\varepsilon^2) = \varepsilon \oint (-f\,dv + g\,dx) + O(\varepsilon^2), \tag{4.71}$$

where the contour integral is taken over the level curve (an iso-energy cycle) of the unperturbed system in the (x, v)-plane. Clearly, ΔE depends on x_0.

For $\Delta E > 0$, the energy increases over time, resulting in an outward spiral in the (x, v)-plane. Conversely, for $\Delta E < 0$, the energy decreases, causing the trajectory to spiral inward toward a stationary point.

In some systems, the sign of ΔE depends on x_0. Consider, for instance, the van der Pol oscillator:

$$\ddot{x} = -x + \varepsilon\dot{x}(1 - x^2).$$

As in Eq. (4.71), we integrate $\frac{dE}{dt}$ over a period, yielding

$$\begin{aligned}
\Delta E &= \varepsilon \int_0^{2\pi} \dot{x}^2(1 - x^2)\, dt + O(\varepsilon^2) \\
&= \varepsilon x_0^2 \int_0^{2\pi} \sin^2 t\, (1 - x_0^2 \cos^2 t)\, dt + O(\varepsilon^2) \\
&= \pi\varepsilon \left(x_0^2 - \frac{x_0^4}{4} \right) + O(\varepsilon^2).
\end{aligned}$$

The $O(\varepsilon)$ term vanishes when $x_0 = 2$, is positive for $x_0 < 2$ and negative for $x_0 > 2$. Thus, if we start with $x_0 < 2$, the system will gain energy, and the maximum amplitude $x(t)$ within a period will approach 2. Conversely, if $x_0 > 2$, the system will lose energy, with the amplitude decreasing until it stabilizes around 2. This behavior, based on the $O(\varepsilon)$ term in ΔE, characterizes a stable limit cycle. For a stable limit cycle, we have

$$\Delta E(x_0) = 0 \quad \text{and} \quad \frac{d}{dx_0}\Delta E(x_0) < 0.$$

In summary, the van der Pol oscillator exhibits singular perturbation behavior, meaning that it behaves fundamentally differently from the unperturbed system. In the unperturbed case, the particle oscillates along orbits determined by initial conditions. In the perturbed case, however, the system's trajectory converges to a stable limit cycle, independent of initial conditions.

Exercise 4.11. Recall the properties of stable and unstable limit cycles:

$$\text{A limit cycle is stable at } x = x_0 \text{ if } \Delta E(x_0) = 0$$
$$\text{and } \frac{d}{dx_0}\Delta E(x_0) < 0.$$
$$\text{A limit cycle is unstable at } x = x_0 \text{ if } \Delta E(x_0) = 0$$
$$\text{and } \frac{d}{dx_0}\Delta E(x_0) > 0.$$

Suggest perturbations f and g in Eq. (4.70) that lead to (a) an unstable limit cycle at $x_0 = 2$ and (b) one stable limit cycle at $x_0 = 2$ and one unstable limit cycle at $x_0 = 4$. Illustrate your proposed perturbations with a computational example.

Consider another example of an ODE:

$$\dot{I} = \varepsilon\left(a + b\cos\left(\frac{\theta}{\omega}\right)\right), \quad \dot{\theta} = \omega, \tag{4.72}$$

where ω, ε, a and b are constants and the ε term in the first equation represents a perturbation. When $\varepsilon = 0$, I becomes an integral of motion, meaning it is constant along the solutions of the ODE. In this unperturbed case, θ increases linearly with time at a rate determined by the frequency ω. The unperturbed system here is equivalent to the one described by Eq. (4.70).

Example 4.6.1.

(a) Show that the unperturbed ($\varepsilon = 0$) version of the system described by Eq. (4.70) can be transformed into the unperturbed version of the system in Eq. (4.72) using the following change of variables (canonical transformation):

$$v = \sqrt{\frac{I}{2}}\cos\left(\frac{\theta}{\omega}\right), \quad x = \sqrt{\frac{I}{2}}\sin\left(\frac{\theta}{\omega}\right).$$

(b) Rewrite Eq. (4.72) in terms of the (x, v) variables.

The transformation described in Example 4.6.1 is a canonical transformation that preserves the Hamiltonian structure of the system. In this case, the Hamiltonian, which is typically a function of the angle θ and the action I, depends only on I. Therefore, we can express the Hamiltonian as $H = \omega I$ and rewrite the unperturbed version of Eq. (4.72) as

$$\dot{\theta} = \frac{\partial H}{\partial I} = \omega, \quad \dot{I} = -\frac{\partial H}{\partial \theta} = 0,$$

thus interpreting θ as the generalized coordinate and I as the generalized momentum.

To analyze the perturbed case ($0 < \varepsilon \ll 1$), we average the perturbed equation (4.72) over one period of the angle θ ($2\pi\omega$), as described in Section 4.6.3. The result is the following expression for the change in I over one period:

$$\Delta I = 2\pi\varepsilon a.$$

For many cycles (i.e., after $2\pi n\omega$ periods), where n is a large integer, we can approximate time t as $2\pi n$ and replace ΔI by J, the action averaged over time t. This leads to the differential equation

$$\dot{J} = \varepsilon a, \tag{4.73}$$

whose solution is

$$J(t) = J_0 + \varepsilon a t.$$

In fact, the system described by Eq. (4.72) can also be solved exactly, yielding

$$I(t) = \varepsilon a t + \varepsilon b \sin t.$$

By comparing this exact solution with the averaged solution from Eq. (4.73), we find that the difference between $J(t)$ and $I(t)$ remains bounded over time:

$$\omega \neq 0: \quad |J(t) - I(t)| \leq O(1)\varepsilon,$$

confirming that the averaged solution remains a good approximation of the exact solution for small ε.

In the general n-dimensional case, we consider the unperturbed system of differential equations:

$$\dot{\boldsymbol{I}} = 0, \quad \dot{\boldsymbol{\theta}} = \boldsymbol{\omega}(\boldsymbol{I}), \quad \boldsymbol{I} = (I_1, \ldots, I_n), \quad \boldsymbol{\theta} = (\theta_1, \ldots, \theta_n), \tag{4.74}$$

where each component I_i is an integral of motion. The perturbed version of Eq. (4.74) takes the form

$$\dot{\boldsymbol{I}} = \varepsilon \boldsymbol{g}(\boldsymbol{I}, \boldsymbol{\theta}, \varepsilon), \quad \dot{\boldsymbol{\theta}} = \boldsymbol{\omega}(\boldsymbol{I}) + \varepsilon \boldsymbol{f}(\boldsymbol{I}, \boldsymbol{\theta}, \varepsilon), \tag{4.75}$$

where $\boldsymbol{f}$ and $\boldsymbol{g}$ are $2\pi\omega$-periodic in each component of $\boldsymbol{\theta}$. Since $\boldsymbol{I}$ changes slowly (due to the small parameter ε), the system can be approximated by a simpler averaged system for the slow (adiabatic) variables $\boldsymbol{J}(t) = \boldsymbol{I}(t) + O(\varepsilon)$:

$$\dot{\boldsymbol{J}} = \varepsilon \boldsymbol{G}(\boldsymbol{J}), \quad \boldsymbol{G}(\boldsymbol{J}) = \frac{\oint \boldsymbol{g}(\boldsymbol{I}, \boldsymbol{\theta}, 0) d\boldsymbol{\theta}}{\oint d\boldsymbol{\theta}},$$

where $\oint$ denotes averaging over one period of the motion in phase space. However, this averaging procedure may fail in higher dimensions ($n > 1$) if there are resonances, i.e., if $\sum_i N_i \omega_i = 0$ for some integers N_i.

If the system is Hamiltonian, where $\boldsymbol{\theta}$ represents generalized coordinates and $\boldsymbol{I}$ represents generalized momenta, the equations in (4.75) simplify to

$$\dot{\boldsymbol{I}} = -\frac{\partial H}{\partial \boldsymbol{\theta}}, \quad \dot{\boldsymbol{\theta}} = \frac{\partial H}{\partial \boldsymbol{I}}.$$

Averaging over $\boldsymbol{\theta}$ in the first equation results in $\dot{\boldsymbol{J}} = 0$, meaning that the slow variables, $\boldsymbol{J}$ (also called adiabatic invariants), remain constant over time. A key challenge in applying this method lies in identifying the appropriate variables that remain conserved in the unperturbed system.

Chapter 5

Partial Differential Equations

A partial differential equation (PDE) is a differential equation that involves one or more unknown multi-variable functions and their partial derivatives. PDEs are essential tools for modeling a wide array of phenomena in both natural and engineered systems, from fluid dynamics to electromagnetic fields and heat conduction. In many cases, PDEs allow for the formulation of physical laws and constraints, describing how quantities evolve over space and time. Their significance lies in their ability to model both static and dynamic behaviors in various fields of science and engineering.

We begin by introducing first-order PDEs, showing how they can be solved using the *method of characteristics*. This powerful method allows us to reduce first-order PDEs to a system of ordinary differential equations (ODEs), providing a direct way to understand the behavior of solutions and offering important intuition that extends to more complex, higher-order problems.

In Section 5.2, we delve into the classification of second-order linear PDEs, distinguishing them as hyperbolic, elliptic, or parabolic. These classifications are critical because they provide insights into the nature of the solutions and the appropriate methods for solving different types of PDEs. Hyperbolic equations typically describe wave propagation, elliptic equations correspond to steady-state phenomena, and parabolic equations are associated with diffusion-like processes. Understanding this classification is fundamental to effectively tackling PDE problems in real-world applications.

Elliptic PDEs, which are commonly linked to static or equilibrium states, are discussed in Section 5.3. We explore various methods for solving elliptic PDEs, including Green's functions, which are instrumental in handling boundary value problems. Examples such as electrostatics, potential flow, and elasticity are used to illustrate the application of these techniques to physical systems.

In Section 5.4, we shift our attention to hyperbolic PDEs, which describe wave-like phenomena. Starting with an intuitive understanding of wave propagation, we examine specific applications such as sound waves and electromagnetic waves. Hyperbolic equations are pivotal in modeling dynamic systems in fields such as acoustics, electromagnetism and fluid mechanics.

Next, in Section 5.5, we focus on parabolic PDEs, using the diffusion or heat equation as a primary example. Parabolic equations describe processes that change over time, such as heat conduction, diffusion of particles and pricing options in financial markets. We extend the analysis of the heat equation to higher dimensions, exploring both the physical and mathematical implications of these equations.

Throughout the chapter, we address boundary value problems, especially in relation to elliptic and parabolic equations. The Fourier method is highlighted as a key tool for solving PDEs with periodic boundary conditions, and Green's functions are introduced as a method to solve linear PDEs, particularly in relation to elliptic boundary value problems.

For readers seeking a deeper and more comprehensive exploration of PDEs, we recommend several key resources. *A Course of Higher Mathematics: Volume IV, Part II: Partial Differential Equations* by Smirnov [12] offers a classical and thorough presentation of PDE theory, especially useful for readers interested in a rigorous mathematical foundation. For a modern perspective, *Introduction to Partial Differential Equations* by Folland [13] provides clear explanations with a focus on functional analysis techniques. *The Boundary Value Problems of Mathematical Physics* by Ladyzhenskaya [14] is highly recommended for its detailed treatment of elliptic PDEs and boundary value problems. For an in-depth discussion on Green's functions and their applications to linear PDEs, we suggest *Green's Functions and Boundary Value Problems* by Stakgold and Holst [15], which offers an extensive treatment of both theoretical and practical aspects of the method. Lastly, *Methods of Mathematical Physics, Vol. 2* by Courant and Hilbert [16] remains an authoritative text, covering a wide range of topics with an emphasis on mathematical physics.

5.1 First-Order PDEs: Method of Characteristics

The method of characteristics provides a way to reduce PDEs to a set of ODEs. This method is particularly useful for solving first-order PDEs, meaning those involving only first-order derivatives, and it applies mainly to equations that are linear in these first-order derivatives.

To introduce this technique, consider a function u of two independent variables (x, y), denoted $u(x, y)$. Suppose that $u(x, y)$ satisfies the following first-order PDE:

$$a(x, y, u)\frac{\partial u}{\partial x} + b(x, y, u)\frac{\partial u}{\partial y} = c(x, y, u), \tag{5.1}$$

where a, b, and c are given functions.

Now, consider an arbitrary differentiable curve $(x(t), y(t))$ and compute the total derivative of u along this curve using the chain rule:

$$\frac{du}{dt} = \frac{dx}{dt}\frac{\partial u}{\partial x} + \frac{dy}{dt}\frac{\partial u}{\partial y}. \tag{5.2}$$

Since the curve $(x(t), y(t))$ was chosen arbitrarily, we can define it in such a way that

$$\frac{dx}{dt} = a(x, y, u), \quad \frac{dy}{dt} = b(x, y, u), \quad \frac{du}{dt} = c(x, y, u). \tag{5.3}$$

Substituting these equations into (5.2) shows that $\frac{du}{dt} = c(x, y, u)$, which is equivalent to the original PDE (5.1). Thus, we have a family of *characteristic curves* that can be used to construct the solution to (5.1). This is a family of curves since we have not specified initial conditions for the system (5.3).

Now, let $u(\boldsymbol{x}) : \mathbb{R}^d \to \mathbb{R}$ be a function of a d-dimensional coordinate vector, $\boldsymbol{x} = (x_1, \ldots, x_d)$. Define the gradient vector

$\boldsymbol{\nabla}_{\boldsymbol{x}} u = (\partial_{x_i} u; i = 1, \ldots, d)$ and consider a first-order PDE that is linear in the gradient:

$$(\boldsymbol{V} \cdot \boldsymbol{\nabla}_{\boldsymbol{x}} u) = f, \tag{5.4}$$

where $\boldsymbol{V}(\boldsymbol{x}) \in \mathbb{R}^d$ is a given velocity field and $f(\boldsymbol{x}) \in \mathbb{R}$ is a forcing function.

First, consider the homogeneous version of Eq. (5.4):

$$(\boldsymbol{V} \cdot \boldsymbol{\nabla}_{\boldsymbol{x}} u) = 0. \tag{5.5}$$

Introduce an auxiliary parameter $t \in \mathbb{R}$, often referred to as time, and define the *characteristic equations*:

$$\frac{d\boldsymbol{x}(t)}{dt} = \boldsymbol{V}(\boldsymbol{x}(t)), \tag{5.6}$$

which describe the evolution of the characteristic trajectory $\boldsymbol{x}(t)$ over time according to the velocity field $\boldsymbol{V}$. A *first integral* is a function, $F(\boldsymbol{x})$, that remains constant along each characteristic curve, satisfying $\frac{d}{dt}F(\boldsymbol{x}(t)) = 0$. Any first integral of Eq. (5.6) provides a solution to Eq. (5.5), and any function of these first integrals, $g(F_1, \ldots, F_k)$, is also a solution.

To see this, let $u = g(F_1, \ldots, F_k)$. Then,

$$(\boldsymbol{V} \cdot \boldsymbol{\nabla}_{\boldsymbol{x}} g) = \sum_{i=1}^{k} \frac{\partial g}{\partial F_i} \sum_{j=1}^{d} \frac{\partial F_i}{\partial x_j} V_j = \sum_{i=1}^{k} \frac{\partial g}{\partial F_i} \frac{d}{dt} F_i = 0,$$

which confirms that $u = g(F_1, \ldots, F_k)$ is indeed a solution to Eq. (5.5).

For a d-dimensional system, Eq. (5.6) typically has $d-1$ independent first integrals (since the system has one degree of freedom in time). Therefore, the general solution to Eq. (5.5) can be expressed as

$$u(\boldsymbol{x}(t)) = g\left(F_1(\boldsymbol{x}(t)), \ldots, F_{d-1}(\boldsymbol{x}(t))\right),$$

where g is a sufficiently smooth function (at least twice differentiable with respect to the first integrals).

Equation (5.5) has a natural geometric interpretation. If we consider $\boldsymbol{V}$ as a velocity field in d-dimensional space, then Eq. (5.5) implies that the derivative of u in the direction of $\boldsymbol{V}$ is zero. In other words, u is constant along the characteristic curves generated by $\boldsymbol{V}$.

To solve a first-order PDE using the method of characteristics, we find the integral curves of the vector field $\boldsymbol{V}(\boldsymbol{x})$, which are tangent to the vector field at every point in space. Along these curves, the solution $u(\boldsymbol{x})$ remains constant. If we locally change variables around each point from $\boldsymbol{x}$ to $(t, F_1, \ldots, F_{d-1})$, where t is a parameter along the integral curve and the transformation is well-defined (meaning that the Jacobian of the transformation is nonzero), then Eq. (5.5) reduces to $du/dt = 0$ along each characteristic.

Let us illustrate the method of characteristics with the following example of a homogeneous first-order PDE:

$$\partial_x u(x, y) + y\, \partial_y u(x, y) = 0.$$

The characteristic equations associated with this PDE are $\frac{dx}{dt} = 1$ and $\frac{dy}{dt} = y$, with general solutions $x(t) = t + c_1$ and $y(t) = c_2 e^t$. The only first integral of the characteristic equations is $F(x, y) = ye^{-x}$, so the general solution to the PDE can be written as $u(x, y) = g(F(x, y)) = g(ye^{-x})$, where g is an arbitrary function. It is useful to visualize the flow along the characteristic curves in the (x, y)-plane.

Example 5.1.1. Find the characteristic curves for each of the following PDEs and use them to determine the general solutions. Verify your solutions by direct substitution:

(a) $\partial_x u - y^2\, \partial_y u = 0$,
(b) $x\, \partial_x u - y\, \partial_y u = 0$,
(c) $y\, \partial_x u - x\, \partial_y u = 0$.

Visualize the characteristic curves in the (x, y)-plane.

Solution. (a) To find the characteristic curves, we parameterize by t and express the left-hand side of the PDE as a total derivative, giving $\frac{d}{dt}u(x(t), y(t)) = 0$. By the chain rule, this condition is equivalent to $\partial_x u\, \dot{x}(t) + \partial_y u\, \dot{y}(t) = 0$, which matches our PDE if we set $\dot{x}(t) = 1$ and $\dot{y}(t) = -y^2$. Solving these characteristic equations gives $x(t) = t + c_1$ and $y(t) = (t + c_2)^{-1}$. Eliminating t yields the first integral $F(x, y) = y^{-1} - x$. Therefore, the general solution to the PDE is $u(x, y) = g(y^{-1} - x)$, where $g : \mathbb{R} \to \mathbb{R}$ is an arbitrary function.

(b) Again, we parameterize by t and use the chain rule to express the PDE as $\frac{d}{dt}u(x(t), y(t)) = 0$. Setting $\dot{x}(t) = x$ and $\dot{y}(t) = -y$

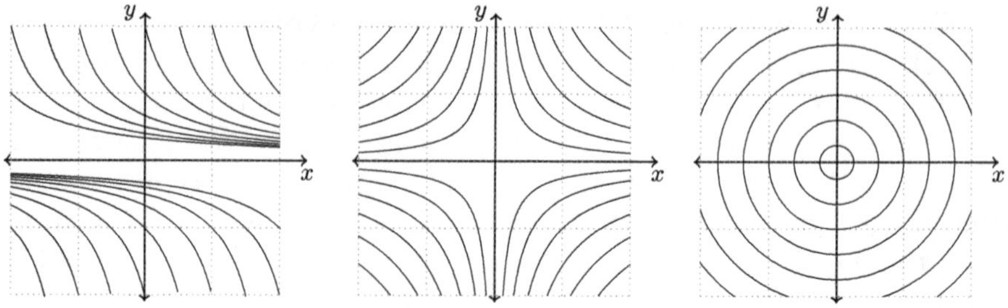

(a) $y^{-1} - x = \text{const}$ (b) $xy = \text{const}$ (c) $x^2 + y^2 = \text{const}$

Figure 5.1. Characteristic curves for the PDEs in Example 5.1.1.

satisfies the PDE. Solving these characteristic equations gives $x(t) = c_1 e^t$ and $y(t) = c_2 e^{-t}$. Eliminating t results in the first integral $F(x, y) = xy$. Thus, the general solution to the PDE is $u(x, y) = g(xy)$, where $g : \mathbb{R} \to \mathbb{R}$ is an arbitrary function.

(c) Parameterizing by t as before, we rewrite the PDE as $\frac{d}{dt}u(x(t), y(t)) = 0$ and set $\dot{x}(t) = y$ and $\dot{y}(t) = -x$, which gives the characteristic equations. Solving these yields $x(t) = c\cos(t)$ and $y(t) = c\sin(t)$. Eliminating t gives the first integral $F(x, y) = x^2 + y^2$. Hence, the general solution to the PDE is $u(x, y) = g(x^2 + y^2)$, where $g : \mathbb{R} \to \mathbb{R}$ is an arbitrary function. □

Consider the following initial value (Cauchy) problem: Solve Eq. (5.5) subject to the boundary condition

$$u(\boldsymbol{x})|_{\boldsymbol{x}_0 \in S} = \vartheta(\boldsymbol{x}_0), \tag{5.7}$$

where S is a surface of dimension $d-1$ (a hypersurface). This Cauchy problem has a well-defined solution in at least some neighborhood of S, provided that S is not tangent to any characteristic curve of Eq. (5.5). As discussed previously, the solution to Eq. (5.5), with the boundary condition (5.7), can be understood as a change of variables along the characteristic curves. The three cases – (a), (b) and (c) – are illustrated in Fig. 5.1.

Example 5.1.2. Let us illustrate the solution of this Cauchy problem with the following example:

$$\partial_x u = y \partial_y u, \quad u(0, y) = \cos(y).$$

Solution. The characteristic equations, $\dot{x} = 1$ and $\dot{y} = -y$, have solutions $x(t) = t + t_1$ and $y(t) = \exp(t_2 - t)$. The first integral is given by

$$F(x, y) = y \exp(x) = \text{constant}.$$

Thus, the general solution to the PDE is

$$u(x, y) = g\left(y \exp(x)\right),$$

where g is an arbitrary function. Now, using the boundary condition $u(0, y) = \cos(y)$ at $x = 0$ (a straight line which is not tangent to any characteristic curve), we find the specific form of g:

$$u(0, y) = g(y) = \cos(y).$$

Therefore, the solution to the Cauchy problem is

$$u(x, y) = \cos\left(y \exp(x)\right). \qquad \square$$

Exercise 5.1. (a) Solve the equation

$$y\partial_x u - x\partial_y u = 0,$$

with the initial condition $u(0, y) = y^2$.

(b) Explain why the same problem with the initial condition $u(0, y) = y$ is ill-posed.

(c) Determine whether the same problem with the initial condition $u(1, y) = y^2$ is ill-posed.

Example 5.1.3. Let $(\boldsymbol{q}, \boldsymbol{p}) = (q_1, \ldots, q_n, p_1, \ldots, p_n)$ represent canonical coordinates for a Hamiltonian system with Hamiltonian $H(\boldsymbol{q}, \boldsymbol{p})$, and let $f = f(t, \boldsymbol{q}, \boldsymbol{p})$ be any function of time t, the coordinates $\boldsymbol{q}$ and momenta $\boldsymbol{p}$. Liouville's theorem states that

$$\partial_t f + \{f, H\} = 0, \quad \text{where } \{f, H\} := \sum_{i=1}^{n} \left(\frac{\partial f}{\partial q_i} \frac{\partial H}{\partial p_i} - \frac{\partial f}{\partial p_i} \frac{\partial H}{\partial q_i} \right),$$

and $\{f, H\}$ is the Poisson bracket of f and H. Find the characteristics of Liouville's PDE.

Solution. We seek a vector field $\boldsymbol{V}(t, \boldsymbol{q}, \boldsymbol{p})$ such that the left-hand side of the PDE can be expressed as $\boldsymbol{V} \cdot \boldsymbol{\nabla} f$. This holds for $\boldsymbol{V} = (1, \partial_{\boldsymbol{p}} H, -\partial_{\boldsymbol{q}} H)$.

We interpret $\boldsymbol{V}$ as the vector field

$$\boldsymbol{V} = \left(\frac{dt}{ds}, \frac{dq}{ds}, \frac{dp}{ds} \right).$$

Introducing this vector field defines the characteristic curves $(t(s), \boldsymbol{q}(s), \boldsymbol{p}(s))$. After some algebra, the characteristic equations reduce to

$$\frac{d}{dt} q_i = \frac{\partial H}{\partial p_i}, \quad \frac{d}{dt} p_i = -\frac{\partial H}{\partial q_i}, \quad \forall i = 1, \ldots, n.$$

Interpretation: In the following chapter, we will explore how Hamilton's equations, $d\boldsymbol{q}/dt = \partial_{\boldsymbol{p}} H$ and $d\boldsymbol{p}/dt = -\partial_{\boldsymbol{q}} H$, describe the evolution of a system's state in phase space. Since the characteristic curves coincide with the solutions to Hamilton's equations and reduce Liouville's PDE to $\frac{df}{dt} = 0$, we conclude that any function of the system's state variables $(\boldsymbol{q}, \boldsymbol{p})$ remains unchanged as the system evolves in time. □

Returning now to the inhomogeneous equation (5.4), we observe that, as with most linear equations, the general solution of the inhomogeneous equation is the sum of a particular solution and the general solution to the corresponding homogeneous equation. To find a particular solution, we use the method of characteristics. Along the characteristic curves, Eq. (5.4) becomes

$$(\boldsymbol{V} \cdot \boldsymbol{\nabla}_{\boldsymbol{x}})\, u = (\dot{\boldsymbol{x}} \cdot \boldsymbol{\nabla}_{\boldsymbol{x}})\, u = \frac{du}{dt} = f(\boldsymbol{x}(t)),$$

which can be integrated along the characteristics to give the desired particular solution:

$$u_{\text{inh}} = \int_{t_0}^{t} f(\boldsymbol{x}(s))\, ds,$$

where $\boldsymbol{x}(s)$ satisfies $(\boldsymbol{V} \cdot \boldsymbol{\nabla}_{\boldsymbol{x}})u = (\dot{\boldsymbol{x}} \cdot \boldsymbol{\nabla}_{\boldsymbol{x}})u$. Note that this solution varies along the characteristics and is not constant.

Example 5.1.4. Solve the Cauchy problem for the following inhomogeneous equation:

$$\partial_x u - y\partial_y u = y, \quad u(0, y) = \sin(y).$$

The method of characteristics can also be extended to quasi-linear first-order PDEs, where the coefficients $\boldsymbol{V}$ and f depend on both the spatial variables $\boldsymbol{x}$ and the unknown function $u(\boldsymbol{x})$. In such cases, the characteristic equations become

$$\frac{d\boldsymbol{x}}{dt} = \boldsymbol{V}(\boldsymbol{x}, u), \quad \frac{du}{dt} = f(\boldsymbol{x}, u). \tag{5.8}$$

The general solution to a quasi-linear PDE is given by $g(F_1, F_2, \ldots, F_n) = 0$, where g is an arbitrary function of the n first integrals of Eq. (5.8).

Consider, for instance, the Hopf equation in $d = 1$:

$$\partial_t u + u\partial_x u = 0, \tag{5.9}$$

which describes the one-dimensional flow of non-interacting particles, where $u(t, x)$ represents the velocity of a particle at position x and time t. The characteristic equations are

$$\dot{x} = u, \quad \dot{u} = 0, \quad x(0) = x_0, \quad u(0) = u_0(x_0).$$

Integrating, we obtain $x = u_0(x_0)t + x_0$, leading to the following implicit solution for u:

$$u = u_0(x - ut).$$

For the specific case of $u_0(x) = c(1 - \tanh(x))$, this results in the implicit equation $u = c(1 - \tanh(x - ut))$. Computing the partial derivative, we find $\partial_x u = -\frac{c}{\cosh^2(x-ut)-ct}$, which shows that u develops a singularity in finite time, $t_* = 1/c$, at $x = ut$. This phenomenon is known as *wave breaking*, where faster particles overtake slower ones, leading to a steepening velocity profile and eventual breakdown. This singularity is non-physical, meaning that the model becomes invalid at the singularity. By introducing a small diffusion term, $\kappa\partial_x^2 u$, to the right-hand side of Eq. (5.9), the breakdown is regularized, and shock formation is explained. The regularized equation is known as *Burgers' equation.*

5.2 Classification of Linear Second-Order PDEs

Consider the most general linear second-order PDE in two independent variables:

$$a_{11}\partial_x^2 u + 2a_{12}\partial_x\partial_y u + a_{22}\partial_y^2 u + b_1\partial_x u + b_2\partial_y u + cu + f = 0, \tag{5.10}$$

where all the coefficients may depend on the independent variables x and y.

The method of characteristics, which applies to first-order PDEs (for example, when $a_{11} = a_{12} = a_{22} = c = 0$ in Eq. (5.10)), can also guide us in the analysis of second-order PDEs. To illustrate this, let us momentarily consider the first-order PDE

$$b_1\partial_x u + b_2\partial_y u + f = 0 \tag{5.11}$$

and interpret its solution as a variable transformation from the (x, y) coordinates to new coordinates $(\eta(x, y), \xi(x, y))$, assuming that the Jacobian of the transformation is nonzero and finite within the domain of interest, i.e.,

$$J = \det\begin{pmatrix} \partial_x\eta & \partial_y\eta \\ \partial_x\xi & \partial_y\xi \end{pmatrix} \neq 0, \infty.$$

Substituting $u = w(\eta(x, y), \xi(x, y))$ into the first derivative terms of Eq. (5.11), we get

$$\begin{aligned} b_1\partial_x u + b_2\partial_y u &= b_1\left(\partial_x\eta\partial_\eta w + \partial_x\xi\partial_\xi w\right) + b_2\left(\partial_y\eta\partial_\eta w + \partial_y\xi\partial_\xi w\right) \\ &= \left(b_1\partial_x\eta + b_2\partial_y\eta\right)\partial_\eta w + \left(b_1\partial_x\xi + b_2\partial_y\xi\right)\partial_\xi w. \end{aligned} \tag{5.12}$$

To simplify this, we require that the second term in Eq. (5.12) vanishes. This happens when $\xi(y(x))$, meaning ξ depends only on $y(x)$ and not explicitly on x. In such cases, the characteristic equation $b_1 dy/dx + b_2 = 0$ describes the characteristic curves.

Now, let us apply the same reasoning to the second-order terms in Eq. (5.10). We derive

$$a_{11}\partial_x^2 u + 2a_{12}\partial_x\partial_y u + a_{22}\partial_y^2 u = \left(A\partial_\xi^2 + 2B\partial_\xi\partial_\eta + C\partial_\eta^2\right) w, \tag{5.13}$$

where

$$\begin{aligned}
A &:= a_{11}(\partial_x \xi)^2 + 2a_{12}(\partial_x \xi)(\partial_y \xi) + a_{22}(\partial_y \xi)^2, \\
B &:= a_{11}(\partial_x \xi)(\partial_x \eta) + a_{12}(\partial_x \xi \partial_y \eta + \partial_y \xi \partial_x \eta) + a_{22}(\partial_y \xi)(\partial_y \eta), \\
C &:= a_{11}(\partial_x \eta)^2 + 2a_{12}(\partial_x \eta)(\partial_y \eta) + a_{22}(\partial_y \eta)^2.
\end{aligned}$$

By analogy with the first-order case, we attempt to simplify the equation by forcing the first and last terms on the right-hand side of Eq. (5.13) to vanish, i.e., $A = C = 0$. This is achieved if we require that $\xi = \psi_+(x, y)$ and $\eta = \psi_-(x, y)$, where

$$\frac{dy_\pm}{dx} = \frac{a_{12} \pm \sqrt{D}}{a_{11}}, \quad \text{where } D := a_{12}^2 - a_{11}a_{22},$$

and D is called the *discriminant*. In the general case, these equations yield distinct integrals $\psi_\pm(x, y) = \text{const}$. We can then choose the new variables $\xi = \psi_+(x, y)$ and $\eta = \psi_-(x, y)$.

The classification of second-order PDEs depends on the discriminant D:

Hyperbolic PDEs: If $D > 0$, Eq. (5.10) is called a *hyperbolic* PDE. The characteristics are real, and any real pair (x, y) can be mapped to a real pair (η, ξ). The equation reduces to the following canonical form:

$$\partial_\xi \partial_\eta u + \tilde{b}_1 \partial_\xi u + \tilde{b}_2 \partial_\eta u + \tilde{c}u + \tilde{f} = 0.$$

Another canonical form can be obtained by changing variables further to $(\alpha, \beta) := \left(\frac{\eta+\xi}{2}, \frac{\xi-\eta}{2}\right)$, resulting in

$$\partial_\alpha^2 u - \partial_\beta^2 u + \tilde{b}_1^{(2)} \partial_\alpha u + \tilde{b}_2^{(2)} \partial_\beta u + \tilde{c}^{(2)} u + \tilde{f}^{(2)} = 0. \tag{5.14}$$

Elliptic PDEs: If $D < 0$, Eq. (5.10) is called an *elliptic* PDE. In this case, the characteristic equations have complex conjugate solutions. To keep the transformation real, we define new variables $\alpha = \Re(\psi_+)$ and $\beta = \Im(\psi_+)$. The canonical form for the elliptic equation is

$$\partial_\alpha^2 u + \partial_\beta^2 u + b_1^{(e)} \partial_\alpha u + b_2^{(e)} \partial_\beta u + c^{(e)} u + f^{(e)} = 0. \tag{5.15}$$

Parabolic PDEs: If $D = 0$, Eq. (5.10) is called a *parabolic* PDE. In this degenerate case, $\psi_+ = \psi_-$, and we choose $\beta = \psi_+(x, y)$ and

$\alpha = \varphi(x, y)$, where φ is an independent function of x and y. The equation reduces to the following parabolic form:

$$\partial_\alpha^2 u + b_1^{(p)} \partial_\alpha u + b_2^{(p)} \partial_\beta u + c^{(p)} u + f^{(p)} = 0. \tag{5.16}$$

Example 5.2.1. Classify the following PDEs and transform them into their respective canonical forms:

(a) $\partial_x^2 u + \partial_x \partial_y u - 2\partial_y^2 u - 3\partial_x u - 15\partial_y u + 27x = 0$,
(b) $\partial_x^2 u + 2\partial_x \partial_y u + 5\partial_y^2 u - 32u = 0$,
(c) $\partial_x^2 u - 2\partial_x \partial_y u + \partial_y^2 u + \partial_x u + \partial_y u - u = 0$.

Solution. (a) The second-order coefficients are $a_{11} = 1$, $a_{12} = 1/2$ and $a_{22} = -2$. The discriminant is $D = a_{12}^2 - a_{11}a_{22} = 9/4$. Since $D > 0$, the equation is hyperbolic. The characteristic equations are $dy/dx = 2$ and $dy/dx = -1$, leading to new variables $\xi = y - 2x$ and $\eta = y + x$. The equation becomes

$$\partial_\xi \partial_\eta u + \partial_\xi u + 2\partial_\eta u + 3(\eta - \xi) = 0.$$

(b) The second-order coefficients are $a_{11} = 1$, $a_{12} = 1$ and $a_{22} = 5$. The discriminant is $D = -4$, so the equation is elliptic. The characteristic equations are $dy/dx = 1 \pm 2i$, leading to new variables $\xi = y - x$ and $\eta = 2x$. The equation becomes

$$\partial_\xi^2 u + \partial_\eta^2 u - 8u = 0.$$

(c) The second-order coefficients are $a_{11} = 1$, $a_{12} = -1$ and $a_{22} = 1$. Since $D = a_{12}^2 - a_{11}a_{22} = 0$, the equation is parabolic. The characteristic equation is $dy/dx = -1$, so we use $\xi = x + y$ and choose $\eta = x$. The equation becomes

$$\partial_\eta^2 u + 2\partial_\xi u + \partial_\eta u - u = 0.$$

Note that the condition for functional independence requires that the Jacobian of the transformation be nonzero. Any other choice of a second independent variable would yield a different canonical form. □

5.3 Elliptic PDEs: The Method of Green's Functions

Elliptic PDEs typically arise in the modeling of static phenomena in two or more dimensions.

Let us first address the generalization of elliptic PDEs to higher dimensions. We extend Eq. (5.15) to the following form:

$$\sum_{i,j=1}^{d} a_{ij}\partial_{x_i}\partial_{x_j}u(\boldsymbol{x}) + (\text{lower-order terms}) = 0, \tag{5.17}$$

where we assume that it is not possible to eliminate all second derivative terms in a well-posed Cauchy problem. Note that in $d > 2$, Eq. (5.17) cannot generally be reduced to a canonical form as in the two-dimensional case (as discussed in the previous section).

Our primary focus here is on cases where $a_{ij} = \delta_{ij}$ in Eq. (5.17) for $d \geq 2$ and on solving inhomogeneous equations driven by a nonzero source term. A common method to solve such equations is to use Green's functions.

In Section 4.4.1, we solved the one-dimensional Poisson equation using Green's functions. Now, we generalize this approach to the Poisson equation in $d \geq 2$:

$$\boldsymbol{\nabla}_{\boldsymbol{r}}^2 u = \phi(\boldsymbol{r}), \tag{5.18}$$

where $\boldsymbol{\nabla}_{\boldsymbol{r}}^2 := \Delta_{\boldsymbol{r}}$ is the Laplacian. For $d = 2$, $\boldsymbol{r} = (x, y) \in \mathbb{R}^2$, and $\Delta_{\boldsymbol{r}} = \partial_x^2 + \partial_y^2$. For $d = 3$, $\boldsymbol{r} = (x, y, z) \in \mathbb{R}^3$, and $\Delta_{\boldsymbol{r}} = \partial_x^2 + \partial_y^2 + \partial_z^2$. The Poisson equation (5.18) has various applications. For example, its solution $u(\boldsymbol{r})$ can describe the electrostatic potential of a charge distribution in $\mathbb{R}^d$ with density $\rho(\boldsymbol{r})$, where $\phi(\boldsymbol{r}) = -4\pi\rho(\boldsymbol{r})$.

The Poisson equation in $\mathbb{R}^d$ can be solved using Green's functions. The Green function $G(\boldsymbol{r} - \boldsymbol{r}')$ is the solution to the inhomogeneous equation with a point source on the right-hand side:

$$\boldsymbol{\nabla}_{\boldsymbol{r}}^2 G = \delta(\boldsymbol{r}), \tag{5.19}$$

where $\delta(\boldsymbol{r})$ is the Dirac delta function. Then, the solution to Eq. (5.18) is given by the convolution of the Green function with the source term:

$$u(\boldsymbol{r}) = \int d\boldsymbol{r}' G(\boldsymbol{r} - \boldsymbol{r}')\phi(\boldsymbol{r}'). \tag{5.20}$$

To find $G(\boldsymbol{r})$, we apply the Fourier transform to Eq. (5.19), leading to the algebraic equation $k^2\hat{G}(\boldsymbol{k}) = -1$, where $k = |\boldsymbol{k}|$. Solving this and applying the inverse Fourier transform, we obtain for $d = 3$,

$$\begin{aligned} G(\boldsymbol{r}) &= -\int \frac{d^3\boldsymbol{k}}{(2\pi)^3} \frac{\exp(i\boldsymbol{k}\cdot\boldsymbol{r})}{k^2} \\ &= -\int \frac{d^2k_\perp}{(2\pi)^3} \int_{-\infty}^{\infty} \frac{dk_\parallel \exp(ik_\parallel r)}{k_\parallel^2 + k_\perp^2} \\ &= -\int \frac{d^2k_\perp}{(2\pi)^3} \frac{\pi}{k_\perp} \exp(-k_\perp r) = -\frac{1}{4\pi r}, \end{aligned} \tag{5.21}$$

where we switched from Cartesian to cylindrical coordinates, with $\boldsymbol{k} = (k_\parallel, \boldsymbol{k}_\perp)$, where $k_\parallel = (\boldsymbol{k}\cdot\boldsymbol{r})/r$ is along $\boldsymbol{r}$ and $\boldsymbol{k}_\perp$ is perpendicular to $\boldsymbol{r}$. Substituting Eq. (5.21) into Eq. (5.20) yields the solution for the electrostatic potential:

$$u(\boldsymbol{r}) = \int d^3\boldsymbol{r}' \frac{\rho(\boldsymbol{r}')}{|\boldsymbol{r} - \boldsymbol{r}'|},$$

which describes the potential due to a charge density $\rho(\boldsymbol{r})$ in space.

For $d = 2$, we typically write $\phi(\boldsymbol{r}) = -2\pi\rho(\boldsymbol{r})$, and the Green function is $G(\boldsymbol{r} - \boldsymbol{r}') = \ln(|\boldsymbol{r} - \boldsymbol{r}'|)$.

The homogeneous case of $\phi = 0$ is known as the Laplace equation. We refer to the two cases as the inhomogeneous and homogeneous Laplace equations, respectively.

Functions that satisfy the homogeneous Laplace equation are called harmonic functions. Notably, there are no nonzero harmonic functions defined over the entirety of $\mathbb{R}^d$ that vanish as $|\boldsymbol{r}| \to \infty$. This can be seen by applying the Fourier transform to the homogeneous Laplace equation, which yields $k^2\hat{u}(\boldsymbol{k}) = 0$, implying $\hat{u}(\boldsymbol{k}) \sim \delta(\boldsymbol{k})$. Taking the inverse Fourier transform results in $u(\boldsymbol{r}) = \text{const}$, and if $u \to 0$ as $r \to \infty$, the constant must be zero. This argument extends to all dimensions and also applies to the homogeneous Debye equation (discussed later).

Consequently, nonzero harmonic functions must be defined in a bounded domain. In many applications, the homogeneous Laplace equation is supplemented with boundary conditions, where u or its normal derivative $\boldsymbol{\nabla}u \cdot \boldsymbol{n}$ is specified at the boundary.

Example 5.3.1. Find the Green function for the Laplace equation in the region outside a sphere of radius R, with zero boundary condition on the sphere. Specifically, solve

$$\nabla_{\boldsymbol{r}}^2 G(\boldsymbol{r}; \boldsymbol{r}') = \delta(\boldsymbol{r} - \boldsymbol{r}'),$$

for $\boldsymbol{r}$ such that $r, r' \geq R$, with the boundary condition $G(\boldsymbol{r}; \boldsymbol{r}') = 0$ when $r = R \leq r'$.

Solution. The Green function can be constructed by using the fact that $|\boldsymbol{r} - \boldsymbol{r}''|^{-1}$ satisfies $\nabla_{\boldsymbol{r}}^2 G(\boldsymbol{r}; \boldsymbol{r}') = 0$ for all $\boldsymbol{r} \neq \boldsymbol{r}''$. The goal is to find, for each $\boldsymbol{r}' \in \mathcal{D}$, a fictitious image point $\boldsymbol{r}'' \notin \mathcal{D}$ such that $G(\boldsymbol{r}; \boldsymbol{r}') = 0$ whenever $|\boldsymbol{r}| = R$. The problem then reduces to determining the correct position and strength of this image point to satisfy the boundary condition on the sphere.

By symmetry, we know that $\boldsymbol{r}'$ and $\boldsymbol{r}''$ must lie along the same radial line. Thus, we can write $\boldsymbol{r}'' = \alpha \boldsymbol{r}'$ for some scalar α. We then seek constants A and α such that

$$G(\boldsymbol{r}; \boldsymbol{r}') := -\frac{1}{4\pi|\boldsymbol{r} - \boldsymbol{r}'|} + \frac{A}{4\pi|\boldsymbol{r} - \alpha\boldsymbol{r}'|} = 0 \quad \text{for } |\boldsymbol{r}| = R.$$

For any $\boldsymbol{r}$ on the boundary and $\boldsymbol{r}' \in \mathcal{D}$, we have $|\boldsymbol{r} - \boldsymbol{r}'| = \sqrt{R^2 + |\boldsymbol{r}'|^2 - 2|\boldsymbol{r}'|R\cos\theta}$, where θ is the angle between $\boldsymbol{r}$ and $\boldsymbol{r}'$. Similarly, $|\boldsymbol{r} - \boldsymbol{r}''| = \sqrt{R^2 + \alpha^2|\boldsymbol{r}'|^2 - 2\alpha|\boldsymbol{r}'|R\cos\theta}$. (See the black and grey triangles in Fig. 5.2.) To enforce the boundary condition, we require that these terms cancel, yielding

$$-\frac{1}{4\pi\sqrt{|\boldsymbol{r}'|^2 + R^2 - 2|\boldsymbol{r}'|R\cos\theta}} + \frac{A}{4\pi\sqrt{\alpha^2|\boldsymbol{r}'|^2 + R^2 - 2\alpha|\boldsymbol{r}'|R\cos\theta}} = 0.$$

To satisfy this condition independently of θ, we solve for A and α. After some algebra, we find that $A = \frac{R}{|\boldsymbol{r}'|}$ and $\alpha = \left(\frac{R}{|\boldsymbol{r}'|}\right)^2$. Therefore, the Green function is given by

$$G(\boldsymbol{r}; \boldsymbol{r}') := -\frac{1}{4\pi|\boldsymbol{r} - \boldsymbol{r}'|} + \frac{R/|\boldsymbol{r}'|}{4\pi|\boldsymbol{r} - \boldsymbol{r}''|},$$

where $\boldsymbol{r}'' = \left(\frac{R}{|\boldsymbol{r}'|}\right)^2 \boldsymbol{r}'$. □

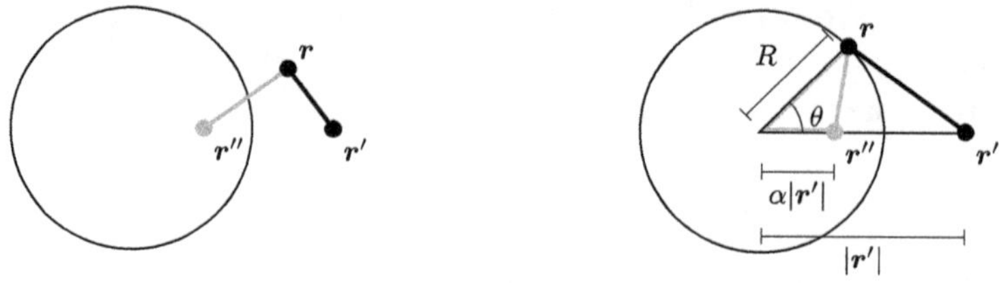

(a) For each $\boldsymbol{r}' \in \mathcal{D}$, identify the associated image point $\boldsymbol{r}'' \notin \mathcal{D}$.

(b) Contributions from $\boldsymbol{r}'$ and $\boldsymbol{r}''$ must cancel on the boundary.

Figure 5.2. Method of images applied to the exterior of a sphere in Example 5.3.1.

Exercise 5.2. Find the general solution to the inhomogeneous Debye equation

$$\left(\nabla_{\boldsymbol{r}}^2 - \kappa^2\right) u = -4\pi\rho(\boldsymbol{r}),$$

where the charge density $\rho(\boldsymbol{r})$ depends only on the distance from the origin, $\rho(r = |\boldsymbol{r}|)$.

Hint: Follow an analogous process to the one used for the Laplace equation.

5.4 Waves in Homogeneous Media: Hyperbolic PDEs*

Although hyperbolic PDEs are typically associated with wave phenomena, we will take a slightly unconventional approach by first discussing the properties of wave solutions and then connecting them back to the PDEs that describe these waves. This approach will help build intuition that generalizes to broader classes of equations, including integro-differential equations.

Consider wave propagation in homogeneous media, such as electromagnetic waves, sound waves, spin waves, surface waves, or electro-mechanical waves in power systems. Despite the variety of physical contexts, these phenomena share a universal description.

*This is an auxiliary section that may be skipped on a first reading. Material from this section will not be included in the midterm or final exams.

The wave process at a general position in d-dimensional space $\boldsymbol{r}$ and time t can be written as an integral over the wave vector $\boldsymbol{k}$:

$$u(t;\boldsymbol{r}) = \int \frac{d\boldsymbol{k}}{(2\pi)^k} \exp\left(i(\boldsymbol{k}\cdot\boldsymbol{r})\right) \psi_{\boldsymbol{k}}(t)\hat{u}(\boldsymbol{k}), \quad \psi_{\boldsymbol{k}}(t) := \exp\left(-i\omega(\boldsymbol{k})t\right), \tag{5.22}$$

where $\omega(\boldsymbol{k})$ is the dispersion relation and $\hat{u}(\boldsymbol{k})$ is the wave amplitude for each wave vector $\boldsymbol{k}$. This equation is analogous to the Fourier integral, with some key differences.

In Eq. (5.22), $\psi_{\boldsymbol{k}}(t)$ satisfies the first-order linear ODE

$$\left(\frac{d}{dt} + i\omega(\boldsymbol{k})\right)\psi_{\boldsymbol{k}} = 0, \tag{5.23}$$

or, equivalently, the second-order ODE

$$\left(\frac{d^2}{dt^2} + \omega(\boldsymbol{k})^2\right)\psi_{\boldsymbol{k}} = 0. \tag{5.24}$$

These are called wave equations in Fourier space. The linearity of these equations is key, as wave dynamics are often linearized in physical systems. Although nonlinear wave interactions can occur, they are beyond the scope of this discussion, where we focus on linear waves. For a comprehensive discussion of the mathematical theory underlying the nonlinear interaction of waves and the resulting phenomena, particularly wave turbulence, we refer the interested reader to Ref. [17].

Dispersion relations

One of the simplest and most important cases occurs when the dispersion relation is $\omega_k = c|\boldsymbol{k}|$, where c is a constant with the dimensions of velocity. The inverse Fourier transform of Eq. (5.24) then yields the wave equation in real space:

$$\left(\frac{d^2}{dt^2} - c^2\boldsymbol{\nabla}_{\boldsymbol{r}}^2\right)\psi(t;\boldsymbol{r}) = 0. \tag{5.25}$$

Here, the time and space derivatives have opposite signs, leading naturally to a hyperbolic PDE, which generalizes Eq. (5.14) to d-dimensional space, where $d \geq 1$.

Equation (5.25), with constant c, describes various physical phenomena, such as the propagation of sound in homogeneous media (e.g., gas, liquid or crystal). In this case, ψ represents the displacement of a material element from its equilibrium position, and c is the speed of sound in the medium.[a]

Another example involves electromagnetic waves, which are described by Maxwell's equations for the electric field $\boldsymbol{E}$ and magnetic field $\boldsymbol{B}$:

$$\partial_t \boldsymbol{E} = c\boldsymbol{\nabla}_{\boldsymbol{r}} \times \boldsymbol{B}, \quad \partial_t \boldsymbol{B} = -c\boldsymbol{\nabla}_{\boldsymbol{r}} \times \boldsymbol{E}, \tag{5.26}$$

along with the divergence-free conditions:

$$\boldsymbol{\nabla}_{\boldsymbol{r}} \cdot \boldsymbol{E} = 0, \quad \boldsymbol{\nabla}_{\boldsymbol{r}} \cdot \boldsymbol{B} = 0,$$

where c is the speed of light in the medium. Differentiating the first equation in Eq. (5.26) with respect to time and using the second equation, along with the fact that $\boldsymbol{E}$ is divergence-free, we again obtain Eq. (5.25) for the components of $\boldsymbol{E}$.

In both sound and electromagnetic waves, the dispersion relation is linear: $\omega(\boldsymbol{k}) = \pm c|\boldsymbol{k}|$. However, more complex dispersion relations arise in other contexts. For example, surface waves on water exhibit the following dispersion relation:

$$\omega(\boldsymbol{k}) = \sqrt{gk + \left(\frac{\sigma}{\rho}\right) k^3},$$

where g is the gravitational constant, σ is the surface tension and ρ is the fluid density. This dispersion relation reflects both gravitational and capillary effects. At large wavelengths (small k), gravitational waves dominate, and $\omega(\boldsymbol{k}) \sim \sqrt{gk}$. At short wavelengths (large k), capillary waves dominate, with $\omega(\boldsymbol{k}) \sim \left(\frac{\sigma}{\rho}\right)^{1/2} k^{3/2}$.

Note that Eq. (5.23) or Eq. (5.24) is written in Fourier space. When the dispersion relation is nonlinear (as in the case of surface

[a]In gases and liquids, there is a unique speed of sound. In three-dimensional crystals, however, three distinct waves with different speeds (corresponding to different polarizations) can propagate. In isotropic crystals, these waves may be longitudinal (parallel to the displacement) or transverse (perpendicular to the displacement).

waves), transforming back to real space does not yield a PDE but rather an integro-differential equation, reflecting the nonlocal nature of the problem in real space.

In general, wave propagation in homogeneous media is characterized by a dispersion relation that depends only on the magnitude of the wave vector, $k = |\boldsymbol{k}|$. The phase velocity $\omega(k)/k$ and group velocity $d\omega(k)/dk$, both with dimensions of speed, describe how the wave propagates.

Example 5.4.1. Solve the Cauchy (initial value) problem for spin waves, which satisfy the following PDE:

$$\partial_t^2 \psi = -(\Omega - b\boldsymbol{\nabla}_{\boldsymbol{r}}^2)^2 \psi, \tag{5.27}$$

in $d = 3$, with the initial conditions $\psi(t = 0; \boldsymbol{r}) = \exp(-r^2)$ and $d\psi/dt(t = 0; \boldsymbol{r}) = 0$.

Solution. Applying the Fourier transform to Eq. (5.27) results in the wave equation (5.23) with the dispersion relation

$$\omega(k) = \Omega + bk^2.$$

The Fourier transform of the initial condition is $\hat{\psi}(t = 0; \boldsymbol{k}) = \pi^{3/2}\exp(-k^2/4)$. Since $d\psi/dt(t = 0; \boldsymbol{r}) = 0$, we also have $d\hat{\psi}/dt$ $(t = 0; \boldsymbol{k}) = 0$. The solution to Eq. (5.23) is

$$\hat{\psi}(t; \boldsymbol{k}) = \pi^{3/2}\exp(-k^2/4)\cos\left((\Omega + bk^2)t\right).$$

Taking the inverse Fourier transform, we find

$$\begin{aligned}
\psi(t; \boldsymbol{r}) &= \pi^{3/2} \int \frac{d^3k}{(2\pi)^3} e^{-k^2/4} \cos\left((\Omega + bk^2)t\right) \exp\left(i(\boldsymbol{k}\cdot\boldsymbol{r})\right) \\
&= \int_0^\infty \frac{kdk}{2\pi^{1/2} r} e^{-k^2/4} \cos\left((\Omega + bk^2)t\right) \sin(kr) \\
&= \operatorname{Re}\left(\frac{\exp\left(i\Omega t - \frac{r^2}{1-4ibt}\right)}{(1-4ibt)^{3/2}}\right).
\end{aligned}$$

□

Exercise 5.3. Solve the Cauchy (initial value) problem for the wave Eq. (5.25) in $d = 3$, where $\psi(t = 0; \boldsymbol{r}) = \exp(-r^2)$ and $d\psi/dt(t = 0;$ $\boldsymbol{r}) = 0$.

Stimulated waves: Radiation

So far, we have focused on the free propagation of waves. Now, consider an inhomogeneous wave equation with a source term $\chi(t;\boldsymbol{r})$ on the right-hand side, generalizing Eq. (5.24):

$$\left(\frac{d^2}{dt^2}+(\omega(-i\boldsymbol{\nabla}_{\boldsymbol{r}}))^2\right)\psi(t;\boldsymbol{r})=\chi(t;\boldsymbol{r}), \tag{5.28}$$

where the operator $\omega(-i\boldsymbol{\nabla}_{\boldsymbol{r}})$ is well defined as a Taylor series in $\boldsymbol{k}$, assuming $\omega(k)$ is a continuous function of the wave vector magnitude $k=|\boldsymbol{k}|$.

The Green function for this PDE is defined as the solution to the equation

$$\left(\frac{d^2}{dt^2}+(\omega(-i\boldsymbol{\nabla}_{\boldsymbol{r}}))^2\right)G(t;\boldsymbol{r})=\delta(t)\delta(\boldsymbol{r}). \tag{5.29}$$

Once the Green function is known, the solution to the inhomogeneous equation, Eq. (5.28), can be expressed as a convolution of the source term $\chi(t_1;\boldsymbol{r}_1)$ with the Green function $G(t;\boldsymbol{r})$:

$$\psi(t;\boldsymbol{r})=\int dt_1\, d\boldsymbol{r}_1\, G(t-t_1;\boldsymbol{r}-\boldsymbol{r}_1)\chi(t_1;\boldsymbol{r}_1). \tag{5.30}$$

The general solution to Eq. (5.28) is the sum of the forced solution from Eq. (5.30) and a homogeneous solution (the solution to Eq. (5.28) with the right-hand side set to zero).

To solve for the Green function $G(t;\boldsymbol{r})$, we first take the Fourier transform of Eq. (5.29):

$$\left(\frac{d^2}{dt^2}+(\omega(\boldsymbol{k}))^2\right)\hat{G}(t;\boldsymbol{k})=\delta(t), \tag{5.31}$$

where $\omega(\boldsymbol{k})$ is the dispersion relation in $\boldsymbol{k}$-space. This inhomogeneous ODE was solved earlier in the course, and its solution is given by Eq. (4.28). Applying the inverse Fourier transform to the solution of Eq. (4.28), we obtain the Green function in real space:

$$G(t;\boldsymbol{r})=\theta(t)\int\frac{d^3k}{(2\pi)^3}\frac{\sin(\omega(k)t)}{\omega(k)}\exp\left(i(\boldsymbol{k}\cdot\boldsymbol{r})\right), \tag{5.32}$$

where $\theta(t)$ is the Heaviside step function.

Example 5.4.2. Show that for a linear dispersion relation $\omega(\boldsymbol{k}) = ck$, the Green function in Eq. (5.32) becomes

$$G(t;\boldsymbol{r}) = \frac{\theta(t)}{4\pi cr}\left(\delta(r-ct) - \delta(r+ct)\right), \tag{5.33}$$

where $r = |\boldsymbol{r}|$.

Solution. The linear dispersion relation implies $\omega(\boldsymbol{k}) = ck$, where $k = |\boldsymbol{k}|$. Substituting into Eq. (5.32), we have

$$G(t;\boldsymbol{r}) = \theta(t)\int_{\mathbb{R}^3} \frac{\sin(ckt)}{ck}\exp(i\boldsymbol{k}\cdot\boldsymbol{r})\frac{d^3\boldsymbol{k}}{(2\pi)^3}.$$

We compute this integral by rotating the coordinate system so that $\boldsymbol{r}$ points along the z-axis and then switch to spherical coordinates. The scalar product $\boldsymbol{r}\cdot\boldsymbol{k}$ becomes $rk\cos(\theta)$:

$$\begin{aligned}
G(t;\boldsymbol{r}) &= \frac{\theta(t)}{(2\pi)^3}\int_0^{2\pi}\int_0^{\pi}\int_0^{\infty}\frac{\sin(ckt)}{ck}\exp(irk\cos\theta)k^2\sin\theta\, dk\, d\theta\, d\phi \\
&= \frac{\theta(t)}{(2\pi)^2 cr}\int_0^{\infty} k\left(e^{ik(r-ct)} - e^{ik(r+ct)}\right)dk \\
&= \frac{\theta(t)}{4\pi cr}\left(\delta(r-ct) - \delta(r+ct)\right),
\end{aligned}$$

which matches the desired expression. □

Substituting Eq. (5.33) into Eq. (5.30), we obtain the following expression for radiation from a source with linear dispersion (such as light or sound):

$$\psi(t;\boldsymbol{r}) = \frac{1}{4\pi c^2}\int \frac{d\boldsymbol{r}_1}{R}\chi\left(t - \frac{R}{c};\boldsymbol{r}_1\right),$$

where $R = |\boldsymbol{r} - \boldsymbol{r}_1|$. This solution shows that the source action is delayed by the time R/c, corresponding to the time taken for light or sound to travel from the source to the observation point.

Example 5.4.3. Solve the radiation problem described by Eq. (5.28) for a point harmonic source, $\chi(t;\boldsymbol{r}) = \cos(\omega t)\delta(\boldsymbol{r})$.

Solution. The solution to Eq. (5.30) is given by

$$\begin{aligned}\psi(t;\boldsymbol{r}) &= \int dt_1 d\boldsymbol{r}_1 G(t-t_1;\boldsymbol{r}-\boldsymbol{r}_1)\chi(t_1;\boldsymbol{r}_1)\\ &= \frac{1}{4\pi c^2}\int \frac{d\boldsymbol{r}_1}{r}\cos\left(\omega\left(t-\frac{r}{c}\right)\right)\delta(\boldsymbol{r}_1)\\ &= \frac{1}{4\pi r c^2}\cos\left(\omega\left(t-\frac{r}{c}\right)\right).\end{aligned}$$

□

5.5 Diffusion Equation

The homogeneous diffusion equation is one of the most common examples of a multi-dimensional generalization of the parabolic equation Eq. (5.16). It is given by

$$\partial_t u = \kappa \nabla_{\boldsymbol{r}}^2 u, \tag{5.34}$$

where κ is the diffusion coefficient. This equation appears in numerous applications, such as describing the evolution of particle density, temperature variation over space and even stochastic processes like Brownian motion.

Consider the Cauchy problem, where the initial condition $u(t=0;\boldsymbol{r})$ is provided. Taking the Fourier transform of $u(t;\boldsymbol{r})$ over the spatial coordinates $\boldsymbol{r}\in\mathbb{R}^d$ yields

$$\hat{u}(t;\boldsymbol{k}) = \int d^d y \exp\left(i\boldsymbol{k}\cdot\boldsymbol{y}\right) u(t;\boldsymbol{y}).$$

Applying the Fourier transform to Eq. (5.34), we obtain

$$\partial_t \hat{u}(t;\boldsymbol{k}) = -k^2 \hat{u}(t;\boldsymbol{k}),$$

where $k=|\boldsymbol{k}|$. Solving this differential equation in time gives

$$\hat{u}(t;\boldsymbol{k}) = \exp(-k^2\kappa t)\hat{u}(0;\boldsymbol{k}),$$

which is the solution in Fourier space. Inverting the Fourier transform results in the following integral expression for $u(t;\boldsymbol{x})$ in real space:

$$u(t;\boldsymbol{x}) = \int \frac{d^d y}{(4\pi\kappa t)^{d/2}} \exp\left(-\frac{(\boldsymbol{x}-\boldsymbol{y})^2}{4\kappa t}\right) u(0;\boldsymbol{y}). \tag{5.35}$$

If the initial field $u(0;\boldsymbol{x})$ is localized around some point, say $\boldsymbol{x}=0$, meaning it decays rapidly as $|\boldsymbol{x}|$ increases, then a universal asymptotic behavior of $u(t;\boldsymbol{x})$ emerges at large times ($t \gg l^2$), where l is the characteristic length scale of the initial distribution. At sufficiently large times, the dominant contribution to the integral in Eq. (5.35) comes from the region $|\boldsymbol{y}| \sim l$. Hence, in the leading order, we can approximate the diffusive kernel in Eq. (5.35) by ignoring the dependence of $u(0;\boldsymbol{y})$ on $\boldsymbol{y}$. This yields

$$u(t;\boldsymbol{x}) \approx \frac{A}{(4\pi\kappa t)^{d/2}} \exp\left(-\frac{x^2}{4\kappa t}\right), \quad A = \int u(0;\boldsymbol{y})\, d^d y.$$

This approximation corresponds to replacing $u(0;\boldsymbol{y})$ with $A\delta(\boldsymbol{y})$ in Eq. (5.35).

Alternatively, this result can be interpreted as expanding the term $\exp\left(-\frac{(\boldsymbol{x}-\boldsymbol{y})^2}{4\kappa t}\right)$ in a Taylor series about $\boldsymbol{y} = 0$ and keeping only the leading order term. If $A = 0$, we need to account for the $O(y^1)$ term in the expansion, which gives the following asymptotic behavior:

$$u(t;\boldsymbol{x}) \approx \frac{(\boldsymbol{B}\cdot\boldsymbol{x})}{(4\pi\kappa t)^{d/2+1}} \exp\left(-\frac{x^2}{4\kappa t}\right), \quad \boldsymbol{B} = 2\pi \int \boldsymbol{y} u(0;\boldsymbol{y})\, d^d y.$$

Exercise 5.4. Find the long-time asymptotic behavior of the one-dimensional diffusion equation for the following initial conditions:

(a) $u(0;x) = x \exp\left(-\frac{x^2}{2l^2}\right)$,

(b) $u(0;x) = \exp\left(-\frac{|x|}{l}\right)$,

(c) $u(0;x) = x \exp\left(-\frac{|x|}{l}\right)$,

(d) $u(0;x) = \frac{1}{x^2+l^2}$,

(e) $u(0;x) = \frac{x}{(x^2+l^2)^2}$.

Hint: Expand the diffusion kernel in the integrand of Eq. (5.35) as a series in y.

Next, we seek the Green function of the diffusion equation, which solves

$$\partial_t G - \kappa \nabla_{\boldsymbol{r}}^2 G = \delta(t)\delta(\boldsymbol{x}).$$

In fact, we have already solved this problem. Setting $u(0;\boldsymbol{y}) = G(0;\boldsymbol{x}) = \delta(\boldsymbol{x})$ in Eq. (5.35) yields

$$G(t;\boldsymbol{x}) = \frac{1}{(4\pi\kappa t)^{d/2}} \exp\left(-\frac{\boldsymbol{x}^2}{4\kappa t}\right). \tag{5.36}$$

This Green function can be used to solve the inhomogeneous diffusion equation

$$\partial_t u - \kappa \nabla_{\boldsymbol{x}}^2 u = \phi(t;\boldsymbol{x}), \tag{5.37}$$

where the solution is expressed as a convolution of the Green function with the source term:

$$u(t;\boldsymbol{x}) = \int_{-\infty}^{t} dt' \int d^d y\, G(t-t';\boldsymbol{x}-\boldsymbol{y})\phi(t';\boldsymbol{y}),$$

assuming $u(\infty;\boldsymbol{x}) = 0$.

Exercise 5.5. Solve Eq. (5.37) for $\phi(t;\boldsymbol{x}) = \theta(t)\exp\left(-\frac{x^2}{2l^2}\right)$ in $d = 4$ dimensions.

5.6 Boundary Value Problems: Fourier Method

Consider the boundary value problem describing sound waves:

$$\begin{aligned} &\partial_t^2 u(t,x) - c^2 \partial_x^2 u(t,x) = 0 \quad \text{s.t.} \\ &0 \le x \le L, \quad u(t,0) = u(t,L) = 0, \\ &u(0,x) = \varphi(x), \quad \partial_t u(0,x) = \psi(x). \end{aligned} \tag{5.38}$$

This problem can be solved using the Fourier method (also called the method of separation of variables), which proceeds in two main steps.

We first look for a particular solution that satisfies the boundary conditions for x. Assume that the solution can be written in the separable form $u(t,x) = X(x)T(t)$. Substituting this ansatz into Eq. (5.38) gives

$$\frac{X''(x)}{X(x)} = \frac{T''(t)}{T(t)} = -\lambda,$$

where λ is a separation constant. The general solution for $X(x)$ is

$$X(x) = A\cos(\sqrt{\lambda}x) + B\sin(\sqrt{\lambda}x).$$

To satisfy the boundary conditions $X(0) = X(L) = 0$, we require that $A = 0$ and that $\sqrt{\lambda}L = n\pi$ for some integer $n = 1, 2, \ldots$. Thus, we obtain

$$\lambda_n = \left(\frac{n\pi}{L}\right)^2, \quad X_n(x) = \sin\left(\frac{n\pi x}{L}\right).$$

Substituting λ_n into the equation for $T(t)$, we solve for $T_n(t)$:

$$T_n(t) = A_n \cos\left(\frac{n\pi ct}{L}\right) + B_n \sin\left(\frac{n\pi ct}{L}\right),$$

where A_n and B_n are constants to be determined.

The functions $X_n(x)$ form a complete basis, so the general solution is given by a linear combination of the basis solutions:

$$u(t,x) = \sum_{n=1}^{\infty} X_n(x)T_n(t).$$

To satisfy the initial conditions, we expand $\varphi(x)$ and $\psi(x)$ in terms of the basis functions $X_n(x)$:

$$\varphi(x) = \sum_{n=1}^{\infty} A_n X_n(x), \quad \psi(x) = \sum_{n=1}^{\infty} \lambda_n B_n X_n(x). \tag{5.39}$$

The eigenfunctions $X_n(x)$ are orthonormal:

$$\int_0^L X_n(x)X_m(x)\,dx = \frac{L}{2}\delta_{nm}.$$

Multiplying both sides of Eq. (5.39) by $X_m(x)$, integrating from 0 to L and using the orthonormality condition, we obtain

$$A_m = \frac{2}{L}\int_0^L \varphi(x)X_m(x)\,dx, \quad B_m = \frac{2}{\lambda_m L}\int_0^L \psi(x)X_m(x)\,dx.$$

Exercise 5.6. The equation describing the deviation of a string from the straight line $u(t,x)$ is $\partial_t^2 u - c^2\partial_x^2 u = 0$, where x is the position along the string, t is time and c is the speed of sound. Assume that the string has an initial parabolic shape, $u(0,x) = 4hx(L-x)/L^2$, with both ends attached at $x=0$ and $x=L$. Further, assume that the string's velocity is zero at $t=0$, i.e., $\forall x \in [0,L]$, $\partial_t u(0,x) = 0$. Find the time evolution of the string deviation $u(t,x)$ at any point $x \in [0,L]$.

Now, let us consider the following parabolic boundary value problem over $x \in [0,L]$:

$$\partial_t u = a^2\partial_x^2 u, \quad u(t,0) = u(t,L) = 0, \quad u(0,x) = \begin{cases} x, & x < L/2, \\ L - x, & x > L/2. \end{cases}$$

We again use the Fourier method. The spectral part of the solution remains the same as in the hyperbolic case, but the time-dependent part is different. For the time dependence, we solve $T_n'(t) = -\lambda_n T_n(t)$, which yields the decaying solution

$$T_n(t) = A_n \exp\left(-\left(\frac{n\pi}{L}\right)^2 a^2 t\right).$$

The Fourier expansion of the initial condition is the same as in the hyperbolic case, leading to the solution

$$\begin{aligned} u(t,x) = \frac{4L}{\pi^2}\sum_{n=0}^{\infty} \frac{(-1)^n}{(2n+1)^2} \exp\left(-\left(\frac{(2n+1)\pi}{L}\right)^2 a^2 t\right) \\ \times \sin\left(\frac{(2n+1)\pi x}{L}\right). \end{aligned}$$

This solution is symmetric with respect to the middle of the interval, i.e., $u(t,x) = u(t,L-x)$, a symmetry inherited from the initial condition.

Exercise 5.7. Solve the following boundary value problem:

$$\partial_t u = a^2\partial_x^2 u - \beta u, \quad u(t,0) = u(t,L) = 0, \quad u(0,x) = \sin\left(\frac{2\pi x}{L}\right).$$

5.7 Case Study: Burgers' Equation*

Burgers' equation is an extension of the Hopf equation, Eq. (5.9), which we discussed earlier to illustrate the method of characteristics. Recall that the Hopf equation leads to wave breaking, which produces a non-physical, multi-valued solution. To address this issue, the Hopf equation can be modified by adding dissipation, resulting in Burgers' equation:

$$\partial_t u + u\partial_x u = \partial_x^2 u. \tag{5.40}$$

At first glance, Burgers' equation appears to be a typical nonlinear PDE that is difficult to solve. However, it is a special case because it allows for the Hopf [18]–Cole [19] transformation, which reduces the nonlinear equation to a simpler form. This transformation introduces a new function $\Psi(t;x)$, related to $u(t;x)$ via

$$u(t;x) = -2\frac{\partial_x \Psi(t;x)}{\Psi(t;x)}.$$

Substituting this transformation into Burgers' equation reduces it to the linear diffusion equation

$$\partial_t \Psi = \partial_x^2 \Psi. \tag{5.41}$$

The solution to the Cauchy problem for Eq. (5.41) can be expressed as a convolution of the initial profile $\Psi(0;x)$ with the Green function of the diffusion equation, given in Eq. (5.36):

$$\Psi(t;x) = \int \frac{dy}{\sqrt{4\pi t}} \exp\left(-\frac{(x-y)^2}{4t}\right) \Psi(0;y). \tag{5.42}$$

This expression allows us to construct some exact solutions to Burgers' equation. For example, consider the initial condition $\Psi(0;x) = \cosh(ax)$. Substituting this into Eq. (5.42) and performing the integration over y yields

$$\Psi(t;x) = \cosh(ax)\exp(a^2 t).$$

According to the Hopf–Cole transformation, this results in a stationary (time-independent) solution to Burgers' equation, commonly

*This auxiliary section can be skipped on the first reading. Material from this section will not be included in midterm or final exams.

referred to as a standing shock:

$$u(t;x) = -2a\tanh(ax).$$

A more general solution to Burgers' equation corresponds to a shock moving with a constant speed u_0, given by

$$u(t;x) = u_0 - 2a\tanh(a(x - x_0 - u_0 t)).$$

Exercise 5.8. Solve the diffusion equation, Eq. (5.41), with the initial condition $\Psi(0;x) = \cosh(ax) + B\cosh(bx)$. Reconstruct the corresponding $u(t;x)$ by solving Burgers' equation, Eq. (5.40). Analyze the result in the regime where $b > a$ and $B \gg 1$. Additionally, verify through a computational experiment that the resulting spatio-temporal dynamics corresponds to a large shock "engulfing" a smaller shock.

Part III

Functional Optimization

Chapter 6

Calculus of Variations

The central theme of this chapter is the connection between differential equations and principles of minimization. In simple terms, minimizing a function $S(q)$ involves solving the condition $S'(q) = 0$. For example, consider a quadratic function $S(q) = \frac{1}{2}q^T K q - q^T g$, where K is positive definite. The minimum of $S(q)$ occurs at q_*, which satisfies the equation $S'(q_*) = Kq_* - g = 0$.

In this example, q is an n-dimensional vector in finite-dimensional space, $q \in \mathbb{R}^n$. Now, let's extend this finite-dimensional optimization problem to an infinite-dimensional, continuous setting, where $q(x)$ is a function, say $q(x) : \mathbb{R} \to \mathbb{R}$, and $S\{q(x)\}$ is a functional. Typically, this functional is expressed as an integral whose integrand depends on $q(x)$ and its derivative, $q'(x)$, for instance:

$$S\{q(x)\} = \int dx \left(\frac{c}{2}(q'(x))^2 - g(x)q(x)\right).$$

The derivative of a functional with respect to $q(x)$ is known as the variational derivative. By analogy with the finite-dimensional case, the Euler–Lagrange (EL) equation

$$\frac{\delta S\{q\}}{\delta q(x)} = 0$$

provides the condition for minimizing the functional. The purpose of this chapter is to explore the concept of variational derivatives and related ideas of the variational calculus, both in theory and through examples.

Taking a broader view, variational calculus is a powerful mathematical framework used to find the extrema of functionals, which are mappings from a space of functions to the real numbers. It provides the foundation for many fundamental principles in physics, mechanics and optimization, including the derivation of equations of motion, control laws and energy minimization problems.

We begin this chapter in Section 6.1 by presenting a range of applications where variational formulations naturally arise, specifically focusing on the minimization of an action functional. These examples serve to illustrate the breadth and utility of variational principles across different fields. In Section 6.2, we derive and discuss the EL equations, using the aforementioned examples/applications to demonstrate how they provide necessary conditions for minimizing an action functional. We also explore discretization techniques in Section 6.3, highlighting the connection between variational calculus and finite-dimensional optimization problems. Numerical approaches for solving the EL equations are introduced in Section 6.4, where we outline key computational methods for practical implementation.

The EL equations are valuable not only because they provide necessary conditions for minimizing an action functional but also because, in the case of a single-variable functional, they enable us to analyze the action as a function (not a functional) of the boundary conditions – a topic explored in Section 6.5. Section 6.6 applies the core concepts of the single-variable variational calculus to classical mechanics, introducing foundational ideas such as Noether's theorem, the Hamiltonian formulation and the Hamiltonian and Hamilton–Jacobi equations.

The remainder of the chapter delves into advanced techniques useful for solving variational calculus problems in practice. In Section 6.7, we discuss the Legendre–Fenchel transform, a key tool in optimization and duality theory. We then explore second variation techniques in Section 6.8, which are crucial for analyzing the stability of extremals. Finally, in Section 6.9, we address the method of Lagrange multipliers, focusing on its application to variational problems with constraints, both function based and functional based.

For readers seeking a deeper and more comprehensive understanding of the calculus of variations, we recommend several key resources. *Calculus of Variations* by Gelfand and Fomin [20] is a classic text that introduces the subject with clarity, covering the derivation of the EL equations and their applications in physics and engineering. *Calculus of Variations I* by Giaquinta and Hildebrandt [21] provides a more advanced and rigorous treatment of the subject, including modern developments in the theory of variational problems. Lastly, for readers interested in the connection between variational principles and mechanics, *Mathematical Methods of Classical Mechanics* by Arnold [22] is an indispensable resource, offering a geometric perspective on mechanics through the lens of variational calculus and Hamiltonian dynamics.

6.1 Examples

To better understand the calculus of variations, we start by describing four examples.

6.1.1 *Fastest path*

Consider a robot navigating in the (x, y)-plane, where its path is described by a function $y = q(x)$. For a small displacement δx, the arc length of the robot's path from (x, y) to $(x + \delta x, y + \delta y)$ can be approximated using the Pythagorean theorem as $\sqrt{(\delta x)^2 + (\delta y)^2}$, which simplifies to $\sqrt{1 + (q'(x))^2}dx$ for infinitesimal δx. If the terrain varies in ruggedness, the robot's speed may change at each point in the plane. Let $\mu^{-1}(x, y)$ be a positive scalar function that describes the robot's speed at each point. The time it takes for the robot to move from $(x, q(x))$ to $(x + dx, q(x + dx))$ along the path $q(x)$ is given by

$$L(x, q(x), q'(x)) := \mu(x, q(x))\sqrt{1 + (q'(x))^2}dx.$$

The total time required to travel from the initial point $(x_i, y_i) := (0, 0)$ to the terminal point $(x_t, y_t) := (a, b)$, where $a > 0$, is

$$S\{q(x)\} = \int_0^a dx\, L(x, q(x), q'(x)).$$

Given reasonable conditions on $\mu(x, y)$, the calculus of variations provides a way to find the optimal path through the domain, that is, the path that minimizes the functional $S\{q(x)\}$, subject to $q(0) = 0$ and $q(a) = b$.

6.1.2 *Minimal surface*

Consider the problem of forming a three-dimensional bubble by dipping a wire loop into soapy water. A natural question arises: Is there an optimal shape that minimizes the surface area of the bubble for a given loop? Physics suggests that the soap film will take the shape that minimizes surface area.

We formalize the problem as follows. The surface of the bubble is described by a continuously differentiable function, $q(x)$, where $x = (x_1, x_2) \in \mathcal{D}$ and $\mathcal{D} \subset \mathbb{R}^2$ is a bounded domain. On the boundary of $\mathcal{D}$, denoted $\partial\mathcal{D}$, the function $q(x)$ is fixed – this represents the shape of the wire loop. Specifically, $q(\partial\mathcal{D}) = g(\partial\mathcal{D})$, where $g(\partial\mathcal{D})$ describes the wire's coordinate along the third dimension. The optimal bubble shape minimizes the following functional:

$$S\{q(x)\} = \int_{\mathcal{D}} dx \sqrt{1 + |\nabla_x q(x)|^2}, \tag{6.1}$$

subject to the boundary condition $q(\partial\mathcal{D}) = g(\partial\mathcal{D})$.

Example 6.1.1. Show that Eq. (6.1) represents a general formula for the surface area of the graph of a continuously differentiable function $q(x)$, where $x = (x_1, x_2) \in \mathcal{D} \subset \mathbb{R}^2$ and $\mathcal{D}$ is a region in the (x_1, x_2)-plane with a smooth boundary.

Solution. It suffices to derive the differential version of Eq. (6.1), showing that the area of an infinitesimal surface element around a point $x^{(0)} = (x_1^{(0)}, x_2^{(0)}) \in \mathcal{D}$ is described by

$$\sqrt{1 + |\nabla_x q(x)|^2} dx = \sqrt{1 + (\partial_{x_1} q(x_1, x_2))^2 + (\partial_{x_2} q(x_1, x_2))^2} dx_1 dx_2,$$

with $\sqrt{1 + |\nabla_x q(x)|^2}$ evaluated at $x = x^{(0)}$. In this infinitesimal case, we can represent the surface element locally by the plane

$$x_3 - x_3^{(0)} = a(x_1 - x_1^{(0)}) + b(x_2 - x_2^{(0)}),$$

in $\mathbb{R}^3$, where $x_3^{(0)} = q(x^{(0)})$, $a = \partial_{x_1} q|_{x=x^{(0)}}$ and $b = \partial_{x_2} q|_{x=x^{(0)}}$. This plane can be described using three points in space (see Fig. 6.1 for

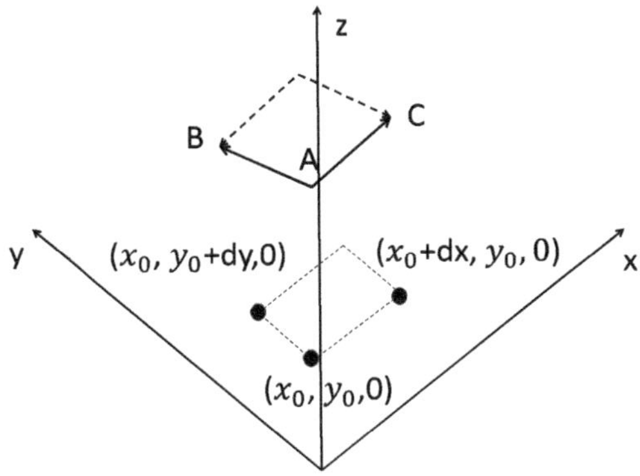

Figure 6.1. Illustration of an infinitesimal surface element in Example (6.1.1).

an illustration):

$$\begin{aligned}\mathrm{A} &= \left(x_1^{(0)}, x_2^{(0)}, x_3^{(0)}\right),\\ \mathrm{B} &= \left(x_1^{(0)} + dx_1, x_2^{(0)}, x_3^{(0)} + adx_1\right),\\ \mathrm{C} &= \left(x_1^{(0)}, x_2^{(0)} + dx_2, x_3^{(0)} + bdx_2\right).\end{aligned}$$

The area of the infinitesimal surface element is given by the absolute value of the cross product of two infinitesimal vectors:

$$\begin{aligned}|u \times v| &= |(\mathrm{B} - \mathrm{A}) \times (\mathrm{C} - \mathrm{A})|\\ &= |(-adx_1dx_2, -bdx_1dx_2, dx_1dx_2)| = dx_1dx_2\sqrt{1 + a^2 + b^2},\end{aligned}$$

where we use the standard vector cross product formula in three dimensions, $(y \times z)_i = \sum_{j,k=1,2,3} \varepsilon_{ijk} y_j z_k$, with ε_{ijk} as the Levi–Civita tensor. □

6.1.3 *Image restoration*

A grayscale image can be represented by a function, $q(x) : [0, 1]^2 \to [0, 1]$, mapping each location x within the square domain $[0, 1]^2 \subset \mathbb{R}^2$ to a real value between 0 (representing white) and 1 (representing black). However, in practice, the true image is often corrupted by noise, so the observed image is a noisy approximation of the true image. The goal of image restoration is to recover the original image from this noisy observation.

The total variation (TV) restoration method [23] is based on the hypothesis that the true image can be reconstructed by minimizing the following functional:

$$S\{q(x)\} = \int_{U=[0,1]^2} dx\, \left((q(x) - f(x))^2 + \lambda |\nabla_x q(x)| \right), \tag{6.2}$$

where $f(x)$ is the noisy image and λ is a positive parameter controlling the amount of smoothing. The minimization is subject to the Neumann boundary condition $\boldsymbol{n} \cdot \nabla_x q(x) = 0$ for all $x \in \delta U$, where $\boldsymbol{n}$ is the unit normal vector to the boundary δU of the domain U.

6.1.4 *Classical mechanics*

Classical mechanics is governed by the evolution of a system's spatial coordinates over time. These coordinates are represented by a function, $q(t) : \mathbb{R} \to \mathbb{R}^d$, mapping time $t \in \mathbb{R}$ to a d-dimensional real-valued spatial coordinate $q(t) \in \mathbb{R}^d$. The trajectory of the system in time is determined by the principle of least action, also known as the Hamiltonian principle. This principle states that the system follows the trajectory that minimizes the action, given by the functional

$$S\{q\} := \int_{t_1}^{t_2} dt\, L(t, q(t), \dot{q}(t)), \tag{6.3}$$

where $L(t, q(t), \dot{q}(t))$ is the system's Lagrangian and $\dot{q}(t) = dq(t)/dt$ is the velocity. The values of q are fixed at the initial and final times: $q(t_1) = q_1$ and $q(t_2) = q_2$, respectively.

For example, consider the dynamics of a particle with unit mass moving in a potential $V(q)$. The Lagrangian for this system is

$$L(t, q(t), \dot{q}(t)) = \frac{\dot{q}^2}{2} - V(q). \tag{6.4}$$

This describes the balance between the kinetic energy $\frac{\dot{q}^2}{2}$ and the potential energy $V(q)$, which governs the particle's motion.

6.2 Euler–Lagrange Equations

Each of the previous examples can be formulated as the problem of minimizing a functional

$$S\{q(x)\} = \int_{\mathcal{D} \subseteq \mathbb{R}^n} dx\, L\left(x, q(x), \nabla_x q(x)\right),$$

where the function $q(x)$ is fixed at the boundary, i.e., $q(x) = g(x)$ for $x \in \partial\mathcal{D}$. The domain $\mathcal{D}$ is bounded, and the Lagrangian L is a given function,

$$L : \mathcal{D} \subseteq \mathbb{R}^n \times \mathbb{R}^d \times \mathbb{R}^{d\times n} \to \mathbb{R},$$

which depends on three variables. For convenience, we denote the partial derivatives of L with respect to its arguments as L_x, L_q and $L_{\nabla q}$. Here, $x \in \mathcal{D} \subseteq \mathbb{R}^n$, $q \in \mathbb{R}^d$ and $\nabla_x q \in \mathbb{R}^{d\times n}$. Throughout, we assume that both L and g are smooth functions.

Theorem 6.2.1 (Necessary Condition for Optimality). *Suppose that $q(x)$ is a minimizer of S, i.e.,*

$$S\{\tilde{q}(x)\} \geq S\{q(x)\} \quad \forall x \in \partial\mathcal{D} : \tilde{q}(x) = q(x),$$

where $\tilde{q}(x) \in C^2(\bar{\mathcal{D}} = \mathcal{D} \cup \partial\mathcal{D})$. Then, L satisfies the EL equation:

$$\nabla_x \left(L_{\nabla q}\left(x, q(x), \nabla_x q(x)\right)\right) - L_q\left(x, q(x), \nabla_x q(x)\right) = 0, \quad \forall x \in \mathcal{D}. \tag{6.5}$$

Sketch of Proof. Consider a small perturbation $q(x) \to q(x) + \varepsilon\delta(x) = \tilde{q}(x)$, where $\varepsilon \in \mathbb{R}$ and $\delta(x)$ is sufficiently smooth and vanishes at the boundary, i.e., $\delta(x) = 0$ for all $x \in \partial\mathcal{D}$. The assumption implies that

$$S\{q(x)\} \leq S\{\tilde{q}(x)\} = S\{q(x) + \varepsilon\delta(x)\}, \quad \forall\varepsilon \in \mathbb{R}.$$

Therefore, we must have

$$\left.\frac{d}{d\varepsilon} S\{q(x) + \varepsilon\delta(x)\}\right|_{\varepsilon=0} = 0.$$

Now, substitute $S\{q(x) + \varepsilon\delta(x)\}$ into the integral expression for S:

$$S\{q(x) + \varepsilon\delta(x)\} = \int_{\mathcal{D}} dx\, L\left(x, q(x) + \varepsilon\delta(x), \nabla_x q(x) + \varepsilon\nabla_x\delta(x)\right).$$

Then, by interchanging the order of differentiation and integration, applying the chain rule to the Lagrangian and evaluating one of the resulting integrals by parts while discarding the boundary term (since $\delta(x) = 0$ on $\partial\mathcal{D}$), we obtain

$$\begin{aligned}
\frac{d}{d\varepsilon} S\{q(x) + \varepsilon\delta(x)\}\bigg|_{\varepsilon=0} &= \int_{\mathcal{D}} dx \frac{d}{d\varepsilon} L\left(x, q(x) + \varepsilon\delta(x), \nabla_x q(x)\right. \\
&\quad \left. + \varepsilon\nabla_x \delta(x)\right)\bigg|_{\varepsilon=0} \\
&= \int_{\mathcal{D}} dx \left(L_q\left(x, q(x), \nabla_x q(x)\right) \cdot \delta(x)\right. \\
&\quad \left. + L_p\left(x, q(x), \nabla_x q(x)\right) \cdot \nabla_x \delta(x)\right) \\
&= \int_{\mathcal{D}} dx\, L_q\left(x, q(x), \nabla_x q(x)\right) \cdot \delta(x) \\
&\quad + \int_{\mathcal{D}} dx\, L_p\left(x, q(x), \nabla_x q(x)\right) \cdot \nabla_x \delta(x) \\
&= \int_{\mathcal{D}} dx \left(L_q\left(x, q(x), \nabla_x q(x)\right)\right. \\
&\quad \left. - \nabla_x \cdot L_{\nabla q}\left(x, q(x), \nabla_x q(x)\right)\right) \cdot \delta(x).
\end{aligned}$$

Since this integral must equal zero for any arbitrary $\delta(x)$, we arrive at the desired result. □

Remark. The solutions to the EL equation, Eq. (6.5), are stationary curves of the functional $S\{q(x)\}$, meaning that they could correspond to minima, maxima or saddle points. Therefore, theorem 6.2.1 provides a necessary, but not sufficient, condition for minimizing $S\{q(x)\}$.

Example 6.2.2. Find the EL equations for the following functionals:

(a) $\displaystyle\int dx\, \left((q'(x))^2 + \exp(q(x))\right),$

(b) $\displaystyle\int dx\, q(x)q'(x),$

(c) $\displaystyle\int dx\, x^2(q'(x))^2.$

Solution. In one dimension, the EL equation simplifies to

$$\frac{d}{dx}\left(\frac{\partial L}{\partial q'}\right) - \frac{\partial L}{\partial q} = 0.$$

We derive the respective equations for each case:

(a) Here, $L(x, q, q') = (q')^2 + \exp(q)$:

$$\frac{d}{dx}\left(\frac{\partial L}{\partial q'}\right) = \frac{d}{dx}(2q') = 2q'',$$
$$\frac{\partial L}{\partial q} = \exp(q).$$

Thus, the EL equation is

$$2q'' - \exp(q) = 0.$$

(b) Here, $L(x, q, q') = qq'$:

$$\frac{d}{dx}\left(\frac{\partial L}{\partial q'}\right) = \frac{d}{dx}q = q',$$
$$\frac{\partial L}{\partial q} = q'.$$

Thus, the EL equation is

$$q' - q' = 0.$$

This implies that any function $q(x)$ satisfies the EL equation, meaning that the functional has no minima or maxima.

(c) Here, $L(x, q, q') = x^2(q')^2$:

$$\frac{d}{dx}\left(\frac{\partial L}{\partial q'}\right) = \frac{d}{dx}(2x^2 q') = 2x^2 q'' + 4xq',$$
$$\frac{\partial L}{\partial q} = 0.$$

Thus, the EL equation is

$$x^2 q'' + 2xq' = 0.$$

Example 6.2.3. Consider the shortest path problem discussed in Section 6.1.1, assuming a uniform speed ($g(x, y) = 1$):

$$\min_{q(x)} \int_0^a dx \sqrt{1 + (q'(x))^2}\bigg|_{q(0)=0, q(a)=b}.$$

Find the EL equation for this problem.

Solution. The EL condition for $q(x)$ is

$$0 = \frac{d}{dx}\left(\frac{q'(x)}{\sqrt{1 + (q'(x))^2}}\right) \to \frac{q'(x)}{\sqrt{1 + (q'(x))^2}} = \text{constant}.$$

This implies $q'(x)$ is constant, and thus

$$q(x) = \frac{b}{a}x,$$

where we used the boundary conditions to solve for the constant. The shortest path is therefore a straight line.

Exercise 6.1.

(a) Derive the EL equation for the fastest path problem described in Section 6.1.1.
(b) Provide an example of a speed function, $\mu(x, y)$, that allows a quadratic solution to the EL equation $q(x) = cx^2$, with an arbitrary non-singular c.

6.3 Phase-Space Intuition and Relation to Optimization (Finite Dimensional, Not Functional)

Consider the special case of the fastest path problem from Section 6.1.1, which generalizes the shortest path problem of Example 6.2.3 by allowing the speed $\mu(x)$ to depend on x. In this case, the action is given by

$$S\{q(x)\} = \int_0^a dx\, \mu(x) \sqrt{1 + (q'(x))^2} = \int_0^a ds\, \mu(x),$$

where ds is the infinitesimal arc length along the curve $q(x)$:

$$ds = \sqrt{1 + (q'(x))^2} dx = \sqrt{dx^2 + dq^2}.$$

The Lagrangian for this system is $L(x; q(x); q'(x)) = \mu(x) \sqrt{1 + (q'(x))^2}$, and its partial derivatives are $L_q = 0$ and $L_{q'} = \mu(x)$

$q'/\sqrt{1+(q'(x))^2}$. The corresponding EL equation is

$$\frac{d}{dx}\left(\frac{\mu(x)q'(x)}{\sqrt{1+(q'(x))^2}}\right)=0,$$

which simplifies to

$$\frac{\mu(x)q'(x)}{\sqrt{1+(q'(x))^2}}=\mu(x)\sin(\theta)=\text{constant}, \tag{6.6}$$

where θ is the angle between the tangent to the curve $q(x)$ and the x-axis.

To gain further insight, we can derive Eq. (6.6) without using variational calculus, instead adopting a finite-dimensional optimization perspective. This method involves discretizing the action $S\{q(x)\}$ and applying standard optimization techniques.

First, we **discretize** the action

$$\begin{aligned}S\{q(x)\}\approx S_k(\ldots,q_k,\ldots)&=\sum_k \mu_k\Delta s_k\\&=\sum_k \mu_k\sqrt{1+\left(\frac{q(x_k)-q(x_{k-1})}{\Delta}\right)^2}\Delta,\end{aligned}$$

where $\Delta = x_{k+1}-x_k$ is the step size in x for all k and Δs_k is the length of the kth segment of the discretized curve, as illustrated in Fig. 6.2.

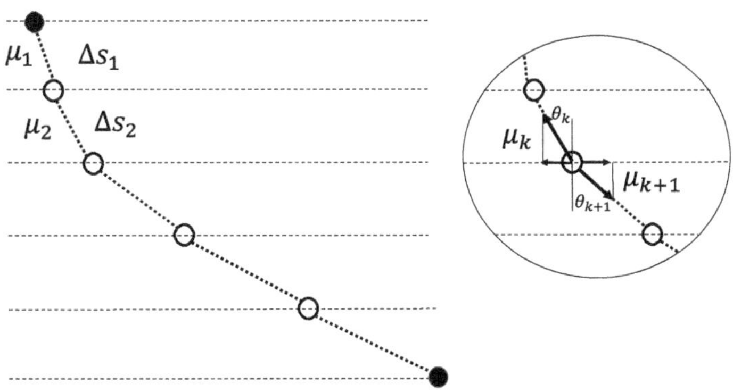

Figure 6.2. Variational calculus via discretization and optimization.

Next, we seek the **extrema** of S_k by varying q_k and setting $\partial_{q_k} S_k = 0$ for all k. This yields a discretized version of the EL equation:

$$\forall k: \quad \frac{\mu_{k+1}(q_{k+1} - q_k)}{\sqrt{1 + \left(\frac{q_{k+1}-q_k}{\Delta}\right)^2}} = \frac{\mu_k(q_k - q_{k-1})}{\sqrt{1 + \left(\frac{q_k - q_{k-1}}{\Delta}\right)^2}}.$$

This can be further simplified to

$$\mu_{k+1} \sin(\theta_{k+1}) = \mu_k \sin(\theta_k),$$

where θ_k is the angle of the kth segment of the discretized path.

6.4 Toward Numerical Solutions of the Euler–Lagrange Equations (*)

In this section,* we explore the image restoration problem introduced in Section 6.1.3. We derive the EL equations and note that solving these equations analytically is challenging. This leads us to discuss the theoretical approach (or philosophy) of solving the EL equations numerically. Following the framework presented in Ref. [24], we first cover gradient descent in this section and later explore the primal-dual method in Section 6.7.

6.4.1 *Smoothing the Lagrangian*

The TV functional in Eq. (6.2) is non-differentiable at $\nabla_x q(x) = 0$, making it difficult to apply variational methods directly. A common workaround is to smooth the Lagrangian by introducing a small regularization parameter, ε, leading to the modified functional

$$S_\varepsilon\{q\} = \int_{[0,1]^2} dx \left(\frac{(q(x) - f(x))^2}{2} + \lambda\sqrt{\varepsilon^2 + (\nabla_x q(x))^2} \right),$$

where $\varepsilon > 0$ is small. The EL equations corresponding to this smoothed functional are

$$q - \lambda \nabla_x \cdot \frac{\nabla_x q}{\sqrt{\varepsilon^2 + (\nabla_x q(x))^2}} = f \quad \text{for all } x \in [0,1]^2, \tag{6.7}$$

*This auxiliary section can be skipped on first reading.

subject to homogeneous Neumann boundary conditions, $\partial q(x)/\partial n = 0$ for all $x \in \partial[0,1]^2$, where n is the unit normal vector to the boundary. Analytical solutions to Eq. (6.7) are generally intractable, motivating the need for numerical techniques, which we now explore.

6.4.2 *Gradient descent and acceleration*

Before diving into numerical methods, we note that this section provides a high-level overview rather than a detailed explanation. The goal here is to highlight connections between numerical methods for solving PDEs and optimization algorithms.

A common numerical approach to solving Eq. (6.7) is gradient descent. To conceptualize gradient descent, we introduce an auxiliary computational "time" dimension. While time will be discrete in actual implementation, we can think of it as continuous for intuition and analysis. The idea is to evolve the function $v(t;x)$ in time according to the following equation:

$$\partial_t v + v - \lambda \nabla_x \cdot \frac{\nabla_x v}{\sqrt{\varepsilon^2 + (\nabla_x v(x))^2}} = f \quad \text{for}\, t > 0,\ x \in [0,1]^2, \tag{6.8}$$

where $v(t;x)$ is the estimate of $q(x)$ at time t, with initial conditions $v(0;x) = f(x)$ and boundary conditions $\partial v(x)/\partial n = 0$ for $x \in \partial[0,1]^2$. Equation (6.8) is a nonlinear heat equation. Near equilibrium, this equation can be linearized.

By discretizing Eq. (6.8) on a spatio-temporal grid with time step Δt and spatial step Δx, we can estimate the relationship between these steps based on numerical stability:

$$\Delta t \sim \frac{\varepsilon(\Delta x)^2}{\lambda}.$$

This shows that Δt must be very small (on the order of the square of Δx) to ensure stability, especially as ε decreases, making the scheme prone to stiffness.

One way to reduce the stiffness of the gradient descent scheme is to replace the diffusion equation (6.8) with a **damped wave equation**:

$$\forall x \in [0,1]^2,\ t > 0: \quad \partial_t^2 v + a\partial_t v + v - \lambda \nabla_x \cdot \frac{\nabla_x v}{\sqrt{\varepsilon^2 + (\nabla_x v(x))^2}} = f, \tag{6.9}$$

where a is the damping coefficient.

To estimate the appropriate spatial discretization step Δx, temporal discretization step Δt and the damping coefficient, we draw an analogy with the diffusive case. Linearizing the nonlinear wave Eq. (6.9) and requiring that the three key terms – temporal oscillation (∂_t^2), damping ($a\partial_t$) and diffusion ($\frac{\lambda}{\varepsilon}\nabla_x^2$) – are balanced, we arrive at the following empirical estimate:

$$(\Delta t)^2 \sim \frac{\Delta t}{a} \sim \frac{\varepsilon(\Delta x)^2}{\lambda},$$

which leads to a much less restrictive, linear scaling of $\Delta t \sim \Delta x$.

This improvement comes from transitioning from overdamped relaxation (pure gradient descent) to a regime where damping and oscillations are balanced. This is the basis of **Polyak's heavy-ball method** [25] and **Nesterov's accelerated gradient descent method** [26] – see also Appendix B – both of which are now widely used in modern optimization, including training of neural networks.

For a deeper exploration of modern, continuous-time interpretations of these acceleration methods and related algorithms, see Ref. [27, 28], as well as Sections 2.3 and 3.6 in Ref. [24].

We will revisit the image restoration problem again in Section 6.7.2, where we will explore an alternative approach using the **primal-dual algorithm**.

6.5 Dependence of the Action on the Endpoints

Consider $x \in \mathbb{R}$ and let the points $A_0 := (x_0, q_0)$ and $A_1 := (x_1, q_1)$ be given, and let $\mathcal{F}$ be the family of continuously differentiable functions on $[x_0, x_1]$ that satisfy $q(x_0) = q_0$ and $q(x_1) = q_1$. For a given Lagrangian $L(x, q, q')$, let $S : \mathcal{F} \to \mathbb{R}$ be the functional

$$S\{q(x)\} = \int_{x_0}^{x_1} L(x, q(x), q'(x))\, dx.$$

In Section 6.2, we showed that if a function $q(x)$ satisfies the EL equations, $\frac{d}{dx}L_{q'} - L_q = 0$, then $q(x)$ is a stationary curve of S and therefore a candidate for a minimizer of S. (See also the discussion of the EL theorem 6.2.1 and the following remark.)

Notation. In the following, we slightly abuse notation and use the symbol S to denote both the action functional

$$S\{q(x)\} = \int_{x_0}^{x_1} L(x, q(x), \dot{q}(x))\, dx$$

and also the action function

$$S(A_0, A_1) = S(x_0, q_0, x_1, q_1) = \min_{q(x)\in\mathcal{F}} \int_{x_0}^{x_1} L(x, q(x), \dot{q}(x))\, dx.$$

Thus, $S(A_0, A_1)$ corresponds to $S\{q(x)\}$ evaluated at the solutions $q(x)$ of the associated EL equations and should be understood as a function of the end points A_0 and A_1 of $q(x)$.

The following statement gives a geometrical interpretation for the derivatives of the action with respect to the end-point parameters:

Theorem 6.5.1 (End-point derivatives of the action).

$$\begin{aligned}
\text{(a)}\quad & \partial_{x_1} S(A_0; A_1) = (L - q' L_{q'})\big|_{x=x_1} \\
& \partial_{x_0} S(A_0; A_1) = \; -(L - q' L_{q'})\big|_{x=x_0}, \\
\text{(b)}\quad & \partial_{q_1} S(A_0; A_1) = L_{q'}\big|_{x=x_1} \\
& \partial_{q_0} S(A_0; A_1) = \; -L_{q'}\big|_{x=x_0}.
\end{aligned}$$

Proof. Here, we sketch the proof. We focus on the part of the theorem concerning derivatives with respect to x_1 and q_1, i.e., the final end point of the critical path.

Let us first fix the final independent variable at x_1 but move the final position by dq, as shown in Fig. 6.3. The trajectory $q(x)$ varies by $\delta q(x)$, where $\delta q(x_0) = 0$ and $\delta q(x_1) = dq$.

Using a first-order Taylor expansion, we can estimate the value of $L(x, q + \delta q, q' + \delta q')$ at each point x:

$$L(x, q+\delta q, q'+\delta q') \approx L(x, q, q') + L_q(x, q, q')\cdot \delta q + L_{q'}(x, q, q')\cdot \delta q'.$$

Thus, the variation of the action becomes

$$dS = \int_{x_0}^{x_1} \left(L_{q'}\delta q' + L_q \delta q\right) dx.$$

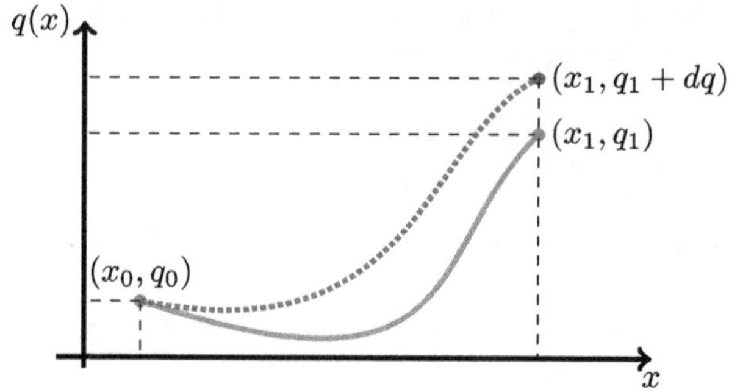

Figure 6.3. Critical curves from (x_0, q_0) to (x_1, q_1) (solid) and to $(x_1, q_1 + dq)$ (dashed).

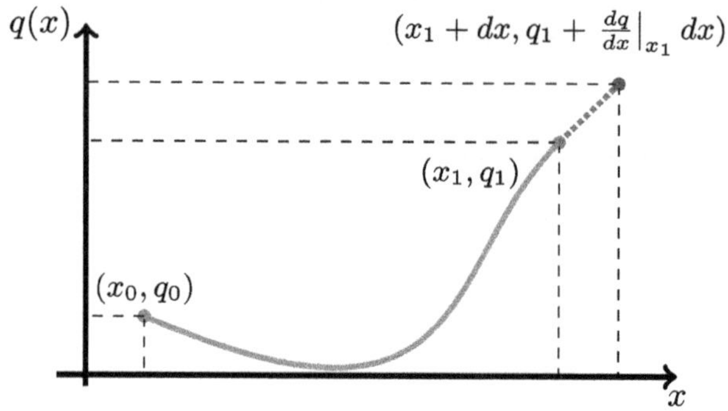

Figure 6.4. Critical curves from (x_0, q_0) to (x_1, q_1) (solid) and to $(x_1 + dx, q_1 + dq)$ (dashed), where $dq = \frac{dq}{dx}\big|_{x_1} dx$.

Substituting $\delta q' = \frac{d}{dx}\delta q$ and using the EL equations (6.5), we get

$$dS = \int_{x_0}^{x_1} \frac{d}{dx}\left(L_{q'}\delta q\right) dx = L_{q'}\big|_{x_1} dq.$$

Thus, keeping the final independent variable fixed, we have $dS = \partial_{q_1} S \, dq$, and we arrive at the desired result:

$$\frac{\partial S}{\partial q_1} = L_{q'}\big|_{x_1}.$$

Next, consider extending the action by changing the final point from (x_1, q_1) to $(x_1 + dx, q_1 + q'(x_1)dx)$, as shown in Fig. 6.4:

$$dS = L\, dx = \frac{\partial S}{\partial x_1} dx + \frac{\partial S}{\partial q_1} q'(x_1) dx = \left(\frac{\partial S}{\partial x_1} + q' L_{q'}\right) dx.$$

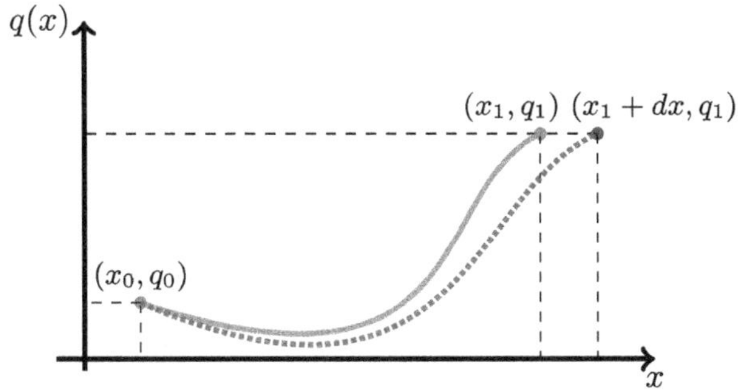

Figure 6.5. Critical curves from (x_0, q_0) to (x_1, q_1) (solid) and to $(x_1 + dx, q_1)$ (dashed).

Finally, from Fig. 6.5, we derive the remaining result:

$$\frac{\partial S}{\partial x_1} = \left(L - q' L_{q'}\right)\Big|_{x_1}. \qquad \square$$

Example 6.5.2. Find the minimizers of the functional

$$S\{q(x)\} = \int_0^1 \left(q'^2 + xq\right) dx$$

for the following cases: (a) $q(0) = 1$ and $q(1) = 0$ and (b) $q(0) = 1$ and $q(1)$ is free.

Solution. The stationary curve satisfies the EL equations:

$$\frac{d}{dx}L_{q'} - L_q = 0 \Rightarrow \frac{d}{dx}2q' - x = 0 \Rightarrow q'' = x.$$

Thus, $q(x) = \frac{1}{6}x^3 + c_1 x + c_2$. For case (a), the conditions $q(0) = 1$ and $q(1) = 0$ yield $c_1 = \frac{7}{6}$ and $c_2 = 1$. In case (b), where $q(1)$ is free, we impose the optimality condition $\partial_{q_1} S = 0$:

$$0 = \frac{\partial S}{\partial q_1} = 2q'\Big|_{x=1}.$$

Thus, $q'(1) = 0$, which gives $c_1 = -\frac{1}{2}$ and $c_2 = 1$. $\square$

Exercise 6.2. Find the critical function(s) $q(x)$ of the functional

$$S\{q(x)\} = \int_0^a \left(q^2 + 2qq' + (q')^2 + 1\right) dx,$$

subject to $q(0) = 0$ and the boundary condition $q(a) = 1$, where a is itself subject to optimization.

6.6 Variational Principle of Classical Mechanics

In this section, we apply the principle of minimal action (also known as the variational principle, or Hamiltonian principle) to classical mechanics, as highlighted in Section 6.1.4. (For further details, see also Ref. [29], which we follow in this section.)

6.6.1 *Conservation of energy and Noether's theorem*

In the context of classical mechanics introduced in Section 6.1.4, the EL equations take the form

$$\frac{d}{dt}L_{\dot{q}} - L_q = 0, \tag{6.10}$$

where the Lagrangian is a function $L(t, q(t), \dot{q}(t))\colon \mathbb{R} \times \mathbb{R}^d \times \mathbb{R}^d \to \mathbb{R}$. Let us consider the special case where the Lagrangian does not depend explicitly on time. (It may still depend implicitly on time through $q(t)$ and $\dot{q}(t)$, i.e., $L(q(t), \dot{q}(t))$.) Remarkably, in this case, the EL equation can be rewritten as a conservation law. Specifically,

$$\frac{d}{dt}\left(\dot{q} \cdot L_{\dot{q}} - L\right) = \ddot{q} \cdot L_{\dot{q}} + \dot{q} \cdot \frac{d}{dt}L_{\dot{q}} - L_q \cdot \dot{q} - L_{\dot{q}} \cdot \ddot{q} = \dot{q} \cdot \left(\frac{d}{dt}L_{\dot{q}} - L_q\right) = 0,$$

where the last equality follows from Eq. (6.10).

Here, we have introduced the Hamiltonian $H = \dot{q} \cdot L_{\dot{q}} - L$, which represents the total energy of the mechanical system. We have also shown that, if the Lagrangian (and thus the Hamiltonian) does not explicitly depend on time, then the Hamiltonian (energy) is conserved. This result is a particular case of Noether's theorem.

Note that invariance under a parametric continuous transformation, such as the time-shift symmetry just explored, implies a conserved quantity – a general feature of Noether's theorem.

To express these invariances more generally, we introduce the following definition:

Definition 6.6.1 (Invariance of the Lagrangian). Consider a family of transformations of $\mathbb{R}^d$, $h_s(q)\colon \mathbb{R}^d \to \mathbb{R}^d$, where $s \in \mathbb{R}$ and $h_s(q)$ is continuous in both q and the parameter s, with $h_0(q) = q$. We say that a time-independent Lagrangian $L(q(t), \dot{q}(t))\colon \mathbb{R}^n \times \mathbb{R}^n \to \mathbb{R}$ is invariant under the action of the family of transformations

$h_s(q(t))\colon \mathbb{R}^n \to \mathbb{R}^d$ if $L(q,\dot{q})$ remains unchanged when $q(t)$ is replaced by $h_s(q(t))$. In other words, for any function $q(t)$,

$$L(h_s(q(t)), \frac{d}{dt}h_s(q(t))) = L(q(t), \frac{d}{dt}q(t)).$$

Common examples of $h_s(q(t))$ in classical mechanics include the following:

- **Translational invariance:** $h_s(q(t)) = q(t) + se$, where e is a unit vector in $\mathbb{R}^n$ and s is the displacement.
- **Rotational invariance:** $h_s(q(t)) = R_e(s)q(t)$, a rotation around the line through the origin defined by the unit vector e.
- **Helical symmetry:** $h_s(q(t)) = aes + R_e(s)q(t)$, where a is a constant.

Theorem 6.6.2 (Noether's theorem (1915)). *If the Lagrangian L is invariant under the action of a one-parameter family of transformations $h_s(q(t))$, then the quantity*

$$I(q(t), \dot{q}(t)) := L_{\dot{q}} \cdot \frac{d}{ds}\left(h_s(q(t))\right)_{s=0} \tag{6.11}$$

is constant along any solution of the EL Eq. (6.10). *Such a constant quantity is called an integral of motion.*

Proof. Following the discussion in Section 6.5, consider the action function $S(t_0, q_0, t_1, q_1)$, i.e., the minimum value of the action functional, analyzed as a function of the end points (t_0, q_0) and (t_1, q_1). Theorem 6.5.1, applied to the case of classical mechanics, gives

$$\begin{aligned}
\text{(a)}\quad & \partial_{t_1} S(A_0; A_1) = (L - \dot{q}L_{\dot{q}})_{t=t_1} \\
& \partial_{t_0} S(A_0; A_1) = -(L - \dot{q}L_{\dot{q}})_{t=t_0}, \\
\text{(b)}\quad & \partial_{q_1} S(A_0; A_1) = L_{\dot{q}}\big|_{t=t_1} \\
& \partial_{q_0} S(A_0; A_1) = -L_{\dot{q}}\big|_{t=t_0}.
\end{aligned}$$

Noether's theorem assumes that the Lagrangian is invariant under the transformation $q(t) \to h_s(q(t))$, implying that the action is invariant under this transformation:

$$S(t_0, h_s(q_0); t_1, h_s(q_1)) = S(t_0, q_0; t_1, q_1), \quad \forall s.$$

Differentiating this with respect to s, applying Theorem 6.5.1 and evaluating at $s = 0$, we find

$$\begin{aligned} 0 &= \partial_{q_0} S \cdot \frac{d}{ds} h_s(q(t_0))\Big|_{s=0} + \partial_{q_1} S \cdot \frac{d}{ds} h_s(q(t_1))\Big|_{s=0} \\ &= -L_{\dot{q}}(q(t_0), \dot{q}(t_0)) \cdot \frac{d}{ds} h_s(q(t_0))\Big|_{s=0} \\ &\quad + L_{\dot{q}}(q(t_1), \dot{q}(t_1)) \cdot \frac{d}{ds} h_s(q(t_1))\Big|_{s=0}. \end{aligned}$$

Since t_1 is arbitrary, this proves that Eq. (6.11) is constant along the solution of the EL Eq. (6.10). □

Exercise 6.3. For $q(t) \in \mathbb{R}^3$, where $t \in [t_0, t_1]$, and for each of the following transformations, find the explicit form of the conserved quantity as given by Noether's theorem, assuming that the respective invariance of the Lagrangian holds:

(a) space translation in the direction of e: $h_s(q(t)) = q(t) + se$;
(b) rotation through an angle s around the vector $e \in \mathbb{R}^3$: $h_s(q(t)) = R_e(s)q(t)$;
(c) helical symmetry: $h_s(q(t)) = aes + R_e(s)q(t)$, where a is a constant.

6.6.2 *Hamiltonian and Hamilton's equations: The case of classical mechanics*

Let us now utilize the specific structure of the classical mechanics Lagrangian, which, as per Eq. (6.4), is split into the difference between kinetic energy, $\frac{1}{2}\dot{q}^2$, and potential energy, $V(q)$. We begin by making an important observation: The minimum of the functional

$$\int dt\, \tfrac{1}{2} (\dot{q} - p)^2$$

over $\{p(t)\}$ is achieved when $\forall t : \dot{q} = p$. Now, we can express the kinetic term of the classical mechanics action (the first term in Eq. (6.4)) in terms of an auxiliary optimization:

$$\int dt \frac{\dot{q}^2}{2} = \max_{\{p(t)\}} \int dt \left(p\dot{q} - \frac{p^2}{2} \right). \tag{6.12}$$

Substituting this result into Eqs. (6.3) and (6.4) yields an alternative variational formulation of classical mechanics:

$$\min_{\{q(t)\}} \max_{\{p(t)\}} \int dt\, (p\dot{q} - H(q;p)), \quad H(q;p) := \frac{p^2}{2} + V(q), \tag{6.13}$$

where p and H are defined as the momentum and the Hamiltonian of the system, respectively. Turning this (Hamiltonian) principle of classical mechanics into equations (which, similar to the EL equations, are sufficient conditions of optimality), we arrive at the so-called Hamilton's equations:

$$\dot{q} = \frac{\partial H(q;p)}{\partial p}, \quad \dot{p} = -\frac{\partial H(q;p)}{\partial q}. \tag{6.14}$$

Example 6.6.3.

(a) **[Conservation of Energy]** Show that, in the case of a time-independent Hamiltonian (i.e., for $H(q;p)$ as defined above), the Hamiltonian H represents the energy of the system, which is conserved along the solutions of Hamilton's equations (6.14).
(b) **[Conservation of Momentum]** Show that if the Lagrangian does not explicitly depend on one of the coordinates, say $q^{(1)}$, where $q = (q^{(1)}, \ldots)$, then the corresponding momentum $\partial L/\partial \dot{q}^{(1)}$ is constant along the physical trajectory, as given by the solutions of either the EL or Hamilton's equations.

Solution. (a) The full time derivative of the Hamiltonian is

$$\frac{dH}{dt} = \frac{\partial H}{\partial q}\dot{q} + \frac{\partial H}{\partial p}\dot{p}.$$

Using Hamilton's equations,

$$\dot{q} = \frac{\partial H}{\partial p}, \quad \dot{p} = -\frac{\partial H}{\partial q},$$

we can rewrite the derivative of H as

$$\frac{dH}{dt} = \frac{\partial H}{\partial q}\dot{q} + \frac{\partial H}{\partial p}\dot{p} = -\dot{p}\dot{q} + \dot{q}\dot{p} = 0,$$

which shows that H is conserved.

(b) Suppose that the Lagrangian does not depend on $q^{(1)}$, so that $\partial L/\partial q^{(1)} = 0$. Then, from the EL equations,

$$0 = \frac{d}{dt}\left(\frac{\partial L}{\partial \dot{q}^{(1)}}\right) - \frac{\partial L}{\partial q^{(1)}} = \frac{d}{dt}\left(\frac{\partial L}{\partial \dot{q}^{(1)}}\right),$$

we conclude that $\frac{\partial L}{\partial \dot{q}^{(1)}}$ is constant over time. □

Note that the Hamiltonian system of Eqs. (6.14) becomes more elegant in vector form:

$$\dot{z} = -J\nabla_z H(z) = -\nabla_z\,(JH(z)), \quad z := \begin{pmatrix} q \\ p \end{pmatrix}, \quad J := \begin{pmatrix} 0 & 1 \\ -1 & 0 \end{pmatrix},$$

where 2×2 matrix J represents a clockwise rotation in the (q,p)-space.

6.6.3 *Hamilton–Jacobi equation*

Let us further explore the critical/optimal trajectory, $\{q(t); t \in [t_0 = 0, t_1]\}$, which solves the EL Eqs. (6.10). We begin by setting $t_0 = 0$ and fixing the initial position at $q(0) = q_0$ and then analyze how the action depends on the final time, t_1, and position, q_1. This builds on Theorem 6.5.1 by considering the action function as a function of $A_1 = (t_1, q_1)$, i.e., the final position of the critical path.

We now re-derive the key results of Theorem 6.5.1 in a slightly different, but equivalent, form. By assuming the action function is sufficiently smooth in t_1 and q_1, we introduce and interpret the derivatives of the action with respect to these variables. First, we compute the derivative of the action function with respect to t_1:

$$\begin{aligned}
\mathcal{S}_{t_1} &:= \partial_{t_1}\mathcal{S}(t_1; q_1) = \partial_{t_1}\int_0^{t_1} dt\, L(q(t), \dot{q}(t)) \\
&= L|_{t=t_1} + \int_0^{t_1} dt\,(L_q \partial_{t_1} q(t) + L_{\dot{q}}\partial_{t_1}\dot{q}(t)) \\
&= L|_{t=t_1} + \int_0^{t_1} dt\,\partial_{t_1} q(t)\left(L_q - \frac{d}{dt}L_{\dot{q}}\right) - L_{\dot{q}}\partial_t q(t)|_0^{t_1} \\
&= (L - L_{\dot{q}}\dot{q})_{t=t_1},
\end{aligned} \tag{6.15}$$

where we integrated by parts, used $\partial_t q(t)|_{t=0} = 0$, $\partial_{t_1} q(t)|_{t=t_1} = \dot{q}(t_1)$ and applied the EL equations, $L_q - \frac{d}{dt} L_{\dot{q}} = 0$, for all $t \in [0, t_1]$.

Next, we compute the derivative of the action function with respect to q_1, the final position:

$$\begin{aligned}
\mathcal{S}_{q_1} := \partial_{q_1} \mathcal{S}(t_1; q_1) &= \partial_{q_1} \int_0^{t_1} dt\, L(q(t), \dot{q}(t)) \\
&= \int_0^{t_1} dt\, (L_q \partial_{q_1} q(t) + L_{\dot{q}} \partial_{q_1} \dot{q}(t)) \\
&= \int_0^{t_1} dt\, \partial_{q_1} q(t) \left(L_q - \frac{d}{dt} L_{\dot{q}} \right) + L_{\dot{q}} \partial_{q_1} q(t)|_0^{t_1} \\
&= L_{\dot{q}}\big|_{t=t_1} .
\end{aligned} \tag{6.16}$$

In classical mechanics, where the Lagrangian is factored as the difference between kinetic and potential energy terms, the right-hand side of Eq. (6.15) becomes the negative Hamiltonian, as defined in Eq. (6.13), and the right-hand side of Eq. (6.16) becomes the momentum, $p = \dot{q}$. In a more general context, where the Lagrangian is not factorizable, we use the right-hand sides of Eqs. (6.15) and (6.16) to define the Hamiltonian and momentum:

$$\forall t:\ p := L_{\dot{q}}, \quad H(t; q; p) := L_{\dot{q}} \dot{q} - L, \tag{6.17}$$

where H is now a function of time, $q(t)$ and $p(t)$.

Combining Eqs. (6.15), (6.16) and (6.17), i.e., parts (a) and (b) of Theorem 6.5.1, and using the definitions of momentum and Hamiltonian, we obtain the Hamilton–Jacobi (HJ) equation:

$$\mathcal{S}_{t_1} + H(t_1; q_1; \partial_{q_1} \mathcal{S}) = 0, \tag{6.18}$$

which provides a first-order nonlinear PDE representation of classical mechanics.

Importantly, if we know the initial values of the action function S, its derivative $\partial_q S$ and the Hamiltonian expressed in terms of time, coordinate and momentum at all moments, Eq. (6.18), combined with appropriate initial conditions, forms a well-posed Cauchy problem, leading to a unique solution of the HJ equation. This is a powerful result with many important consequences, as it guarantees a unique solution to the optimization problem, despite the fact that solutions to the EL equations are not necessarily unique. We will explore these consequences in later sections, including the HJB equation, which generalizes the HJ equation in the context of optimal control and dynamic programming (DP).

To summarize, the schematic derivation of the HJ equation revealed the meaning of the action derivative with respect to final time and position. We learned that $\partial_{t_1} S$ is the negative Hamiltonian and $\partial_{q_1} S$ is the momentum $p_1 = p(t_1)$, which, under our unit mass convention, is also equal to the velocity.

Alternative derivation: We can also derive the HJ equation using differentials. Transforming the action from a functional of $\{q(t); t \in [0, t_1]\}$ into a function of t_1 and q_1, $\mathcal{S}\{q(t)\} \to \mathcal{S}(t_1; q_1)$, we rewrite Eqs. (6.3) as

$$\mathcal{S} = \int p\, dq - \int H\, dt,$$

which implies the differential form

$$d\mathcal{S} = \frac{\partial \mathcal{S}}{\partial t_1} dt_1 + \frac{\partial \mathcal{S}}{\partial q_1} dq_1,$$

so that

$$\partial_{t_1} \mathcal{S} = -H(t_1; q_1; p_1), \quad \partial_{q_1} \mathcal{S} = p_1,$$

leading to the HJ equation (6.18).

Once the HJ equations are established, we may return to a simpler notation and use t for both current and final times, where appropriate.

Example 6.6.4. Find and solve the HJ equation for a free particle.

In this case, the Hamiltonian is $H = \frac{p^2}{2}$. The HJ equation becomes

$$\frac{(\partial_q \mathcal{S})^2}{2} = -\partial_t \mathcal{S}.$$

Looking for a solution of the form $\mathcal{S} = f(q) - Et$, we find that $f(q) = \sqrt{2E} q - c$, so the general solution of the HJ equation is

$$S(t; q) = \sqrt{2E} q - Et - c.$$

Exercise 6.4. Find and solve the HJ equation for a two-dimensional oscillator (unit mass and unit elasticity) in spherical coordinates. The action functional is

$$\mathcal{S}\{r(t), \varphi(t)\} = \int dt \left(\frac{1}{2} \left(\dot{r}^2 + r^2 \dot{\varphi}^2 \right) - \frac{1}{2} r^2 \right).$$

We conclude this brief discussion of classical/Hamiltonian mechanics by noting that, in addition to its relevance for optimal control and DP (to be discussed in Section 7), the HJ equations are instrumental in establishing transformations from coordinate-momentum variables (q, p) to canonical variables, where paths of motion reduce to single points, and the Hamiltonian becomes zero.

6.7 Legendre–Fenchel Transform (*)

This section* is dedicated to the Legendre–Fenchel (LF) transform, which has already been used in its simpler, infinite-dimensional form in Eq. (6.12). Given its importance in variational calculus(as we have seen) and its relevance to finite-dimensional optimization (see Appendix B), we have allocated this section to a thorough treatment of the LF transform and its key applications. Toward the end of this section, we will explore two of its practical uses: (a) solving the image restoration problem using a primal-dual algorithm, and (b) estimating integrals via the Laplace method.

Definition 6.7.1 (Legendre–Fenchel Transform). The LF transform of a function $f : \mathbb{R}^n \to \mathbb{R}$ is defined as

$$f^*(k) := \sup_{x \in \mathbb{R}^n} (x \cdot k - f(x)). \tag{6.19}$$

The LF transform is often referred to as a "dual" transform, with $f^*(k)$ representing the dual of $f(x)$.

Example 6.7.2. Find the LF transform of the quadratic function $f(x) = \frac{1}{2} x \cdot A \cdot x - b \cdot x$, where A is a symmetric positive definite matrix, $A \succ 0$.

*This auxiliary section can be skipped on the first reading.

Solution. The LF transform of a positively defined quadratic function results in another positively defined quadratic function, as shown by the following steps:

$$\begin{aligned}\sup_x \left(x \cdot k - \frac{1}{2} x \cdot A \cdot x + b \cdot x \right) &= \sup_x \Big(-\frac{1}{2} \left(x - A^{-1}(k+b)\right) \\ &\quad \cdot A \cdot \left(x - A^{-1}(k+b)\right) \\ &\quad + \frac{1}{2}(k+b) \cdot A^{-1} \cdot (k+b) \Big) \\ &= \frac{1}{2}(k+b) \cdot A^{-1} \cdot (k+b), \end{aligned} \tag{6.20}$$

where the maximum is attained at $x_* = A^{-1}(k+b)$.

Definition 6.7.3 (Convex Function on $\mathbb{R}^n$). A function $u : \mathbb{R}^n \to \mathbb{R}$ is convex if

$$\forall x, y \in \mathbb{R}^n,\ \lambda \in (0,1): \quad u(\lambda x + (1-\lambda) y) \le \lambda u(x) + (1-\lambda) u(y). \tag{6.21}$$

The combination of these two concepts – the LF transform and convexity – leads to the following theorem (proof is beyond the scope of this book).

Theorem 6.7.4 (Convexity and Involution of the LF Transform). *The LF transform of a convex function is itself convex, and it exhibits involution, meaning that* $(f^*)^* = f$.

6.7.1 *Geometric interpretation: Supporting lines, duality and convexity*

After establishing the formal definitions, let us explore the one-dimensional case ($n = 1$) to build intuition about the LF transform and convexity. In one dimension, the LF transform has a straightforward geometric interpretation, often explained using the concept of supporting lines (see, e.g., Ref. [30]).

Definition 6.7.5 (Supporting Lines). A function $f : \mathbb{R} \to \mathbb{R}$ has a supporting line at $x \in \mathbb{R}$ if

$$\forall x' \in \mathbb{R}: \quad f(x') \ge f(x) + \alpha(x' - x).$$

If the inequality is strict for all $x' \neq x$, the line is called *strictly supporting*.

Note that supporting lines are defined locally at a particular point x, rather than globally for all $x \in \mathbb{R}$.

Example 6.7.6. Find $f^*(k)$ and the supporting line(s) for $f(x) = ax + b$.

Solution. Since $f(x)$ is a straight line, we cannot draw any other line that doesn't cross it unless it has the same slope. Hence, $f(x)$ serves as its own supporting line. The LF transform of a straight line is finite only at the point corresponding to the slope of the line, i.e., $k = a$:

$$f^*(k) = \begin{cases} -b, & k = a, \\ \infty, & \text{otherwise.} \end{cases}$$ □

Example 6.7.7. Consider the quadratic function $f(x) = \frac{a}{2}x^2 - bx$. Find $f^*(k)$, the supporting lines for $f(x)$ and the supporting lines for $f^*(k)$.

Solution. The solution, obtained from the one-dimensional version of Eq. (6.20), gives $f^*(k) = \frac{(b+k)^2}{2a}$, where the maximum in the LF transform is achieved at $x_* = \frac{b+k}{a}$. We observe that $f^*(k)$ is well-defined (finite) for all $k \in \mathbb{R}$. The supporting line of $f(x)$ at x is given by its Taylor expansion truncated at the linear term:

$$f_x(x') = f(x) + f'(x)(x' - x) = \frac{a}{2}x^2 - bx + (ax - b)(x' - x).$$

Similarly, for $f^*(k)$,

$$f_k^*(k') = f^*(k) + (f^*)'(k)(k' - k) = \frac{(b+k)^2}{2a} + \frac{b+k}{a}(k' - k).$$ □

This example generalizes into the following key results:

Proposition 6.7.8. *If $f(x)$ admits a supporting line at x and $f'(x)$ exists, then the slope of the supporting line at x is $f'(x)$; in other words, for a differentiable function, the supporting line is always the tangent line.*

Theorem 6.7.9. *If $f(x)$ admits a supporting line at x with slope k, then $f^*(k)$ admits a supporting line at k with slope x.*

Example 6.7.10. Draw the supporting lines for the example of a smooth, non-convex function shown in Fig. 6.6.

Solution. Sketching supporting lines for this smooth, non-convex and bounded function with two local minima leads to the following observations:

- Point a admits a strictly supporting line, as it touches the graph of $f(x)$ only at $x = a$.
- Point b does not admit a supporting line, as any line passing through $(b, f(b))$ crosses $f(x)$ at another point.
- Point c admits a supporting line but not a strictly supporting line, as it also touches $f(x)$ at point d. □

These observations yield further insights into the relationship between supporting lines and convexity, leading to several important results.

Theorem 6.7.11. *The LF transform $f^*(k)$ is always convex in k.*

Corollary 6.7.12. *The double LF transform $f^{**}(x)$ is always convex in x.*

This tells us that $f^{**}(x)$ is not always equal to $f(x)$ because $f^{**}(x)$ is convex, even if $f(x)$ is not. This leads to the following.

Theorem 6.7.13. *$f^{**}(x) = f(x)$ if and only if $f(x)$ admits a supporting line at x.*

Corollary 6.7.14. *If $f(x)$ is convex, then $f^{**}(x) = f(x)$.*

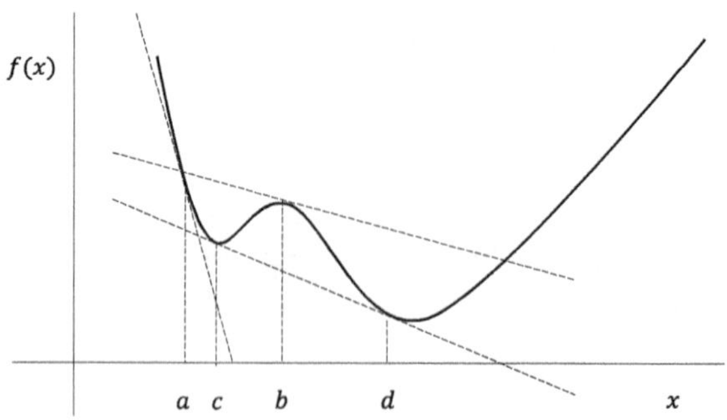

Figure 6.6. Geometric interpretation of supporting lines.

Corollary 6.7.15. *If $f^*(k)$ is differentiable for all k, then $f^{**}(x) = f(x)$.*

Corollary 6.7.16. *A convex function can always be written as the LF transform of another function.*

Theorem 6.7.17. *$f^{**}(x)$ is the largest convex function satisfying $f^{**}(x) \leq f(x)$, also known as the* convex envelope *of $f(x)$.*

In the following, we continue exploring the concepts of supporting lines, convexity and duality with illustrative examples.

Example 6.7.18. Consider a function with a non-differentiable point (cusp), as depicted in Fig. (6.7(a)). Using the concept of supporting lines, construct and interpret $f^*(k)$, the LF transform. Does $f^{**}(x) = f(x)$ hold?

Solution. When analyzing a function with a non-differentiable point, it is helpful to examine the differentiable and non-differentiable regions separately:

- **Differentiable part of $f(x)$:** Each point $(x, f(x))$ in the differentiable regions (labeled as (l) and (r) in Fig. 6.7(a)) admits a unique supporting line with slope $f'(x) = k$. Under the LF transformation, these points map to points $(k, f^*(k))$ that also admit supporting lines with slopes $(f^*)'(k) = x$. This transformation yields the left (l') and right (r') branches shown in Fig. (6.7(b)). Thus, the left (l) and right (r) branches in $f(x)$ correspond to left (l') and right (r') branches in $f^*(k)$.
- **Cusp at $x = x_c$:** The non-differentiable point x_c supports an infinite family of supporting lines with slopes within the range $[k_1, k_2]$.

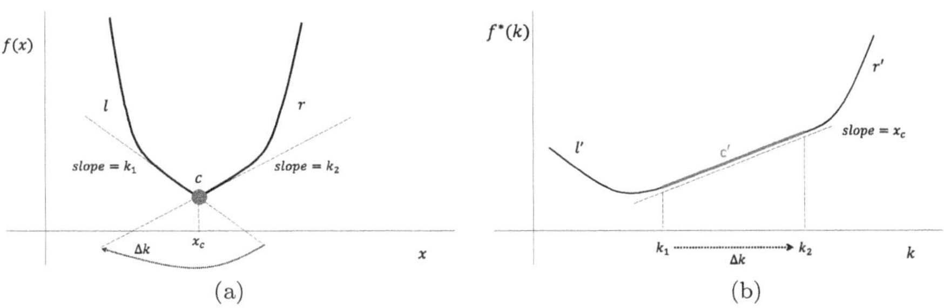

Figure 6.7. Function with a singularity (cusp, (a)) and its LF transform (b).

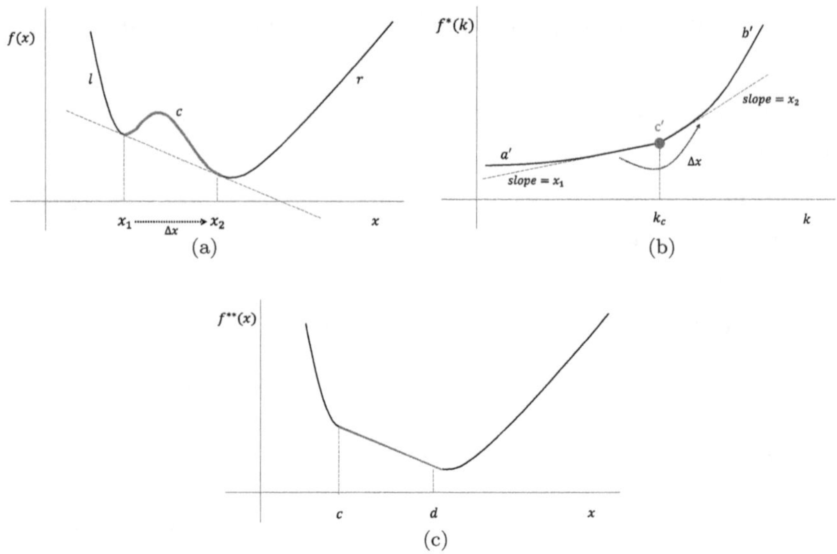

Figure 6.8. (a) An example of a non-convex function $f(x)$; (b) its LF transform $f^*(k)$; (c) the double LF transform $f^{**}(x)$.

Consequently, $f^*(k)$ for $k \in [k_1, k_2]$ must include a supporting line with constant slope x_c, shown as the (c') branch in Fig. 6.7(b). This (c') branch appears as a linear (affine) segment in $f^*(k)$.

Since $f(x)$ is convex, Corollary 6.7.14 confirms that $f^{**}(x) = f(x)$. □

Example 6.7.19. Illustrate $f^*(k)$ and $f^{**}(x)$ for the function $f(x)$ shown in Fig. (6.6).

Solution. We divide the function curve into three sections: left (l), center (c) and right (r). For each section, we construct the LF and double-LF transforms using supporting lines, as follows:

- The left (l) and right (r) branches are strictly convex, allowing strict supporting lines. Their LF transforms are smooth, and the double LF transform precisely reproduces the original function.
- The center (c) branch, which spans from x_1 to x_2, is not convex. Consequently, none of its points admits supporting lines, so these points do not appear in $f^*(k)$. In Fig. (6.8(b)), this non-convex branch collapses to a single point under the LF transform, represented as $(k_c, f^*(k_c))$, where k_c is the slope of the supporting

line connecting the branch's endpoints. The supporting line is not strict, making $f^*(k)$ non-differentiable at this point. In $f^{**}(x)$, the LF transformation extends $(k_c, f^*(k_c))$ into a straight line with slope k_c (shown in bold in Fig. (6.8(c)), effectively creating a convex envelope of $f(x)$ across its non-convex section. □

Exercise 6.5. (a) Find the supporting lines and compute the LF transform for

$$f(x) = \begin{cases} p_1 x + b_1, & x \le x_* \\ p_2 x + b_2, & x \ge x_* \end{cases}$$

where $x_* = \frac{b_2 - b_1}{p_1 - p_2}$, with $b_2 > b_1$ and $p_2 > p_1$. Also, determine $f^{**}(x)$.

(b) Suggest an example of a convex function defined on a bounded domain with diverging (infinite) slopes at the boundary. Illustrate schematically the LF transform $f^*(k)$ and the double-LF transform $f^{**}(x)$ for this function.

6.7.2 *Example of dual optimization in variational calculus*

We now return to the image restoration problem introduced in Section 6.1.3. Our objective is to bypass the ε-smoothing approach discussed in Section 6.4.2 by applying the LF transform. This elegant theoretical step will lead us to a computationally efficient primal-dual algorithm. Using Theorem 6.7.4, we transform the original problem into its dual form.

To generalize, consider the following setup, which extends beyond the image restoration problem. Suppose $f : \mathbb{R}^n \to \mathbb{R}$ is a convex function, and we aim to solve

$$\min_{\{q(x)\}} \int_U dx\, (g(x, q(x)) + f(\nabla_x q(x)))\Big|_{n^T \cdot \nabla_x q = 0,\ \forall x \in \partial U},$$

where $q : U \to \mathbb{R}$ and n denotes the normal vector to the boundary ∂U. To leverage the LF transform, we recast the problem using Theorem 6.7.4:

$$\min_{\{q(x)\}} \max_{\{p(x)\}} \int_U dx\, (g(x, q(x)) + p(x) \cdot \nabla_x q(x) - f^*(p(x)))\Big|_{n^T \cdot \nabla_x q = n^T \cdot p = 0,\ \forall x \in \partial U}, \tag{6.22}$$

where $p : U \to \mathbb{R}^n$ is a vector field aligned with $\nabla_x q$. Here, $q(x)$ is the *primal* variable, while $p(x)$ is the *dual* variable. We "dualize" only the second term in the integrand, as it is non-smooth, while leaving the first term (which is smooth) unchanged. The problem in Eq. (6.22) is often referred to as a *saddle-point* formulation due to its min-max structure.

We can simplify Eq. (6.22) further by applying integration by parts to the middle term:

$$\min_{\{q(x)\}} \max_{\{p(x)\}} \int_U dx \left(g(x, q(x)) - q(x)\nabla \cdot p(x) - f^*(p(x))\right)\Big|_{n^T \cdot p=0,\ \forall x \in \partial U}. \tag{6.23}$$

To solve Eq. (6.22) or Eq. (6.23), one can use a primal-dual method, alternating between the minimization and maximization steps via gradient descent for q and gradient ascent for p.

Let's now apply this approach to the total variation (TV) image restoration problem from Section 6.1.3. The objective is given by

$$g(x,q) = \frac{(q - f(x))^2}{2\lambda}, \quad f(w = \nabla_x q(x)) = |w|,$$

resulting in the following optimization problem:

$$\min_q \int_U dx \left(\frac{(q-f)^2}{2\lambda} + |\nabla_x q|\right)\Big|_{n^T \cdot \nabla_x q=0,\ \forall x \in \partial U}.$$

The function $f(w) = |w|$ is convex, and its LF transform is easily computed as

$$f^*(p) = \sup_{w \in \mathbb{R}^n} (p \cdot w - |w|) = \begin{cases} 0, & |p| \le 1, \\ \infty, & |p| > 1. \end{cases}$$

We can now "invert" this transformation using Theorem 6.7.4:

$$f(w) = |w| = \sup_{|p| \le 1} p \cdot w.$$

Substituting this expression into Eq. (6.23), we obtain the primal-dual formulation:

$$\min_q \max_{|p| \le 1} \int_U dx \left(\frac{(q-f)^2}{2\lambda} - q\nabla_x \cdot p\right)\Big|_{n^T \cdot p=0,\ \forall x \in \partial U}.$$

By the strong convexity theorem (see Appendix), we can swap the min-max order, yielding

$$\max_{|p|\leq 1} \min_{q} \int_U dx \left(\frac{(q-f)^2}{2\lambda} - q\nabla_x \cdot p \right) \Bigg|_{n^T \cdot p = 0,\ \forall x \in \partial U} . \tag{6.24}$$

This trick simplifies the problem, as the quadratic term in q can be minimized explicitly. The optimal q satisfies

$$q = f + \lambda \nabla_x \cdot p.$$

Substituting this back into Eq. (6.24) leads to the dual problem:

$$\max_{|p|\leq 1} \int_U dx \left(f\nabla_x \cdot p - \frac{\lambda}{2}(\nabla_x \cdot p)^2 \right) \Bigg|_{n^T \cdot p = 0,\ \forall x \in \partial U} . \tag{6.25}$$

If we ignore the constraint in Eq. (6.25), the optimum occurs when $\nabla \cdot p = f/\lambda$. To handle the constraint, the projected gradient ascent algorithm [31] can be used:

$$p^{k+1}(x) = \frac{p^k + \tau \nabla_x \left(\nabla_x \cdot p^k - \frac{f}{\lambda} \right)}{1 + \tau |\nabla_x \cdot p^k - \frac{f}{\lambda}|},$$

where p^0 is initialized such that $|p^0| \leq 1$. The iterations converge to the optimal p, and once the optimal p is found, the optimal image q is reconstructed using $q = f + \lambda \nabla_x \cdot p$.

6.7.3 *Geometric interpretation of the Legendre–Fenchel transform*

In this section, we provide additional geometric intuition behind the LF transform, drawing inspiration from Ref. [32]. We focus on the one-dimensional case, which offers a clear visual framework for understanding key concepts.

First, observe that if the function $f : \mathbb{R} \to \mathbb{R}$ is strictly convex, then its derivative, $f'(x)$, is strictly increasing and monotonic with respect to x. This implies a one-to-one relationship between the original variable, x, and the dual variable, $k = f'(x)$. This one-to-one correspondence provides an intuitive explanation for why the LF transform is self-inverse in the case of convex functions (strictly

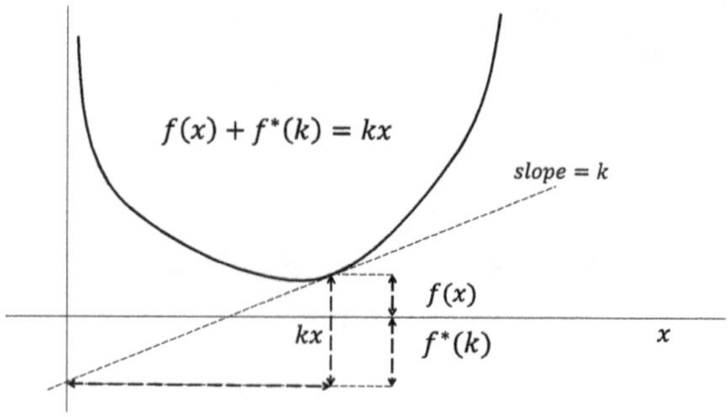

Figure 6.9. Graphical representation of the LF transform.

convex, to be precise, though this holds even in the general convex case).

Next, let's examine the relationship between the original function $f(x)$ and its LF transform $f^*(k)$ at points where the function admits a strict supporting line, as illustrated in Fig. 6.9. The LF transform evaluated at $k = f'(x)$ satisfies the following symmetric relation:

$$kx = f(x) + f^*(k), \quad \text{where } k = f'(x). \tag{6.26}$$

This equation shows that $f^*(k)$ can be interpreted as the extension of $f(x)$ by the term kx, where kx represents the supporting line at x with slope $k = f'(x)$. The geometric meaning becomes clear in Fig. 6.9, where the LF transform captures how $f^*(k)$ adds to the supporting line to fully describe the dual function.

Remarkably, Eq. (6.26) reveals a symmetry between the primal and dual formulations under the exchange of variables: $x \leftrightarrow k$ and $f \leftrightarrow f^*$. However, it's important to remember that the variables x and k are not independent; once one is selected, the other becomes its conjugate through the relationships $k = f'(x)$ or $x = (f^*)'(k)$, depending on which is tracked.

6.7.4 *Hamiltonian-to-Lagrangian duality in classical mechanics*

The LF transform plays a key role in understanding the relationship between the Hamiltonian and Lagrangian formulations of classical mechanics. Let us illustrate this using a simplified example with no explicit dependence on q, i.e., where the Hamiltonian, typically a function of t, q and p, depends only on p.

Consider the case of a free relativistic particle, where the Hamiltonian is given by

$$H(p) = \sqrt{p^2 + m^2},$$

with m representing the particle's mass and the speed of light set to unity ($c = 1$). According to Hamilton's equations, the velocity $\dot{q}$ is derived as

$$\dot{q} = \frac{\partial H}{\partial p} = \frac{dH}{dp} = \frac{p}{\sqrt{p^2 + m^2}}.$$

The Lagrangian, which typically depends on both $\dot{q}$ and q, in this case depends solely on $\dot{q}$ and can be computed as

$$L(\dot{q}) = p\dot{q} - H(p).$$

Rewriting this relation in a symmetric form,

$$p\dot{q} = L(\dot{q}) + H(p),$$

we see a clear resemblance to the LF relation in Eq. (6.26). Here, p and $\dot{q}$, similar to x and k in the LF transform, are conjugate variables. Thus, the Lagrangian, L, can be viewed as the LF transform of the Hamiltonian, i.e., $L = H^*$, and vice versa, $H = L^*$.

For further discussions on applications of the LF transform in physics, such as in statistical thermodynamics (where temperature and energy are conjugate variables and the free energy is the LF dual of the entropy), see Ref. [32].

6.7.5 *Legendre–Fenchel transformation and the Laplace method*

The LF transform also appears in the Laplace approximation method for evaluating integrals. Consider the following integral:

$$F(k, n) = \int_{\mathbb{R}} dx \exp\left(n\left(kx - f(x)\right)\right).$$

When $n \to \infty$, the Laplace method provides an approximation for the logarithm of the integral:

$$\log F(k,n) = n \sup_{x\in\mathbb{R}} (kx - f(x)) + o(n).$$

Here, we recognize that the expression $kx - f(x)$ in the exponent is precisely the structure of the LF transform. Thus, in the limit $n \to \infty$, the dominant contribution to the integral comes from the LF transform of $f(x)$ evaluated at the optimal value of x. This insight links the Laplace method directly to the LF transform and shows its broad applicability in various optimization and approximation techniques.

6.8 Second Variation (*)

Finding the extrema of a function involves more than merely identifying its critical points.* A critical point can be a minimum, maximum or saddle point. To determine the nature of a critical point, one must compute the Hessian matrix of the function. The same principle applies to functionals when characterizing solutions of the EL equations.

We begin by discussing the second variation in the finite-dimensional case. Let $f : U \subset \mathbb{R}^n \to \mathbb{R}$ be a $\mathcal{C}^2$ function (with continuous first and second derivatives). The Hessian matrix of f at a point x is a symmetric bi-linear form on the tangent space $\mathbb{R}^n_x$ to $\mathbb{R}^n$ at x, defined as

$$\forall \epsilon, \eta \in \mathbb{R}^n_x : \quad \text{Hess}_x(\epsilon, \eta) = \left. \frac{\partial^2 f(x + s\epsilon + w\eta)}{\partial s \partial w} \right|_{s=w=0}. \qquad (6.27)$$

If the Hessian is positive definite (i.e., the matrix of second derivatives has only positive eigenvalues), the critical point is a minimum.

Let us generalize the notion of the Hessian to the action, $\mathcal{S} = \int dt\, L(q, \dot{q})$, and the Lagrangian, $L(q, \dot{q})$, where $q(t) : \mathbb{R} \to \mathbb{R}^n$ is a

*This auxiliary section can be skipped on the first reading.

$\mathbb{C}^2$ function. The direct generalization of Eq. (6.27) becomes

$$\begin{aligned}
&\text{Hess}_x\{\epsilon(t), \eta(t)\} \\
&= \left.\frac{\partial^2 \mathcal{S}\{q(t) + s\epsilon(t) + w\eta(t)\}}{\partial s \partial w}\right|_{s=w=0} \\
&= \left(\frac{\partial}{\partial w}\left(\frac{\partial \mathcal{S}\{q(t) + s\epsilon(t) + w\eta(t)\}}{\partial s}\right)_{s=0}\right)_{w=0} \\
&= \int dt \sum_{i=1}^{n} \left.\left(\frac{\partial L(q + s\epsilon, \dot{q} + s\dot{\epsilon})}{\partial q^i} - \frac{d}{dt}\frac{\partial L(q + s\epsilon, \dot{q} + s\dot{\epsilon})}{\partial \dot{q}^i}\right)\eta^i\right|_{s=0} \\
&= \int dt \sum_{i,j=1}^{n} \left(\frac{\partial^2 L}{\partial q^j \partial q^i}\epsilon^j + \frac{\partial^2 L}{\partial \dot{q}^j \partial q^i}\dot{\epsilon}^j - \frac{d}{dt}\left(\frac{\partial^2 L}{\partial q^j \partial \dot{q}^i}\epsilon^j + \frac{\partial^2 L}{\partial \dot{q}^j \partial \dot{q}^i}\dot{\epsilon}^j\right)\right)\eta^i \\
&:= \int dt \sum_{i,j=1}^{n} J_{ij}\epsilon^j \eta^i, \qquad (6.28)
\end{aligned}$$

where $(J_{ij}|i, j = 1, \ldots, n)$ is the matrix of differential operators known as the Jacobi operator. Determining whether the bilinear form is positive definite can be challenging, but in some cases, it can be resolved.

Consider $q : \mathbb{R} \to \mathbb{R}$, with $q \in \mathbb{C}^2$, and the quadratic action

$$\mathcal{S}\{q(t)\} = \int_0^T dt\, (\dot{q}^2 - q^2), \qquad (6.29)$$

subject to zero boundary conditions, $q(0) = q(T) = 0$. To gain insight into the landscape of the action (6.29), let us consider a specific class of functions, such as single-harmonic oscillatory functions:

$$\bar{q}(t) = a \sin\left(n\frac{\pi t}{T}\right), \qquad (6.30)$$

where $a \in \mathbb{R}$ (any real) and $n \in \mathbb{Z} \setminus \{0\}$ (any nonzero integer). Substituting Eq. (6.30) into Eq. (6.29) yields

$$\begin{aligned}
\mathcal{S}\{\bar{q}(t)\} = &\frac{n^2\pi^2a^2}{T^2}\left(\int_0^T dt \cos^2\left(\frac{n\pi t}{T}\right)\right) \\
&-a^2\left(\int_0^T dt \sin^2\left(\frac{n\pi t}{T}\right)\right) = \frac{Ta^2}{2}\left(\frac{n^2\pi^2}{T^2} - 1\right).
\end{aligned}$$

This shows that for $T < \pi$, the action $\mathcal{S}$ is positive for this particular class of functions. However, for $T > \pi$, some functions yield a negative action, indicating that the quadratic form corresponding to the action (6.29) is not positive definite when $T > \pi$.

Thus, this exercise with "probe functions" raises the following question: Could the quadratic form corresponding to the action (6.29) be not positive definite in general? While this initial analysis is inconclusive, it can be shown that the action (6.29) is indeed positive over the entire class of zero-boundary condition, twice-differentiable functions if $T < \pi$.

Example 6.8.1. Prove that the action $\mathcal{S}\{q(t)\}$, given by Eq. (6.29), is positive for $T < \pi$ for any twice-differentiable function $q \in \mathbb{C}^2$ with zero boundary conditions, $q(0) = q(T) = 0$.

Solution. Consider the general Fourier series expansion of $q(t)$:

$$q(t) = \frac{a_0}{2} + \sum_{n=1}^{\infty} \left[a_n \cos\left(\frac{2\pi n t}{T}\right) + b_n \sin\left(\frac{2\pi n t}{T}\right) \right].$$

Calculating $\int_0^T (\dot{q}^2 - q^2)\, dt$ and using the orthogonality of terms, we obtain

$$\int_0^T (\dot{q}^2 - q^2)\, dt = -T\frac{a_0^2}{4} + T\sum_{n=1}^{\infty} \frac{a_n^2 + b_n^2}{2} \left(\frac{4\pi^2 n^2}{T^2} - 1\right).$$

For the "worst-case" scenario of $T = \pi$, we must show that

$$\sum_{n=1}^{\infty} \frac{a_n^2 + b_n^2}{2} (4n^2 - 1) \geq \frac{a_0^2}{4}$$

to ensure positivity of the action. Noting that the boundary conditions imply $a_0/2 + \sum_n a_n = 0$, we can, without loss of generality, scale $q(t)$ such that $a_0 = -2$. It will then suffice to show that

$$\sum_{n=1}^{\infty} \frac{a_n^2}{2} (4n^2 - 1) \geq 1.$$

To verify this, we can construct the dual problem and show that the minimal value of the left-hand side is 1 when varying the a_n.

Specifically, the problem is

$$\min_{a=(a_1,\ldots,a_k)} \sum_{n=1}^{k} \frac{a_n^2}{2}\left(4n^2-1\right), \quad \text{s.t.} \sum_{n=1}^{k} a_n = 1,$$

considering the partial sum case and then taking the $k \to \infty$ limit. The Lagrangian is

$$\mathcal{L}(a,\mu) = \sum_{n=1}^{k} \frac{a_n^2}{2}\left(4n^2-1\right) - \mu\left(\sum_{n=1}^{k} a_n - 1\right).$$

Setting $\nabla_a \mathcal{L} = 0$ gives

$$a_n(4n^2-1) - \mu = 0, \qquad \forall n = 1,\ldots,k,$$

yielding $a_n = \mu/(4n^2-1)$. Enforcing the equality constraint, we find

$$\sum_{n=1}^{k} \frac{\mu}{4n^2-1} = 1,$$

which simplifies to

$$\sum_{n=1}^{k} \frac{\mu}{4n^2-1} = \frac{\mu k}{2k+1},$$

implying $\mu = (2k+1)/k$. Substituting back into the objective function gives

$$\sum_{n=1}^{k} \frac{a_n^2}{2}\left(4n^2-1\right) = \frac{2k+1}{2k} \geq 1.$$

Therefore, in the $k \to \infty$ limit, we remain above 1, proving the positivity of the action. □

6.9 Methods of Lagrange Multipliers

So far, we have only discussed unconstrained variational problems. In this section, we generalize these problems to include constraints, specifically by formulating and solving variational problems with constraints using Lagrange multipliers.

6.9.1 *Functional constraint(s)*

Consider the shortest path problem discussed in Example 6.2.3 but now constrained by a given area, A, under the curve:

$$\min_{\{q(x)\,|\,x\in[0,a]\}} \int_0^a dx\,\sqrt{1+(q'(x))^2},$$

$$\text{s.t.}\quad q(0)=0,\ q(a)=b,\ \int_0^a q(x)\,dx = A.$$

We can incorporate the area constraint into the optimization by adding a term with a Lagrange multiplier, λ, to the objective:

$$\lambda\left(\int_0^a q(x)\,dx - A\right).$$

The EL equations for this "extended" action become

$$\begin{aligned}
0 &= \nabla_x\left(L_{\nabla q}\left(x,q(x),\nabla_x q(x)\right)\right) - L_q\left(x,q(x),\nabla_x q(x)\right) - \lambda\\
&= \frac{d}{dx}\left(\frac{q'(x)}{\sqrt{1+(q'(x))^2}}\right) - \lambda\\
&\to \frac{q'(x)}{\sqrt{1+(q'(x))^2}} = \text{constant} + \lambda x.
\end{aligned}$$

Example 6.9.1. The principle of maximum entropy (also known as the principle of maximum likelihood) selects the probability distribution that maximizes the entropy

$$\mathcal{S} = -\int_D dx\,P(x)\log P(x),$$

subject to the normalization condition $\int_D dx\,P(x) = 1$.

(a) Consider $D\subset\mathbb{R}^n$. Find the optimal distribution $P(x)$.
(b) Consider $D = [a,b]\subset\mathbb{R}$. Find the optimal distribution $P(x)$ assuming that the mean of x is known: $\mathbb{E}_{\{P(x)\}}(x) := \int_D dx\,x\,P(x) = \mu$.

Solution. (a) The effective action is

$$\tilde{\mathcal{S}} = \mathcal{S} + \lambda\left(1 - \int_D dx\, P(x)\right),$$

where λ is the Lagrange multiplier. Varying $\tilde{\mathcal{S}}$ with respect to $P(x)$ yields the EL equation:

$$\frac{\delta \tilde{\mathcal{S}}}{\delta P(x)} = 0 \Rightarrow -\log(P(x)) - 1 - \lambda = 0.$$

Solving this equation and applying the normalization condition, the optimal distribution is the uniform distribution:

$$P(x) = \frac{1}{\|D\|},$$

where $\|D\|$ is the size (volume) of D.

(b) The effective action is

$$\tilde{\mathcal{S}} = \mathcal{S} + \lambda\left(1 - \int_D dx\, P(x)\right) + \lambda_1\left(\mu - \int_D dx\, xP(x)\right),$$

where λ and λ_1 are two Lagrange multipliers. Varying $\tilde{\mathcal{S}}$ with respect to $P(x)$ leads to the EL equation:

$$\frac{\delta \tilde{\mathcal{S}}}{\delta P(x)} = 0 \Rightarrow -\log(P(x)) - 1 - \lambda - \lambda_1 x = 0,$$

which gives

$$P(x) = e^{-1-\lambda} \exp(-\lambda_1 x).$$

The constants λ and λ_1 can be determined by solving the normalization constraint and the constraint on the mean:

$$e^{-1-\lambda}\left[-\frac{\exp(-\lambda_1 x)}{\lambda_1}\right]_a^b = 1,$$

$$e^{-1-\lambda}\left[-\frac{x\exp(-\lambda_1 x)}{\lambda_1} - \frac{\exp(-\lambda_1 x)}{\lambda_1^2}\right]_a^b = \mu. \qquad \square$$

Exercise 6.6. Consider the setting of Example 6.9.1(b) with $a = -\infty$ and $b = \infty$. Assuming that the mean and variance of the probability distribution are known, i.e., $\mathbb{E}_{\{P(x)\}}(x) = \mu$ and $\mathbb{E}_{\{P(x)\}}\left[x^2\right] = \sigma^2 + \mu^2$, find the distribution $P(x)$ that maximizes the entropy.

6.9.2 *Function constraints*

The method of Lagrange multipliers in the calculus of variations can be extended to constrained optimizations where the constraint is not a functional but rather a function. Consider, for example, the standard one-dimensional action functional

$$\mathcal{S}\{q(t)\} = \int dt\, L(t; q(t); \dot{q}(t)),$$

where $q : \mathbb{R} \to \mathbb{R}$. Suppose now that this functional is constrained by the condition

$$\forall t : \quad G(t; q(t); \dot{q}(t)) = 0, \tag{6.31}$$

where $L(t; q; \dot{q})$ and $G(t; q; \dot{q})$ are sufficiently smooth functions of their arguments, including the derivative $\dot{q}$.

The approach is to introduce a new, "modified" action that incorporates the constraint by introducing a Lagrange multiplier function, $\lambda(t)$:

$$\tilde{\mathcal{S}}\{q(t), \lambda(t)\} = \int dt\, (L(t; q(t); \dot{q}(t)) - \lambda(t) G(t; q(t); \dot{q}(t))),$$

which is now a functional of both $q(t)$ and $\lambda(t)$. We then extremize this modified action with respect to both $q(t)$ and $\lambda(t)$. Solutions of the corresponding EL equations, derived by taking variations of the modified action, will provide sufficient conditions for extremizing $\mathcal{S}$ under the constraint given by Eq. (6.31).

To illustrate, let us derive the EL equation for a Lagrangian, $L(q; \dot{q}; \ddot{q})$, that depends on the second derivative of a $\mathbb{C}^3$ function, $q : \mathbb{R} \to \mathbb{R}$, but not explicitly on t. Following the approach outlined above, we define the modified action as

$$\tilde{\mathcal{S}}\{q(t), \lambda(t)\} = \int dt\, (L(q(t); \dot{q}(t); \ddot{q}(t)) - \lambda(t)\, (v(t) - \dot{q}(t))),$$

where $v(t)$ is an auxiliary variable introduced to enforce the constraint $\dot{q}(t) = v(t)$ through the Lagrange multiplier $\lambda(t)$.

The EL equations for this modified action are

$$\frac{\partial L}{\partial q} = \frac{d}{dt}\left(\frac{\partial L}{\partial \dot{q}} + \lambda\right), \quad -\lambda = \frac{d}{dt}\frac{\partial L}{\partial \ddot{q}}, \quad v = \dot{q}.$$

Eliminating λ and v, we arrive at the modified EL equation, expressed solely in terms of the derivatives of the Lagrangian with respect to $q(t)$ and its derivatives:

$$\frac{\partial L}{\partial q} - \frac{d}{dt}\frac{\partial L}{\partial \dot{q}} + \frac{d^2}{dt^2}\frac{\partial L}{\partial \ddot{q}} = 0.$$

Exercise 6.7. Find the extrema of $\mathcal{S}\{q(t)\} = \int_0^1 dt\, \|\dot{q}(t)\|$ for $q : [0,1] \to \mathbb{R}^3$, subject to the norm constraint $\forall t : \|q(t)\|^2 = 1$ and the boundary conditions $q(0) = q_0$ and $q(1) = q_1$, where both boundary points satisfy the norm constraint.

This method for handling function constraints through Lagrange multipliers will reappear in later sections, especially in the context of optimal control.

Chapter 7

Optimal Control and Dynamic Programming

Our treatment of optimal control in this chapter is intentionally brief and selective, aligned with the applied and practical focus of this book. We aim to provide a foundational understanding of the key concepts in optimal control, specifically as they relate to the variational calculus methods discussed in Chapter 6. The primary objective here is to present the basic principles that naturally lead to a more algorithmic approach, culminating in the powerful framework of *dynamic programming* (DP), which will be explored in detail in Section 7.4. Readers seeking a more comprehensive exploration of optimal control theory, beyond the scope of our focused discussion, are encouraged to consult several seminal texts.

The Mathematical Theory of Optimal Processes by Pontryagin *et al.* [33] provides a rigorous and classical introduction to optimal control theory, including Pontryagin's maximum principle – discussed briefly in Section 7.3. For readers interested in the DP approach in more depth, *Dynamic Programming* by Bellman [34] lays the foundational work, introducing the recursive principle of optimality that underpins the DP method. A more modern and comprehensive treatment of DP, with applications to both deterministic and stochastic control problems – to be discussed in Section 9.5 – can be found in Bertsekas' *Dynamic Programming and Optimal Control* [35]. Finally, Boltyanskii's *Mathematical Methods of Optimal Control* [36] offers an alternative approach with a strong focus on mathematical methods for solving control problems, including a variety of examples and theoretical developments relevant to constrained variational calculus – with related finite-dimensional optimization concepts, e.g., convex analysis and duality, briefly discussed in Appendix B.1.

Let's begin with an illustrative optimal control problem.

Example 7.0.1. Consider the trajectory of a particle in one dimension, $\{q(\tau) : [0,t] \to \mathbb{R}\}$, which is subject to a control input $\{u(\tau) : [0,t] \to \mathbb{R}\}$. The control input governs the velocity of the particle, constrained by $\dot{q}(t) = u(t)$, where $|u(t)| \leq 1$. Our goal is to solve the following variational problem:

$$\min_{\{u(\tau),q(\tau)\}} \int_0^t d\tau\,(q(\tau))^2 \quad \text{subject to} \quad \dot{q}(\tau) = u(\tau),\ |u(\tau)| \leq 1, \tag{7.1}$$

with initial condition $q(0) = q_0$ for some fixed q_0 and $t > 0$.

Solution. If $q_0 > 0$, the optimal solution can be found quickly by intuition: Immediately drive q to zero at $\tau = 0^+$ and then remain at $q(\tau) = 0$ for the rest of the time interval. To justify this solution, first drop the control constraint and observe that the minimal solution of the unconstrained problem is $q(\tau) = u(\tau) = 0$ for all $\tau \in (0,t]$. Then, verify that this solution satisfies the control constraint, confirming its optimality. Note that a discontinuity at $\tau = 0$ is allowed, as the problem formulation does not prevent it.

For the case of $q_0 \leq 0$, the analysis is more intricate. We can eliminate the control variable, reducing the pair of constraints in Eq. (7.1) to a single inequality, $\dot{q}(\tau) \leq 1$ for all τ. To solve this, we use *duality theory* and the *Karush–Kuhn–Tucker* (KKT) conditions, which are discussed in Appendix B. This practical example extends the KKT framework from finite-dimensional optimization to variational calculus.

We define the Lagrangian function,

$$L(q(\tau), \mu(\tau)) = q^2 + \mu(\tau)(\dot{q}(\tau) - 1),$$

and write the following KKT conditions:

1. **KKT-1: Primal Feasibility:** $\dot{q}(\tau) \leq 1$ for all $\tau \in (0,t]$.
2. **KKT-2: Dual Feasibility:** $\mu(\tau) \geq 0$ for all $\tau \in (0,t]$.

3. **KKT-3: Stationarity (Euler–Lagrange equation):** $2q(\tau) = \dot{\mu}(\tau)$ for all $\tau \in (0, t]$.
4. **KKT-4: Complementary Slackness:** $\mu(\tau)(\dot{q}(\tau) - 1) = 0$ for all $\tau \in (0, t]$.

We find that the following solution satisfies both the KKT conditions and the initial condition $q(0) = q_0$:

$$q(\tau) = \tau + q_0, \quad \mu(\tau) = \tau^2 + 2q_0\tau + c, \tag{7.2}$$

where c is a constant. However, we should check if there is an alternative solution that satisfies the KKT conditions. Consider a discontinuous control strategy where the particle reaches $q = 0$ with maximum control and then remains at rest:

$$q(\tau) = \begin{cases} q_0 + \tau, & 0 \le \tau \le -q_0, \\ 0, & -q_0 < \tau \le t, \end{cases}$$

$$\mu(\tau) = \begin{cases} \tau^2 + 2q_0\tau + q_0^2, & 0 \le \tau \le -q_0, \\ 0, & -q_0 < \tau \le t. \end{cases} \tag{7.3}$$

For $0 < -q_0 < t$, this solution satisfies the KKT conditions and offers an alternative to the continuous solution in Eq. (7.2). By comparing the objectives in Eq. (7.1) for the two solutions, we conclude that Eq. (7.3) is optimal when $0 < -q_0 < t$, whereas Eq. (7.2) is optimal when $t < -q_0$. □

Exercise 7.1. Solve Example 7.0.1 with the constraint $|u(\tau)| \le 1$ instead of $u(\tau) \le 1$.

7.1 Linear Quadratic Control via Calculus of Variations (*)

This auxiliary section covers linear quadratic (LQ) control and can be skipped on a first reading.* LQ control considers a d-dimensional real vector that represents the system state evolving over time, $\{q(\tau) \in \mathbb{R}^d \mid \tau \in [0, t]\}$, governed by a system of linear ODEs:

$$\forall \tau \in (0, t] : \quad \dot{q}(\tau) = Aq(\tau) + Bu(\tau), \quad q(0) = q_0, \tag{7.4}$$

*This auxiliary section can be dropped at the first reading.

where A and B are constant, time-independent, possibly asymmetric, nonsingular matrices ($A \neq A^T$ and $B \neq B^T$), with $A, B \in \mathbb{R}^{d\times d}$. The control vector $\{u(\tau) \in \mathbb{R}^d \mid \tau \in [0,t]\}$ has the same dimension as q. The objective is to minimize a combined action, often referred to as the *cost-to-go*:

$$\mathcal{S}\{q(\tau), u(\tau)\} := \mathcal{S}_{\text{eff}}\{u(\tau)\} + \mathcal{S}_{\text{des}}\{q(\tau)\} + \mathcal{S}_{\text{fin}}(q(t)),$$

$$\mathcal{S}_{\text{eff}}\{u(\tau)\} := \frac{1}{2}\int_0^t d\tau\, u^T(\tau)Ru(\tau),$$

$$\mathcal{S}_{\text{des}}\{q(\tau)\} := \frac{1}{2}\int_0^t d\tau\, q^T(\tau)Qq(\tau),$$

$$\mathcal{S}_{\text{fin}}(q(t)) := \frac{1}{2}q^T(t)Q_{\text{fin}}q(t),$$

where $\mathcal{S}_{\text{eff}}$ represents the cost of control effort (i.e., energy or force required for control), $\mathcal{S}_{\text{des}}$ is the cost of maintaining a desired system state and $\mathcal{S}_{\text{fin}}$ penalizes deviations from the desired final state at time t. We assume that R, Q and Q_{fin} are symmetric, positive-definite matrices. Our goal is to optimize the *cost-to-go* with respect to $\{q(\tau)\}$ and $\{u(\tau)\}$, subject to the dynamics in Eq. (7.4).

As is standard in variational calculus with constraints, we introduce a Lagrange multiplier $\lambda(\tau)$ (also called the *adjoint vector*) to enforce the ODE constraint (7.4). The extended action becomes

$$\mathcal{S}\{q, u, \lambda\} := \mathcal{S}\{q, u\} + \int_0^t d\tau\, \lambda^T(\tau)\left(-\dot{q}(\tau) + Aq(\tau) + Bu(\tau)\right),$$

where $\lambda(\tau)$ acts as a time-dependent Lagrange multiplier. The Euler–Lagrange (EL) equations and the primal feasibility condition are obtained by varying the effective action with respect to $q(\tau)$, $u(\tau)$ and $\lambda(\tau)$:

$$\frac{\delta \mathcal{S}}{\delta q(\tau)} = 0 \Rightarrow Qq(\tau) + \dot{\lambda}(\tau) + A^T\lambda(\tau) = 0, \quad \forall \tau \in (0, t], \tag{7.5}$$

$$\frac{\delta \mathcal{S}}{\delta u(\tau)} = 0 \Rightarrow Ru(\tau) + B^T\lambda(\tau) = 0, \quad \forall \tau \in [0, t], \tag{7.6}$$

$$\frac{\delta \mathcal{S}}{\delta \lambda(\tau)} = 0 \Rightarrow \dot{q}(\tau) = Aq(\tau) + Bu(\tau) \quad \text{(primal feasibility)}. \tag{7.7}$$

At the final time, we impose the boundary condition

$$\lambda(t) = Q_{\text{fin}} q(t),$$

derived by varying the action with respect to $q(t)$.

7.1.1 *Solving the LQ problem*

Equation (7.6) is algebraic and allows us to express the control vector $u(\tau)$ in terms of the adjoint vector $\lambda(\tau)$:

$$u(\tau) = -R^{-1}B^T\lambda(\tau). \tag{7.8}$$

Substituting this into Eqs. (7.5) and (7.7), we obtain the following coupled system of ODEs:

$$\begin{pmatrix} \dot{q}(\tau) \\ \dot{\lambda}(\tau) \end{pmatrix} = \begin{pmatrix} A & -BR^{-1}B^T \\ -Q & -A^T \end{pmatrix} \begin{pmatrix} q(\tau) \\ \lambda(\tau) \end{pmatrix}, \quad \begin{pmatrix} q(0) \\ \lambda(t) \end{pmatrix} = \begin{pmatrix} q_0 \\ Q_{\text{fin}} q(t) \end{pmatrix}.$$

This system represents a boundary value problem (BVP), which can be solved using the *shooting method.* However, for linear systems such as LQ control, the solution can be found in a more straightforward manner. By integrating the system of ODEs, we get

$$\begin{pmatrix} q(\tau) \\ \lambda(\tau) \end{pmatrix} = W(\tau) \begin{pmatrix} q(0) \\ \lambda(0) \end{pmatrix}, \quad W(\tau) = \exp\left(\tau \begin{pmatrix} A & -BR^{-1}B^T \\ -Q & -A^T \end{pmatrix}\right).$$

Using the boundary condition $\lambda(t) = Q_{\text{fin}} q(t)$, we solve for $\lambda(0)$:

$$\lambda(0) = Mq_0,$$
$$M := -\left(W^{2,2}(t) + Q_{\text{fin}} W^{1,2}(t)\right)^{-1} \left(W^{2,1}(t) + Q_{\text{fin}} W^{1,1}(t)\right).$$

Finally, we substitute $\lambda(0)$ back into Eq. (7.8) to obtain the optimal control in terms of the initial state q_0:

$$u(\tau) = -R^{-1}B^T \left(W^{2,1}(\tau) + W^{2,2}(\tau)M\right) q_0. \tag{7.9}$$

7.1.2 Feedback loop control

The control policy in Eq. (7.9) depends only on the initial state q_0, which is called an *open-loop* control. However, in the presence of uncertainty or disturbances, a *feedback loop* (or *closed-loop*) control is often preferable. To express the control as a feedback law depending on the current state $q(\tau)$, we define the feedback matrix $P(\tau)$, such that

$$\lambda(\tau) = P(\tau)q(\tau).$$

This leads to the feedback control law:

$$u(\tau) = -R^{-1}B^T P(\tau)q(\tau).$$

The matrix $P(\tau)$ satisfies the *Riccati equation*:

$$\dot{P}(\tau) + A^T P(\tau) + P(\tau)A + Q = P(\tau)BR^{-1}B^T P(\tau), \tag{7.10}$$

with the boundary condition $P(t) = Q_{\text{fin}}$.

Example 7.1.1. Consider a one-dimensional unstable system described by

$$\tau \in [0, \infty): \quad \dot{q}(\tau) = Aq(\tau) + u(\tau),$$

where $A > 0$ is a constant. Design an LQ controller that minimizes the cost,

$$\mathcal{S}\{q(\tau), u(\tau)\} = \int_0^\infty d\tau \, \left(q^2(\tau) + Ru^2(\tau)\right),$$

and find the feedback control law, $u(\tau) = -Pq(\tau)/R$.

Solution. Since P is a constant in this case, $\dot{P} = 0$. The Riccati equation simplifies to

$$2AP + 1 = \frac{P^2}{R}.$$

Solving this quadratic equation gives two branches:

$$P = RA \pm \sqrt{R^2A^2 + R}.$$

We select the positive root to ensure stability. As $R \to 0$, we find $P \to \sqrt{R}$, while as $R \to \infty$, we have $P \to 2RA$. □

7.2 From Variational Calculus to the Hamilton–Jacobi–Bellman Equation

In this section, we explore a more general optimal control problem, extending the previous examples to systems governed by nonlinear dynamics in the primal variable $\{q(\tau) : [0,t] \to \mathbb{R}^d\}$, while still retaining linear dependence on the control variable $\{u(\tau) : [0,t] \to \mathbb{R}^d\}$. The dynamics are governed by the following equation:

$$\forall \tau \in [0,t]: \quad \dot{q}(\tau) = f(q(\tau)) + u(\tau), \tag{7.11}$$

where $f(q(\tau))$ describes the nonlinear dynamics of the system. Our goal is to minimize the following objective function, which includes both control effort and potential costs:

$$\int_0^t d\tau \left(\frac{1}{2}u^T(\tau)u(\tau) + V(q(\tau))\right), \tag{7.12}$$

subject to the dynamics in Eq. (7.11). The objective in Eq. (7.12) is composed of two terms: (a) the cost of control, which is quadratic in the control effort, and (b) a potential function $V(q)$, which penalizes or constrains the system's state. The potential can be soft, such as the quadratic potential

$$V(q) = \frac{1}{2}q^T\Lambda q = \frac{1}{2}\sum_{i=1}^{d} q_i\Lambda_{ij}q_j,$$

where Λ is a positive semi-definite matrix, or hard, such as

$$V(q) = \begin{cases} 0, & \text{if } |q| < a, \\ \infty, & \text{if } |q| \geq a, \end{cases}$$

which constrains the system to stay within a ball of radius a around the origin.

Thus, the general optimal control problem can be written as

$$\min_{\{u(\tau),q(\tau)\}} \int_0^t d\tau \left(\frac{u^T(\tau)u(\tau)}{2} + V(q(\tau))\right), \tag{7.13}$$

subject to the dynamics $\dot{q}(\tau) = f(q(\tau)) + u(\tau)$, with initial and final conditions $q(0) = q_0$ and $q(t) = q_t$, respectively.

7.2.1 *Eliminating the control variable*

We now restate Eq. (7.13) as an unconstrained variational problem. First, we eliminate the control variable $u(\tau)$ by solving for $u(\tau)$ directly from the dynamics:

$$u(\tau) = \dot{q}(\tau) - f(q(\tau)).$$

Substituting this into the objective, the problem becomes

$$\min_{\{q(\tau)\}} \int_0^t d\tau \left(\frac{(\dot{q}(\tau) - f(q(\tau)))^T (\dot{q}(\tau) - f(q(\tau)))}{2} + V(q(\tau)) \right),$$

with boundary conditions $q(0) = q_0$ and $q(t) = q_t$.

7.2.2 *Lagrangian and Hamiltonian formulations*

Following the Lagrangian and Hamiltonian methods from the variational calculus section (Section 6), we define the action, Lagrangian, momentum and Hamiltonian for this optimization problem as follows:

$$S\{q(\tau), \dot{q}(\tau)\} = \int_0^t d\tau \left(\frac{(\dot{q} - f(q))^T (\dot{q} - f(q))}{2} + V(q) \right),$$

$$L = \frac{(\dot{q} - f(q))^T (\dot{q} - f(q))}{2} + V(q),$$

$$p = \frac{\partial L}{\partial \dot{q}} = \dot{q} - f(q),$$

$$\begin{aligned} H &= \dot{q}^T \frac{\partial L}{\partial \dot{q}} - L = \frac{\dot{q}^T \dot{q}}{2} - \frac{f(q)^T f(q)}{2} - V(q) \\ &= \frac{p^T p}{2} + p^T f(q) - V(q). \end{aligned}$$

7.2.3 Euler–Lagrange and Hamiltonian equations

The EL equations, derived from the Lagrangian, are

$$\forall i = 1, \ldots, d: \quad \frac{d}{dt}\frac{\partial L}{\partial \dot{q}_i} = \frac{\partial L}{\partial q_i},$$

$$\frac{d}{dt}\left(\dot{q}_i - f_i(q)\right) = -\sum_{j=1}^{d}\left(\dot{q}_j - f_j(q)\right)\frac{\partial f_j(q)}{\partial q_i} + \frac{\partial V(q)}{\partial q_i}.$$

The corresponding Hamiltonian equations are

$$\dot{q}_i = \frac{\partial H}{\partial p_i} = p_i + f_i(q),$$

$$\dot{p}_i = -\frac{\partial H}{\partial q_i} = -p_i\frac{\partial f(q)}{\partial q_i} + \frac{\partial V(q)}{\partial q_i}.$$

7.2.4 Hamilton–Jacobi–Bellman equation

Next, we treat the action S as a function of the final time t and final position q_t. From Hamiltonian mechanics, we know that

$$\frac{\partial S}{\partial t} = -H\Big|_{\tau=t}, \quad \frac{\partial S}{\partial q_t} = p(t),$$

which leads to the Hamilton–Jacobi (HJ) equation:

$$\frac{\partial S}{\partial t} = -\frac{1}{2}\left(\frac{\partial S}{\partial q_t}\right)^T\left(\frac{\partial S}{\partial q_t}\right) - \left(\frac{\partial S}{\partial q_t}\right)^T f(q_t) + V(q_t).$$

Alternatively, if we consider the problem in reverse time, we express the action $\mathcal{S} = \int_\tau^t d\tau' L$ as a function of the starting time τ and $q(\tau)$. This gives a backwards-in-time version of the HJ equation:

$$-\frac{\partial S}{\partial \tau} = \frac{1}{2}\left(\frac{\partial S}{\partial q}\right)^T\left(\frac{\partial S}{\partial q}\right) - \left(\frac{\partial S}{\partial q}\right)^T f(q) + V(q), \tag{7.14}$$

which is derived using $\partial_\tau \mathcal{S} = H\big|_\tau$ and $\partial_q \mathcal{S} = -\partial_{\dot{q}} L\big|_\tau$. This form emphasizes the close connection between the HJ equation and the principles of DP.

The HJ equation in the control context is known as the Hamilton–Jacobi–Bellman (HJB) equation, with the control-part extension added in recognition of Richard Bellman, who pioneered the use of this equation in optimal control.

7.2.5 *Conclusion*

In Section 7.4, we will revisit the HJB equation from the perspective of DP and explore its broader applications.

Exercise 7.2. Solve the following one-dimensional optimal control problem:

$$\{u^*(\tau), q^*(\tau)\} = \underset{\{u(\tau),q(\tau)\in\mathbb{R}\}}{\arg\min} \int_0^t d\tau \frac{(u(\tau))^2 + \beta^2(q(\tau))^2}{2},$$

subject to

$$\dot{q}(\tau) = -\alpha q(\tau) + u(\tau), \quad q(0) = q_0, \quad q(t) = 0.$$

Use a substitution to eliminate $u(\tau)$ and derive the EL equations, Hamiltonian equations and HJ equations. Find the optimal trajectory $q^*(\tau)$ and verify consistency with the three approaches. Reconstruct the optimal control $u^*(\tau)$ in terms of (a) $q^*(\tau)$ (closed-loop control) and (b) q_0 (open-loop control). [*Hint*: You may find it useful to define $\gamma^2 = \beta^2 + \alpha^2$.]

7.3 Pontryagin's Minimum Principle

We now explore a more general formulation of the optimal control problem for a dynamical system with state variables $\{q(\tau) \in \mathbb{R}^d | \tau \in [0,t]\}$, governed by the dynamics

$$\{u^*(\tau), q^*(\tau)\} = \underset{\{u(\tau),q(\tau)\}}{\arg\min} \left(\phi(q(t)) + \int_0^t d\tau\, L(\tau, q(\tau), u(\tau)) \right), \tag{7.15}$$

subject to the following constraints:

$$\dot{q}(\tau) = f(\tau, q(\tau), u(\tau)), \quad u(\tau) \in U \subseteq \mathbb{R}^d, \quad q(0) = q_0,$$

where U is the set of allowable control values and $\phi(q(t))$ is a terminal cost or reward function evaluated at the final time t.

The necessary conditions for optimality, analogous to the EL conditions in variational calculus, are encapsulated by Pontryagin's minimum principle (PMP), named after Lev Pontryagin, who introduced this approach in the context of control theory [37]. For further discussion of the PMP and its historical context, see Ref. [38]. PMP provides the necessary conditions for the optimal control problem formulated in Eq. (7.15).

To derive the PMP conditions, we introduce the following augmented action (cost functional):

$$\tilde{\mathcal{S}} = \phi(q(t)) + \int_0^t d\tau \, (L(\tau, q(\tau), u(\tau)) \\ + \lambda^T(\tau) \, (f(\tau, q(\tau), u(\tau)) - \dot{q}(\tau))), \tag{7.16}$$

where $\lambda(\tau) \in \mathbb{R}^d$ is a Lagrange multiplier function, often referred to as the adjoint or costate variable. By extremizing/minimizing this augmented action over the control variables $\{u(\tau)\}$ and state variables $\{q(\tau)\}$, we obtain the necessary conditions for optimality, known as the Pontryagin system of equations.

The PMP conditions are as follows:

1. *Control minimization condition*:

$$u^*(\tau) = \arg\min_{\tilde{u}} \left(L(\tau, q^*(\tau), \tilde{u}) + \lambda^*(\tau)^T f(\tau, q^*(\tau), \tilde{u}) \right) \\ \forall \tau \in [0, t].$$

   This condition finds the control $u^*(\tau)$ that minimizes the augmented action sequentially for each τ.
2. *Costate/Adjoint equation*:

$$\dot{\lambda}^*(\tau) = -\frac{\partial}{\partial q} \left(L(\tau, q^*(\tau), u^*(\tau)) + \lambda^*(\tau)^T f(\tau, q^*(\tau), u^*(\tau)) \right), \\ \forall \tau \in [0, t].$$

   This describes the evolution of the adjoint variable $\lambda(\tau)$.
3. *Boundary conditions*:

$$\lambda^*(t) = \frac{\partial \phi(q^*(t))}{\partial q^*(t)}.$$

   This boundary condition connects the final state $q(t)$ and the costate variable $\lambda(t)$.

This is a BVP, with conditions specified at both ends of the time interval, making it more challenging to solve analytically in general.

Numerical methods such as the *shooting method* are typically used to solve this system.

Exercise 7.3. Consider a rocket modeled as a particle of constant unit mass, moving in a two-dimensional, zero-gravity space. The rocket is subject to a thrust force $f(\tau)$, and the direction of the thrust can be controlled. The system's dynamics are governed by the following equations of motion:

$$\ddot{q}_1 = f(\tau)\cos u(\tau), \quad \ddot{q}_2 = f(\tau)\sin u(\tau),$$

where $u(\tau)$ is the control variable that specifies the thrust direction at time τ.

(a) Optimal control in terms of a bi-linear tangent law: Assume that the thrust magnitude $f(\tau) > 0$ for all $\tau \in [0, t]$. Show that the optimal control $u^*(\tau)$, which minimizes the cost functional $\phi(q(t))$, takes the following form:

$$\tan(u^*(\tau)) = \frac{a + b\tau}{c + d\tau},$$

where a, b, c and d are constants to be determined.

(b) Maximizing horizontal velocity: Now, assume that the rocket starts at rest at the origin, $q_1(0) = q_2(0) = 0$, and we want to maximize the horizontal velocity $\dot{q}_1(t)$ at a given final time t, while ensuring that the rocket reaches a given height $q_2(t) = q_*$ and that the vertical velocity is zero at the final time ($\dot{q}_2(t) = 0$). Show that the optimal control in this case is reduced to a linear tangent law:

$$\tan(u^*(\tau)) = a + b\tau.$$

7.4 Dynamic Programming in Optimal Control and Beyond

7.4.1 *Discrete time optimal control*

Discretizing the continuous-time optimal control problem from Eq. (7.15) yields the following discrete-time formulation:

$$\min_{u_{0:n-1}, q_{1:n}} \left(\phi(q_n) + \Delta \sum_{k=0}^{n-1} L(\tau_k, q_k, u_k) \right), \tag{7.17}$$

subject to the discrete-time state dynamics

$$q_{k+1} = q_k + \Delta f(\tau_k, q_k, u_k),$$

where $\tau_k = kt/n$, $q_k := q(\tau_k)$, $u_k := u(\tau_k)$ and $\Delta := t/n$. The initial state q_0 is given and fixed.

The core idea behind **DP** is to optimize Eq. (7.17) sequentially – one step at a time – in a greedy, backward fashion, rather than solving for all variables simultaneously. This approach significantly reduces the complexity of the problem.

Step-by-step optimization: We begin by optimizing the final control variable u_{n-1} and the corresponding state q_n. From the dynamics, we can substitute $q_n = q_{n-1} + \Delta f(\tau_{n-1}, q_{n-1}, u_{n-1})$ into the objective function. This leads to

$$S(n, q_n) = \phi(q_n), \tag{7.18}$$

$$u^*_{n-1} = \arg\min_{u_{n-1}\in U} \left[S\left(n, q_{n-1} + \Delta f(\tau_{n-1}, q_{n-1}, u_{n-1})\right)\right.$$
$$\left. + \Delta L(\tau_{n-1}, q_{n-1}, u_{n-1})\right], \tag{7.19}$$

$$S(n-1, q_{n-1}) = S\left(n, q_{n-1} + \Delta f(\tau_{n-1}, q_{n-1}, u^*_{n-1})\right)$$
$$+ \Delta L(\tau_{n-1}, q_{n-1}, u^*_{n-1}), \tag{7.20}$$

where $S(n, q_n)$ represents the **cost-to-go** (or **value function**) at time step n, which is essentially the remaining cost from q_n onward.

Recursive backward optimization: We now apply the same optimization logic recursively, going backward from $k = n - 1$ to $k = 0$, updating the control and state at each step. The recursive equations are

$$u^*_{k-1} = \arg\min_{u_{k-1}\in U} \left[S(k, q_{k-1} + \Delta f(\tau_{k-1}, q_{k-1}, u_{k-1}))\right.$$
$$\left. + \Delta L(\tau_{k-1}, q_{k-1}, u_{k-1})\right], \tag{7.21}$$

$$S(k-1, q_{k-1}) = S(k, q_{k-1} + \Delta f(\tau_{k-1}, q_{k-1}, u^*_{k-1}))$$
$$+ \Delta L(\tau_{k-1}, q_{k-1}, u^*_{k-1}), \tag{7.22}$$

where Eq. (7.18) serves as the terminal condition for the backward time iterations. The value function $S(k, q_k)$ at each step represents the minimum cost-to-go from time step τ_k.

Key notions: $S(k, q_k)$ is the **cost-to-go** or **value function**, which accumulates the cost from time step k to the final time. $L(\tau, q, u)$ is the **incremental cost** (or reward) at each step, while $f(\tau, q, u)$ governs the **state transition** dynamics.

The procedure described forms the basis for the **DP** algorithm, which is summarized in Algorithm 1.

Algorithm 1 Dynamic Programming [Backward in Time Value Iteration]

Input: $L(\tau, q, u), f(\tau, q, u), \quad \forall \tau, q, u.$

$S(n, q) \leftarrow \phi(q)$
for $k = n-1, \ldots, 0$ **do**
$u_k^*(q) \leftarrow \arg\min_u (\Delta L(\tau_k, q, u) + S(k+1, q + \Delta f(\tau_k, q, u))), \ \forall q$
$S(k, q) \leftarrow \Delta L(\tau_k, q, u_k^*(q)) + S(k+1, q + \Delta f(\tau_k, q, u_k^*(q))), \ \forall q$
end for

Output: $u_k^*(q), \quad \forall q, k = 0, \ldots, n-1$

This approach was introduced in the seminal work of Richard Bellman in 1952 [39] and is now known as the **Bellman equation** approach. The Bellman equation allows for a recursive decomposition of the original optimization problem into smaller, more manageable sub-problems solved iteratively.

Optimality of dynamic programming: DP is a **greedy** algorithm, optimizing each time step by making local decisions that appear optimal for that step. Remarkably, in the case of optimal control, this greedy approach leads to a globally optimal solution. We effectively demonstrated this by the sequence of transformations in Eqs. (7.18)–(7.22), showing that DP leads to the exact solution for the discrete-time problem.

7.4.2 *Continuous-time optimal control*

Taking the continuous limit of Eqs. (7.18), (7.21), (7.22), we arrive at the **Bellman equation**, also called the **HJB** equation. This equation is a central tool in continuous-time optimal control and was introduced earlier in Section 7.2, where it was derived in a specific case:

$$-\partial_\tau \mathcal{S}(\tau,q) = \min_{u\in U}\left(L(\tau,q,u) + \partial_q \mathcal{S}(\tau,q)\cdot f(\tau,q,u)\right). \quad (7.23)$$

Here, $\mathcal{S}(\tau,q)$ is the *value function* or *cost-to-go*, $L(\tau,q,u)$ is the running cost or reward and $f(\tau,q,u)$ describes the system's dynamics.

The expression for the **optimal control** – the continuous-time version of step 3 in Algorithm 1 – becomes

$$\forall \tau \in (0,t]: \quad u^*(\tau,q) = \arg\min_{u\in U}\left(L(\tau,q,u) + \partial_q \mathcal{S}(\tau,q)\cdot f(\tau,q,u)\right). \quad (7.24)$$

A special case discussed in Section 7.2, where

$$L(\tau,q,u) \to \frac{u^2}{2} + V(q), \quad f(\tau,q,u) \to f(q) + u,$$

and $U \to \mathbb{R}^d$, leads to the HJ equation derived in Eq. (7.14) after performing quadratic optimization.

Example 7.4.1 (Bang-bang control of an oscillator). Consider a particle of unit mass attached to a spring, subject to bounded control:

$$\tau \in (0,t]: \quad \ddot{x}(\tau) = -x(\tau) + u(\tau), \quad |u(\tau)| \le 1, \quad (7.25)$$

where $x(\tau) \in \mathbb{R}$ is the position of the particle and $u(\tau) \in \mathbb{R}$ is the control force. The goal is to maximize the position of the particle at the final time t, given that initially $x(0) = x_0$ and $\dot{x}(0) = 0$. Find the optimal control $\{u(\tau)\}$, and describe the optimal trajectory for the specific case of $x(0) = 0$ and $t = 2\pi$.

Solution. First, we rewrite the second-order ODE in Eq. (7.25) as a system of two first-order ODEs:

$$\forall \tau \in (0,t]: \quad q = \begin{pmatrix} q_1 \\ q_2 \end{pmatrix} := \begin{pmatrix} x \\ \dot{x} \end{pmatrix}, \quad \dot{q} = Aq + Bu,$$
$$A = \begin{pmatrix} 0 & 1 \\ -1 & 0 \end{pmatrix}, \quad B = \begin{pmatrix} 0 \\ 1 \end{pmatrix}.$$

This reformulation brings us to the standard optimal control problem of Eq. (7.15), where $\phi(q) = C^T q$, with $C^T = (-1, 0)$, $L(\tau,q,u) = 0$, and $f(\tau, q, u) = Aq + Bu$. Substituting this into the HJB equation (7.23), we get

$$\forall \tau \in (0, t]: \quad -\partial_\tau \mathcal{S} = (\partial_q \mathcal{S})^T Aq - \left|(\partial_q \mathcal{S})^T B\right|. \tag{7.26}$$

Here, the absolute value comes from the fact that the optimal control $u(\tau)$ will always be at one of its extremes, either $+1$ or -1, depending on the sign of $(\partial_q \mathcal{S})^T B$.

To solve this, we use the method of separation of variables, assuming the ansatz $\mathcal{S}(\tau, q) = \psi(\tau)^T q + \alpha(\tau)$. Substituting this into Eq. (7.26), we obtain

$$\forall \tau \in (0, t]: \quad \dot{\psi} = -A^T \psi, \quad \dot{\alpha} = |\psi^T B|.$$

The boundary conditions are $\psi(t) = C$ and $\alpha(t) = 0$. Solving the first equation for $\psi(\tau)$ and using the result in Eq. (7.24), we find

$$\forall \tau \in (0, t]: \quad \psi(\tau) = \begin{pmatrix} -\cos(\tau - t) \\ \sin(\tau - t) \end{pmatrix}, \quad u(\tau) = -\text{sign}\left(\sin(\tau - t)\right).$$

Thus, the optimal control depends only on time and alternates between $+1$ and -1.

For the specific case where $x(0) = 0$ and $t = 2\pi$, the optimal control takes the form

$$u(\tau) = \begin{cases} -1, & 0 < \tau < \pi \\ 1, & \pi < \tau < 2\pi \end{cases},$$

and the corresponding optimal trajectory is

$$q^T = (q_1, q_2) = \begin{cases} (\cos(\tau) - 1, -\sin(\tau)), & 0 < \tau < \pi \\ (3\cos(\tau) + 1, -3\sin(\tau)), & \pi < \tau < 2\pi \end{cases}.$$

In this solution, the control first pushes the particle downward ($u = -1$) and then pulls it upward ($u = 1$), both to the maximum extent allowed by the control bounds. This strategy is known as *bang-bang control*, typically observed in cases where there are hard constraints on control but no soft penalties for using control effort.

Exercise 7.4. Consider a **soft version** of the problem discussed in Example 7.4.1:

$$\min_{\{u(\tau)\},\{q(\tau)\}}\left(C^T q(t)+\frac{1}{2}\int_0^t d\tau (u(\tau))^2\right),$$

subject to $\forall \tau \in (0,t]$: $\dot{q}(\tau) = Aq(\tau)+Bu(\tau)$, where $(q(0))^T = (x_0, 0)$ and A, B, C are as defined in Example 7.4.1. Derive the Bellman/ HJB equation, construct a generic solution and illustrate it for the case of $t = 2\pi$ and $q_1(0) = x_0 = 0$. Compare your results with Example 7.4.1.

7.5 Dynamic Programming in Discrete Mathematics

DP is a powerful method in discrete mathematics, particularly in problems related to combinatorics and graph theory. In this section, we explore how DP can be applied to discrete optimization problems, using examples that illustrate its fundamental principles.

7.5.1 *LaTeX Engine Line Breaking Optimization*

Consider a sequence of words of varying lengths, $w_1, \ldots, w_n$, and the task of optimally placing line breaks to minimize the visual discomfort caused by uneven spacing between words. In LaTeX, this is handled by stretching spaces to ensure text aligns with both the left and right margins. Our goal is to minimize the overall "stretching cost," ensuring that the lines are as aesthetically pleasing as possible.[a]

Each word has a length $w_i > 0$ for $i = 1, \ldots, n$. Let $c(i, j)$ denote the cost of placing words w_i–w_j on the same line. The total cost of formatting the paragraph with l lines is given by

$$c(1, j_1) + c(j_1 + 1, j_2) + \cdots + c(j_l + 1, n),$$

where $1 < j_1 < j_2 < \cdots < j_l < n$ denotes the positions where line breaks are introduced. The objective is to find the optimal sequence of line breaks that minimizes the total cost.

[a]This example is adapted from Ref. [40], Section 3.3.1.

The total length of a line is the sum of word lengths plus the spaces between them (i.e., the number of words minus one). If the total length exceeds the maximum allowable line width L, the cost is set to infinity. Otherwise, the cost is defined as a function of the remaining space, typically with a cubic penalty:

$$c(i,j) = \begin{cases} +\infty, & \text{if } L < (j-i) + \sum_{k=i}^{j} w_k, \\ \left(\dfrac{L - (j-i) - \sum_{k=i}^{j} w_k}{j-i} \right)^3, & \text{otherwise.} \end{cases} \tag{7.27}$$

The cubic dependence favors smaller stretching factors, ensuring that tighter spacing is preferred.

At first glance, this problem may seem computationally expensive, requiring 2^{n-1} possible configurations of line breaks. However, DP provides an efficient approach by leveraging the optimal substructure of the problem. Specifically, the solution to the entire problem depends on the solutions to smaller sub-problems involving shorter segments of the paragraph.

Let $f(i)$ denote the minimum cost of formatting the words from position i to the end of the paragraph. Then, the minimum cost of formatting the entire paragraph is

$$f(1) = \min_j \left(c(1,j) + f(j+1) \right),$$

with the recursive relation for partial costs being

$$\forall i: \quad f(i) = \min_{j:i \leq j} \left(c(i,j) + f(j+1) \right),$$

subject to the boundary condition $f(n+1) = 0$, which reflects that no cost is incurred after the last word.

This recursion is the discrete analog of the Bellman equation and is implemented in Algorithm 2.

While Algorithm 2 efficiently reduces the number of potential solutions from exponential to polynomial, it still recomputes values multiple times. By memorizing (storing) previously computed results, the algorithm's time complexity is reduced to $O(n^2)$, as each computation of $f(i)$ involves checking $O(n)$ possible line breaks.

Algorithm 2 Dynamic Programming for LaTeX Line Breaking

Input: $c(i,j), \forall i,j = 1,\ldots,n$, as defined in Eq. (7.27). Boundary condition $f(n+1) = 0$.

for $i = n,\ldots,1$ **do**
  $f(i) = +\infty$
  for $j = i,\ldots,n$ **do**
    $f(i) \leftarrow \min\left(f(i), c(i,j) + f(j+1)\right)$
  end for
end for

Output: $f(i),\ \forall i = 1,\ldots,n$

7.5.2 *Cheapest path on a grid*

Consider another problem: navigating a rectangular grid of size $N \times M$, where each cell contains a number. Starting from the top-left corner, you must reach the bottom-right corner by moving only right or down, and at each step, you "pay" the value of the cell you move into. What is the minimum cost to complete the path?

The minimum cost to reach any cell (i,j) depends on the cost of reaching either the cell directly above $(i-1,j)$ or the one to the left $(i,j-1)$. Let $p(i,j)$ be the minimum cost to reach cell (i,j). The recursive relation is

$$p(i,j) = \min\left(p(i-1,j), p(i,j-1)\right) + a(i,j),$$

where $a(i,j)$ is the value in cell (i,j). To avoid handling boundary conditions explicitly, we can initialize the first row and first column with large values, simulating the effect of an "out-of-bounds" penalty. The final answer will be stored in $p(N,M)$. The solution is implemented in Algorithm 3.

Algorithm performance is illustrated in Fig. (7.1).

Exercise 7.5. Consider a weighted directed acyclic graph (DAG) where each edge has a weight, and the goal is to find the **maximum cost path** from the start node to the end node. Construct an algorithm that propagates the maximum cost path by adjacency, starting from the initial node, and updates the maximum path cost for each node. Your algorithm should avoid arbitrarily computing all possible

Algorithm 3 Dynamic Programming for Minimum Cost Path on a Grid

Input: Costs $a(i,j), \forall i = 1, \ldots, N; \forall j = 1, \ldots, M$. Initialize boundary conditions: $p(i,0) \leftarrow +\infty, \forall i = 1, \ldots, N$ and $p(0,j) \leftarrow +\infty$, $\forall j = 1, \ldots, M$. Set $p(1,1) \leftarrow 0$.

for $t = 2, \ldots, N+M$ **do**
    for $i + j = t, \ i,j \geq 0$ **do**
        $p(i,j) \leftarrow \min\left(p(i-1,j), p(i,j-1)\right) + a(i,j)$
    end for
end for

Output: $p(i,j), \ \forall i = 1, \ldots, N; j = 1, \ldots, M$.

paths. Provide a pseudo-code and test your algorithm on the DAG in Fig. 7.2.

7.5.3 *Dynamic programming for graphical model optimization*

DP can efficiently solve a wide range of combinatorial optimization problems. One notable application is finding the optimal configuration of a system described by a graphical model, such as a chain or tree. Consider the following combinatorial optimization problem over a binary n-dimensional variable x:

$$E := \min_{x \in \{\pm 1\}^n} \sum_{i=1}^{n-1} E_i(x_i, x_{i+1}), \tag{7.28}$$

where the goal is to minimize the total energy, E, over all 2^n possible configurations of x. Here, $E_i(x_i, x_{i+1})$ is a known real-valued function that depends on neighboring variables x_i and x_{i+1}, which are binary. This problem is also known as finding the ground state of the Ising model in statistical physics.

To explain the DP algorithm for this problem, we represent it as a linear graph (chain), where each node corresponds to a variable x_i and the edges represent pairwise interactions between neighboring variables, as shown in Fig. 7.3. The goal is to minimize the

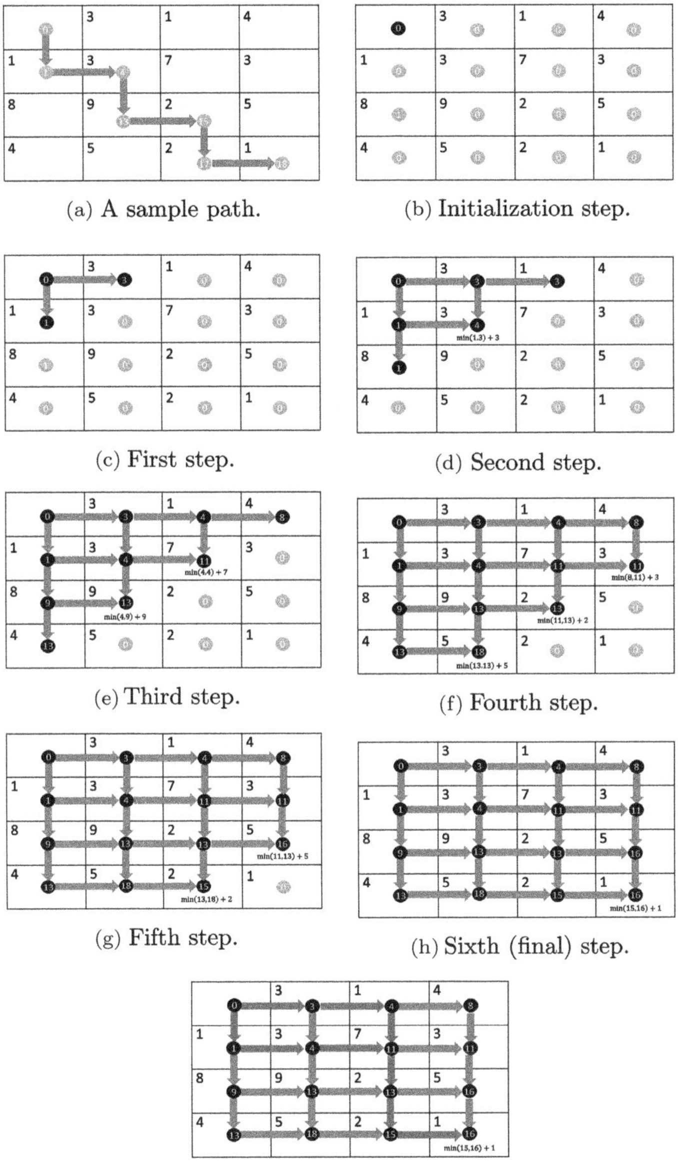

(a) A sample path.

(b) Initialization step.

(c) First step.

(d) Second step.

(e) Third step.

(f) Fourth step.

(g) Fifth step.

(h) Sixth (final) step.

(i) Optimal path(s).

Figure 7.1. Step-by-step illustration of the cheapest-path Algorithm 3 for an exemplary 4×4 grid. Number in the corner of each cell (except cell $(1, 1)$ is respective a_{ij}. Nonzero values in the circles are respective final p_{ij}, corresponding to the cost of the optimal path from $(1, 1)$ to (i, j).

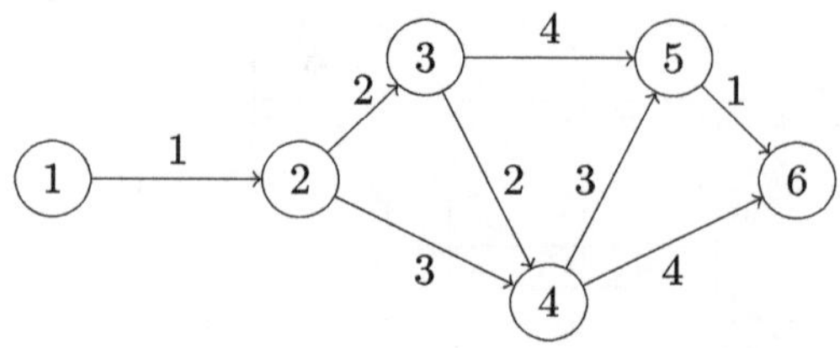

Figure 7.2. A weighted directed acyclic graph.

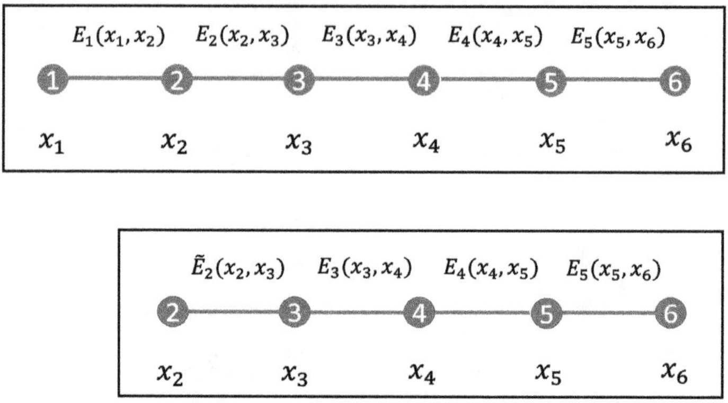

Figure 7.3. Top: Example of a linear graphical model (chain). Bottom: Modified graphical model (shortened chain) after one DP step.

total energy by progressively reducing the size of the problem, which involves renormalizing the pairwise energies.

The key insight of the DP approach is that we can minimize the energy sequentially by processing one node at a time, starting from one end of the chain and moving to the other. Let's demonstrate the process with an example, as shown in Fig. 7.3.

We begin by minimizing over x_1:

$$
\begin{aligned}
E &= \min_{x_2,\ldots,x_n} \left(\min_{x_1} E_1(x_1, x_2) + \sum_{i=2}^{n-1} E_i(x_i, x_{i+1}) \right) \\
&= \min_{x_2,\ldots,x_n} \left(\tilde{E}_2(x_2, x_3) + \sum_{i=3}^{n-1} E_i(x_i, x_{i+1}) \right),
\end{aligned}
$$

where

$$\tilde{E}_2(x_2, x_3) := E_2(x_2, x_3) + \min_{x_1} E_1(x_1, x_2),$$

defines the new "renormalized" energy $\tilde{E}_2(x_2, x_3)$ after minimizing over x_1. This procedure effectively shortens the chain by one node and reduces the problem to a similar form as the original, except with renormalized pairwise energies. We can repeat this process for each node sequentially until we arrive at the final solution.

At each step, we minimize over the current node and update the pairwise energy for the next node. Figure 7.3 illustrates this reduction process step by step. By performing this sequence of greedy steps, we reduce the original problem in n steps, each time reducing the chain by one node and recalculating the energy for the remaining sub-problem.

The DP algorithm for this problem is described in Algorithm 4, where we generalize the approach to allow for variables from any set (often called an *alphabet* in computer science and information theory).

Algorithm 4 DP for Combinatorial Optimization on a Chain

Input: Pairwise energies $E_i(x_i, x_{i+1})$, $\forall i = 1, \ldots, n-1$.

```
for i = 1, ..., n − 2 do
    for x_{i+1}, x_{i+2} ∈ Σ do
        E_{i+1}(x_{i+1}, x_{i+2}) ← E_{i+1}(x_{i+1}, x_{i+2}) + min_{x_i} E_i(x_i, x_{i+1})
    end for
end for
```

Output: $E = \min_{x_{n-1}, x_n} E_{n-1}(x_{n-1}, x_n)$

Now, let's generalize the combinatorial optimization problem to tree structures. Consider a tree $\mathcal{T} = (\mathcal{V}, \mathcal{E})$, where $\mathcal{V}$ represents the set of nodes and $\mathcal{E}$ the set of edges (see Fig. 7.4). The optimization problem over the tree is given by

$$E := \min_{x \in \Sigma^{|\mathcal{V}|}} \sum_{\{i,j\} \in \mathcal{E}} E_{i,j}(x_i, x_j), \tag{7.29}$$

where Σ is the set of possible values (alphabet) for each variable x_i at node $i \in \mathcal{V}$ and $|\mathcal{V}|$ is the number of nodes in the tree.

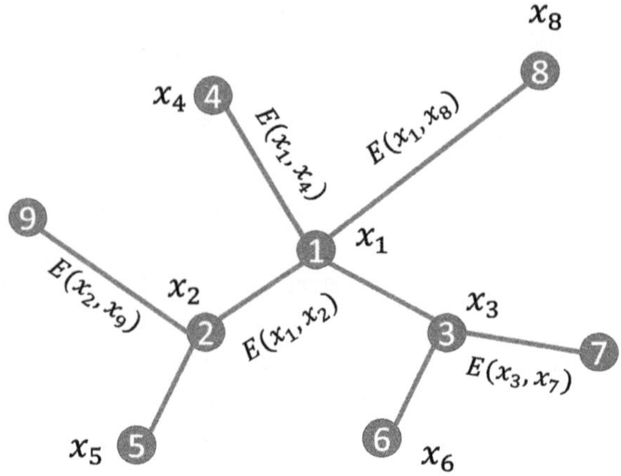

Figure 7.4. Example of a tree-like graphical model.

The approach for solving this optimization problem over a tree is similar to that used for a chain. Starting from any leaf node of the tree, we use a recursive DP process to minimize the energy over the variables associated with each node, updating the pairwise energies along the edges. By systematically eliminating nodes, we eventually reduce the problem to a simpler form, solving it efficiently in linear time with respect to the number of nodes.

Exercise 7.6. Generalize Algorithm 4 to solve the graphical model optimization problem for trees, as defined in Eq. (7.29).

Hint: Begin by processing the leaf nodes, and then use a recursive DP scheme similar to the chain case.

Part IV

Mathematics of Uncertainty

Chapter 8

Basic Concepts from Statistics

This chapter introduces key concepts in probability theory and statistics that will be foundational for later discussions of stochastic processes and uncertainty. We begin with the notion of *random variables* (Section 8.1), both discrete and continuous, introducing basic tools such as probability distributions and density functions. In Section 8.2, we discuss statistical moments and cumulants, which summarize distributional characteristics such as mean, variance, skewness and kurtosis. Probabilistic inequalities are the subject of Section 8.3. Moving forward, we explore the transition from univariate to *multivariate random variables* (Section 8.4), including concepts of marginalization and conditional probability. Lastly, the chapter introduces an *information-theoretic view on randomness*, focusing on entropy and its role in quantifying uncertainty (Section 8.5).

This chapter is accompanied by two Jupyter/Julia notebooks, RandomVariables.ipynb and Entropy.ipynb, which are available on the author's living-book website: https://sites.google.com/site/mchertkov/living-books/applied-math-book.

For further reading and a deeper understanding of the topics covered in this chapter, the following references are suggested: *Probability Theory: The Logic of Science* by Jaynes [41] offers a Bayesian perspective on probability and statistics, with a focus on logical reasoning. For a more classical and comprehensive mathematical treatment, we recommend *A Course of Modern Probability* by Durrett [42], which is well suited for readers seeking a rigorous understanding of probability theory. Additionally, *Information Theory, Inference, and Learning Algorithms* by MacKay [43] covers modern statistical concepts, including an intuitive approach to entropy and information theory.

8.1 Distributions and Random Variables

Consider a system that can exist in a number of different states. State spaces can be either continuous or discrete. An example of a continuous state space is the angle between the hands of a clock, measured clockwise from the hour hand, so $\Sigma = [0, 2\pi)$. An example of a discrete state space is the number showing on the top of a die, $\Sigma = \{1, 2, 3, 4, 5, 6\}$. If the system's state is influenced by a source of randomness, each state, $x \in \Sigma$, is associated with a probability, $P(x)$, which describes the likelihood of observing state x.

8.1.1 *Discrete random variables*

For discrete state spaces, P must satisfy

$$\forall x : \quad 0 \leq P(x) \leq 1, \quad \sum_{x\in\Sigma} P(x) = 1. \tag{8.1}$$

It is often useful to work with quantitative state spaces. For example, the set of possible outcomes of a coin toss, $\{\text{Tail}, \text{Head}\}$, can be mapped to the quantitative state space $\Sigma = \{0, 1\}$ by counting how many heads are observed after one toss. The probability mass function associated with this binary sample space is fully determined by one parameter, β. If $P(1) = \beta$, then $P(0) = 1 - \beta$ (see Fig. 8.1).

Terminology. In the example of the coin toss, we defined a random variable as the number of heads after one toss. If we call this random variable X, then the notation $P(1) = \beta$ and $P(0) = 1-\beta$ is shorthand for $P(X = 1) = \beta$ and $P(X = 0) = 1-\beta$. Another common notation is $P_X(1) = \beta$ and $P_X(0) = 1 - \beta$. All three notations mean the following: "The probability of observing exactly one head after a toss is β."

We can also write

$$P(X = k) = \begin{cases} 1 - \beta, & \text{for } k = 0; \\ \beta, & \text{for } k = 1. \end{cases} \tag{8.2}$$

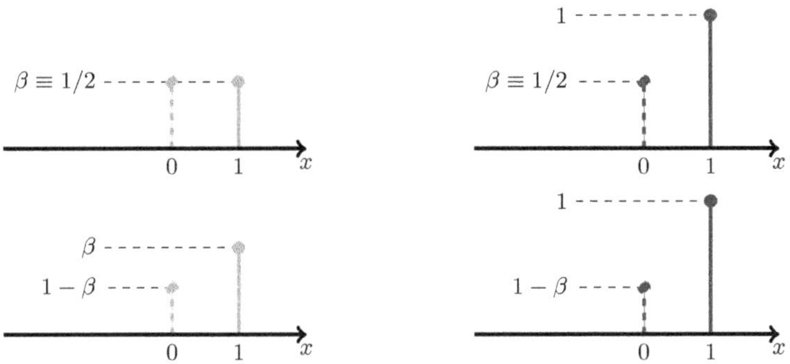

Figure 8.1. Probability mass function (left column) and cumulative distribution function (right column) for a Bernoulli random variable with parameter $\beta = 1/2$ (top) and $\beta > 1/2$ (bottom).

The probability distribution described by Eq. (8.2) is called the *Bernoulli distribution* with parameter β. A random variable X that follows the Bernoulli distribution is called a *Bernoulli random variable*, denoted as $X \sim \text{Bernoulli}(\beta)$.

Equation (8.2) and Fig. 8.1 describe the Bernoulli distribution via its *probability mass function* (PMF), which defines the probability that a random variable takes certain values. Distributions can also be described by their *cumulative distribution function* (CDF), which defines the probability that a random variable is less than or equal to a certain value. For example, the CDF of the Bernoulli distribution is

$$P(X \leq k) = \begin{cases} 1 - \beta, & \text{for } k = 0; \\ 1, & \text{for } k = 1. \end{cases}$$

See Fig. 8.1 for illustration.

Next, consider tossing the coin n times. The set of possible outcomes is the set of sequences of length n consisting of heads and tails. For instance, the sequence $(H, T, H, H, \ldots, T)$ is one possible outcome. If we define the random variable X_i to be the number of heads on the ith toss, the sequence $(X_i)_{i=1}^n$ represents the outcome of n tosses.

In this case, we say that the random variables X_i are *independent* because the outcome of each toss does not depend on previous tosses and are *identically distributed* because the principles governing each toss are identical. Random variables that are both independent and identically distributed are denoted as "i.i.d."

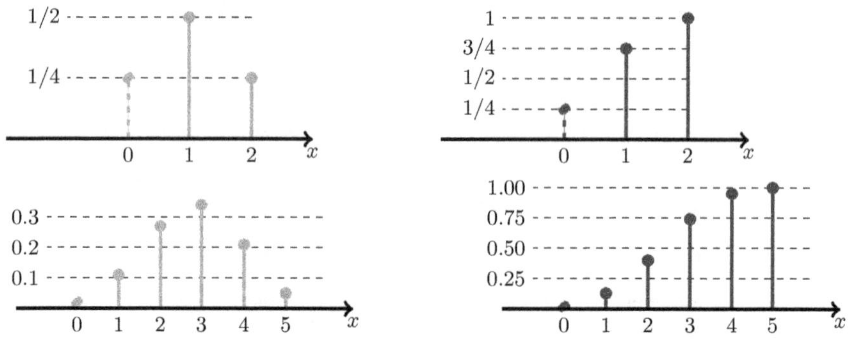

Figure 8.2. Probability mass function (left) and cumulative distribution function (right) for Binomial random variables with parameters $n = 2$, $\beta = 1/2$ (top) and $n = 5$, $\beta > 1/2$ (bottom).

Now, define a new random variable, Y, as the total number of heads after n tosses, so $Y = X_1 + X_2 + \cdots + X_n$. Since the X_i are i.i.d., the probability of observing exactly k heads is given by the binomial formula

$$P(Y = k) = \binom{n}{k} \beta^k (1 - \beta)^{n-k}. \tag{8.3}$$

The probability distribution described by Eq. (8.3) is called the *binomial distribution* with parameters n and β (see Fig. 8.2). A random variable Y that follows the binomial distribution is called a *binomial random variable*, denoted as $Y \sim \mathrm{B}(n, \beta)$ or $Y \sim \mathrm{Binom}(n, \beta)$.

Let us now discuss an unbounded discrete state space example. Consider an event that occurs by chance, such as observing a meteor. Let K be the random variable counting the number of occurrences during a given period. Then, $\Sigma = \{0, 1, 2, \dots\}$ (i.e., it is possible to observe zero, one, two, etc., occurrences). Under certain conditions, K has the probability distribution

$$P(K = k) = \frac{\lambda^k e^{-\lambda}}{k!}, \quad k = 0, 1, 2, \dots. \tag{8.4}$$

The distribution in Eq. (8.4) is called the *Poisson distribution* with parameter λ (see Fig. 8.3). A random variable K that follows the Poisson distribution is called a *Poisson random variable*, denoted as $K \sim \mathrm{Pois}(\lambda)$. (You can verify that the probability defined in Eq. (8.4) satisfies Eq. 8.1.)

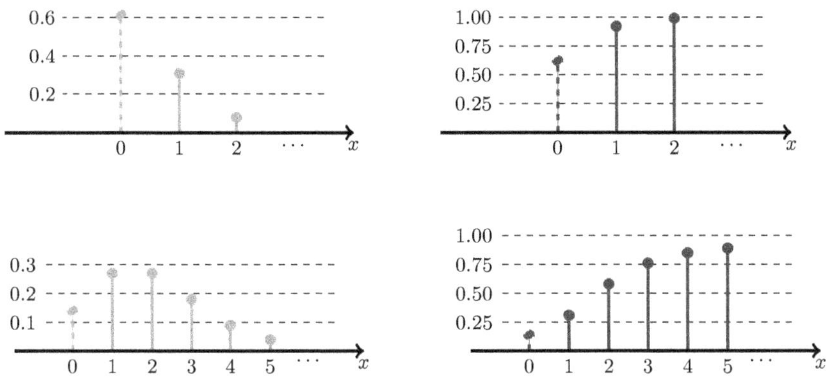

Figure 8.3. Probability mass function (left) and cumulative distribution function (right) for Poisson random variables with $\lambda = 0.5$ (top) and $\lambda = 2$ (bottom).

Common real-world examples of Poisson processes include the number of phone calls received in an hour, the number of customers arriving at a bank and the number of typing errors on a page.

Example 8.1.1. Are the Bernoulli and Poisson distributions related? Can a Poisson process be derived from a Bernoulli process?

Solution. Consider repeating the Bernoulli process n times independently, drawing a sequence of zeros and ones from a Binomial distribution. Focus on the ones and record the times (indices) of their occurrence. Analyzing the distribution of k arrivals in n steps as $n \to \infty$ and assuming $n\beta$ converges to a constant, $n\beta \to \lambda$, recovers a Poisson distribution. This result is known as the Poisson Limit Theorem, and we will explore it further in later lectures. □

8.1.2 *Continuous random variables*

The state space Σ can also be continuous. Random variables on continuous state spaces are associated with a probability density function (PDF), which must satisfy

$$\forall x \in \Sigma : \quad p(x) \geq 0, \quad \int_{\Sigma} dx\, p(x) = 1. \tag{8.5}$$

It is customary to use lowercase p for the PDF and uppercase P to denote actual probabilities. The PDF provides a way to compute the probability that an outcome lies within a given set or interval.

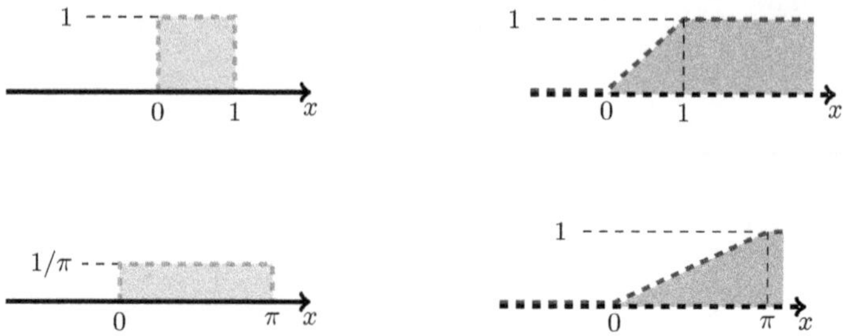

Figure 8.4. PDF (left column) and CDF (right column) for uniform random variables on $(0,1)$ (top) and $(0,\pi)$ (bottom).

For example, for $\mathcal{A} \subset \Sigma$, the probability of observing an outcome in the set $\mathcal{A}$ is given by

$$P(\mathcal{A}) = \int_{\mathcal{A}} p(x)\, dx.$$

Consider the probability that a real-valued random variable X takes a value less than or equal to x:

$$P(X \leq x) = \int_{-\infty}^{x} p(x')\, dx'. \tag{8.6}$$

Equation (8.6) extends the concept of the CDF to the continuous case, as introduced earlier in Section 8.1.1.

This framework can also apply to finite intervals. For example, the uniform distribution on the interval $[a, b]$ is a distribution on a bounded continuous state space, defined by

$$\forall x \in [a, b]: \quad p(x) = \frac{1}{b-a}. \tag{8.7}$$

A random variable X with a probability distribution given by Eq. (8.7) is denoted as $X \sim \text{Unif}(a, b)$. Figure 8.4 shows the PDF and CDF of $\text{Unif}(a, b)$.

The Gaussian distribution is perhaps the most widely known and important continuous distribution:

$$\forall x \in \mathbb{R}: \quad p(x|\sigma, \mu) = \frac{1}{\sigma\sqrt{2\pi}} \exp\left(-\frac{(x-\mu)^2}{2\sigma^2}\right). \tag{8.8}$$

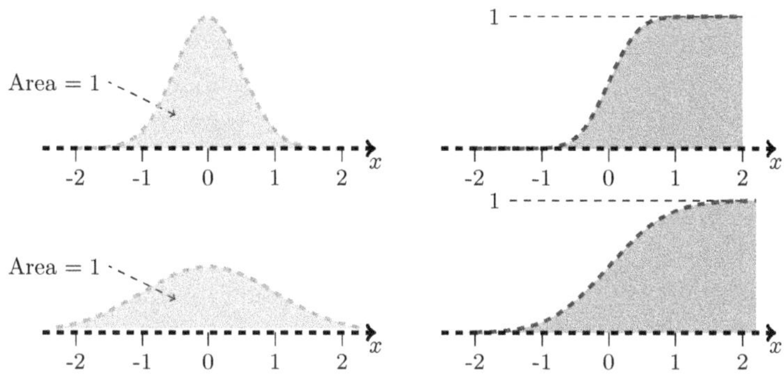

Figure 8.5. PDF (left column) and CDF (right column) for normally distributed random variables $\mathcal{N}(0, 1)$ (top) and $\mathcal{N}(1, 0.5^2)$ (bottom).

Another possible notation for the Gaussian PDF is $p_{\sigma,\mu}(x)$. This distribution is parameterized by its mean, μ, and variance, σ^2. The standard notation for the Gaussian (or normal) distribution is $\mathcal{N}(\mu, \sigma^2)$ or $N(\mu, \sigma^2)$. Figure 8.5 illustrates the PDF and CDF of $\mathcal{N}(\mu, \sigma^2)$.

The probability distribution in Eq. (8.8) is also referred to as a *normal distribution.* The term "normal" reflects the fact that the Gaussian distribution naturally arises from the sum of many independent random variables, regardless of the individual distributions – a result of the central limit theorem, which we will explore in later sections.

A brief note on notation: We may write $P(X = x)$, or use the shorthand $P(x)$. Sometimes the notation $P_X(x)$ is also used. By convention, uppercase letters represent random variables. When a random variable takes a specific observed value (after sampling), this particular value is denoted with a lowercase letter, e.g., x. The notation $X \sim P(x)$ indicates that the random variable X is drawn from the distribution $P(x)$. Additionally, to streamline notation (sometimes at the cost of precision), we may use lowercase letters for both random variables and their realized values when it causes no confusion.

8.1.3 *Sampling and histograms*

Random Process Generation. A random process is generated or sampled using a random number generator, which is a fundamental feature in any computational package or software. These generators (often pseudo-random) are critical, and designing a good random number generator is important. However, in this course, we will primarily use existing pseudo-random number generators.

Histograms. To visualize distributions graphically, one can create a histogram by "binning" data within a defined domain. This is a convenient way to represent $p(\sigma)$. For an illustration, see RandomVariables.ipynb, available at https://sites.google.com/site/mchertkov/living-books/applied-math-book.

8.2 Moments and Cumulants

8.2.1 *Expectation and variance*

It is often useful to summarize a probability distribution with as few numbers as possible, while still capturing its key characteristics.

Cumulants are a common set of descriptors used for probability distributions. The first two cumulants are widely known: The *mean* describes the central tendency of a distribution, while the *variance* measures the spread around the mean. The third and fourth cumulants are less familiar: *Skewness* measures asymmetry around the mean, and *kurtosis* indicates whether extreme values are unusually rare or common (see Fig. 8.6).

Cumulants are often derived from a distribution's moments, which can be computed via summation or integration. Alternatively, cumulants can be found using the moment-generating function or the characteristic function of a distribution, though this topic is more theoretical and beyond the scope of this course.

For a random variable X (discrete or continuous), the *expectation of* X is defined as

$$\mathbb{E}[X] := \sum_{x\in\Sigma} x\, P(x) \quad \text{(discrete)},$$

$$\mathbb{E}[X] := \int_{x\in\Sigma} dx\, x\, p(x) \quad \text{(continuous)}.$$

Notation. Common notations for the expectation of a random variable X include $\mathbb{E}[X]$, $\mathbb{E}_P[X]$, $\langle X\rangle$ and $\langle X\rangle_P$.

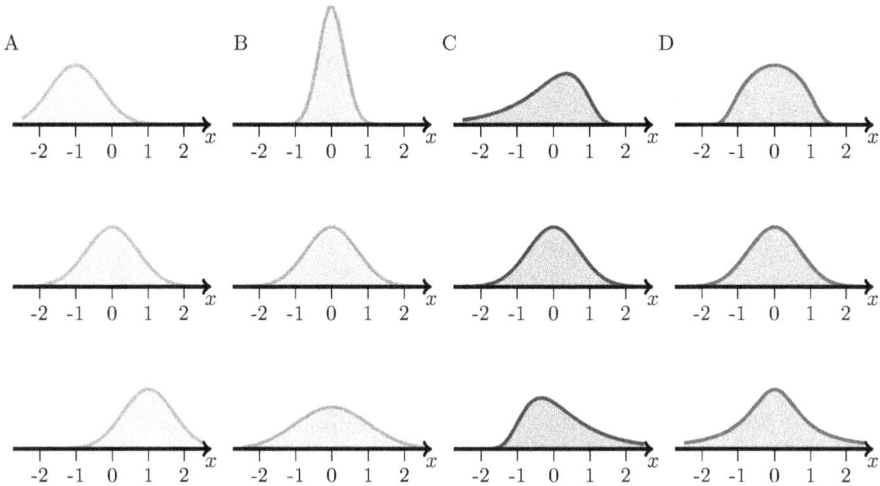

Figure 8.6. The first cumulant describes the central tendency (mean) of a probability distribution (A). The second cumulant describes the spread about the mean (B). The third cumulant describes the asymmetry of the distribution (C). The fourth cumulant describes whether extreme values are unusually rare or common (D).

Example 8.2.1. Consider tossing a pair of fair coins. The set of possible outcomes is $\{(T,T),(T,H),(H,T),(H,H)\}$, each occurring with equal probability. Define the random variable X as the number of heads observed. Then, $X \sim B(2,1/2)$, and

$$P(X=x) = \begin{cases} 1/4, & \text{for } x=0; \\ 1/2, & \text{for } x=1; \\ 1/4, & \text{for } x=2. \end{cases}$$

The expected number of heads is

$$\mathbb{E}[X] = \sum_{x=\{0,1,2\}} x\,P(x) = 0\cdot 1/4 + 1\cdot 1/2 + 2\cdot 1/4 = 1.$$

The expectation can also be extended to functions of random variables. For a function $f(x)$, the expectation over the probability distribution $P(x)$ is

$$\mathbb{E}_P\left[f(x)\right] = \langle f(x)\rangle_P = \sum_{x\in\Sigma} f(x)P(x), \quad \text{(discrete)},$$

$$\mathbb{E}_p\left[f(x)\right] = \langle f(x)\rangle_p = \int_{x\in\Sigma} f(x)\,p(x)\,dx, \quad \text{(continuous)}.$$

Example 8.2.2. Consider a scenario where a gambler wins \$200 for tossing a pair of heads but loses \$100 for any other outcome. If $f : \Sigma \to \mathbb{R}$ represents the earnings, then the expected value of f is

$$\mathbb{E}[f] = -100 \cdot 3/4 + 200 \cdot 1/4 = -25.$$

The *variance* of a random variable measures its expected spread about the mean and is defined as

$$\text{Var}[X] := \mathbb{E}\left[(X - \mathbb{E}[X])^2\right].$$

Since variance and mean have different units (variance is in squared units of the mean), it is often easier to interpret the *standard deviation*, σ, which is the square root of the variance.

Example 8.2.3. Compute the variance and standard deviation of X and f from Examples 8.2.1 and 8.2.2.

Solution. For X,

$$\begin{aligned}\text{Var}[X] &= (1-0)^2 \cdot 1/4 + (1-1)^2 \cdot 1/2 + (1-2)^2 \cdot 1/4 = 1/2,\\ \sigma &= \sqrt{1/2} \approx 0.707.\end{aligned}$$

For $f(X)$,

$$\begin{aligned}\text{Var}[f(X)] &= (-100+25)^2 \cdot 3/4 + (200+25)^2 \cdot 1/4 = 4275,\\ \sigma &= \sqrt{4275} \approx 65.4.\end{aligned}$$

□

Example 8.2.4. The Cauchy distribution is important in physics, as it describes resonance behavior (e.g., the shape of a spectral width of a laser). The PDF of a Cauchy distribution with parameters $a \in \mathbb{R}$ and $\gamma > 0$ is

$$p(x|a,\gamma) = \frac{1}{\pi}\frac{\gamma}{(x-a)^2+\gamma^2}, \quad -\infty < x < +\infty. \tag{8.9}$$

Show that this distribution is properly normalized and find its mean. What can you say about its variance?

Solution. To verify that Eq. (8.9) is properly normalized, we must show that the integral of the probability density equals 1. This can be done using the trigonometric substitution $\tan(\theta) = (x - a)/\gamma$.

The mean of the Cauchy distribution is

$$\text{mean} = \frac{\gamma}{\pi} \int_{-\infty}^{+\infty} \frac{x\,dx}{(x-a)^2 + \gamma^2} = a.$$

This result is computed using a principal value integral. However, the variance is unbounded:

$$\text{variance} = \frac{\gamma}{\pi} \int_{-\infty}^{+\infty} \frac{(x-a)^2\,dx}{(x-a)^2 + \gamma^2},$$

indicating that the variance of the Cauchy distribution does not exist (it is infinite). □

8.2.2 *Higher moments*

The concepts of expectation and variance can be generalized to the *moments* of a distribution. For a discrete random variable X with probability distribution $P(x)$, the *moments* of $P(x)$ are defined as

$$k = 0, 1, 2, \ldots \qquad \mu_k := \mathbb{E}_P\left[X^k\right] = \langle X^k \rangle_P = \sum_{x \in \Sigma} x^k P(x).$$

For a continuous random variable X with probability density $p(x) = p_X(x)$, the moments of X are given by

$$k = 0, 1, 2, \ldots \qquad \mu_k := \mathbb{E}_p\left[X^k\right] = \langle X^k \rangle_p = \int_\Sigma dx\, x^k p(x).$$

From these definitions, it follows that the first moment of a random variable corresponds to its mean, $\mathbb{E}[X] = \mu_1$, and the second moment is related to the variance by $\text{Var}(X) = \mu_2 - \mu_1^2$.

Example 8.2.5. Give a closed-form expression for the moments of a Bernoulli distribution with parameter β. Use the first and second moments to find the mean and variance of the Bernoulli distribution

$$p(x) = \beta\delta(1 - x) + (1 - \beta)\delta(x). \tag{8.10}$$

Solution. The moments of the Bernoulli distribution are

$$k = 1, 2, \ldots \qquad \mu_k = \langle X^k \rangle = \int_{-\infty}^{\infty} x^k\, p(x)\, dx = \beta. \tag{8.11}$$

The mean of the distribution is its first moment: $\mu_1 = \beta$. The variance is given by $\sigma^2 = \mu_2 - \mu_1^2$. In this case, $\sigma^2 = \beta - \beta^2 = \beta(1-\beta)$. □

Example 8.2.6. What is the mean number of events in a Poisson process Pois(λ)? What is the variance of the Poisson distribution?

Solution. To compute the mean and variance of the Poisson distribution, we first calculate the first and second moments:

$$\begin{aligned}
\mu_1 &= \sum_{k=0}^{\infty} k\, P(k) = \sum_{k=0}^{\infty} \frac{k\lambda^k}{k!} e^{-\lambda} = \sum_{k=1}^{\infty} \frac{\lambda^k}{(k-1)!} e^{-\lambda} \\
&= \lambda \sum_{n=0}^{\infty} \frac{\lambda^n}{n!} e^{-\lambda} = \lambda, \\
\mu_2 &= \sum_{k=0}^{\infty} k^2\, P(k) = \sum_{k=0}^{\infty} \frac{k^2\lambda^k}{k!} e^{-\lambda} = \sum_{k=1}^{\infty} \frac{k\lambda^k}{(k-1)!} e^{-\lambda} \\
&= \lambda \sum_{n=0}^{\infty} \frac{(n+1)\lambda^n}{n!} e^{-\lambda} = \lambda(\lambda+1).
\end{aligned}$$

Therefore, the mean and variance of the number of events are

$$\mu := \mu_1 = \lambda \quad \text{and} \quad \sigma^2 := \mu_2 - \mu_1^2 = \lambda.$$

Note that both the mean and the variance of the Poisson distribution are equal to λ, which is a unique property of the Poisson distribution. □

8.2.3 *Moment-generating functions*

The *moment-generating function* (MGF) of a random variable X is defined as

$$M_X(t) = \mathbb{E}\left[\exp(tX)\right] = \int_{-\infty}^{\infty} dx\, p(x) \exp(tx), \tag{8.12}$$

where $t \in \mathbb{R}$ and all integrals are assumed to be well-defined. When $\exp(tx)$ is expanded as a Taylor series, the MGF can be written as an infinite sum involving the moments μ_k of the random variable

$$M_X(t) = \int_{-\infty}^{\infty} dx\, p(x) \sum_{k=0}^{\infty} \frac{(tx)^k}{k!} = \sum_{k=0}^{\infty} \frac{\mu_k t^k}{k!}.$$

The name "moment-generating function" comes from the fact that differentiating $M_X(t)$ k times and evaluating the result at $t = 0$ yields the kth moment of X:

$$\left. \frac{d^k}{dt^k} M_X(t) \right|_{t=0} = \mu_k.$$

Example 8.2.7. Consider the standard Boltzmann distribution from statistical mechanics, where the probability density $p(x)$ of a random state X is given by

$$p(x) = \frac{1}{Z} e^{-\beta E(x)}, \quad Z(\beta) = \sum_x e^{-\beta E(x)},$$

where $\beta = 1/T$ is the inverse temperature, $E(x)$ is the energy of state x and Z is the *partition function*. Suppose we know $Z(\beta)$ as a function of β. Compute the expected mean value and variance of the energy.

Solution. The mean value of the energy is

$$\langle E(X) \rangle = \sum_x p(x)E(x) = \frac{1}{Z} \sum_x E(x) e^{-\beta E(x)} = -\frac{1}{Z}\frac{\partial Z}{\partial \beta} = -\frac{\partial \ln Z}{\partial \beta}.$$

The variance of the energy (energy fluctuations) is

$$\mathrm{Var}\left[E(X)\right] = \langle (E(X) - \langle E(X) \rangle)^2 \rangle = \frac{\partial^2 \ln Z}{\partial \beta^2}.$$

Note that up to a sign inversion, the partition function $Z(\beta)$ is equivalent to the MGF $M_{E(X)}(-\beta)$. □

8.2.4 *Characteristic functions*

The *characteristic function* of a random variable is defined as the Fourier transform of its PDF:

$$G(t) := \mathbb{E}_p\left[\exp(itX)\right] = \int_{-\infty}^{+\infty} p(x)\exp(itx)\,dx, \tag{8.13}$$

where $i^2 = -1$. The characteristic function exists for any real t and satisfies the following properties:

$$G(0) = 1, \quad |G(t)| \leq 1.$$

The characteristic function encodes information about all the moments μ_k of the distribution. It can be expanded in a Taylor series in terms of the moments:

$$G(t) = \sum_{k=0}^{\infty} \frac{(it)^k}{k!}\langle X^k\rangle. \tag{8.14}$$

From this, we obtain

$$\langle X^k\rangle = \frac{1}{i^k}\frac{\partial^k}{\partial t^k}G(t)\bigg|_{t=0}.$$

This shows that the derivatives of $G(t)$ at $t = 0$ give the moments of the distribution as long as the corresponding moments exist.

Example 8.2.8. Find the characteristic function of a Bernoulli distribution with parameter β.

Solution. Substituting Eq. (8.10) into Eq. (8.13), we derive

$$G(t) = (1-\beta) + \beta e^{it},$$

and the moments are given by

$$\mu_k = \frac{\partial^k}{i^k\partial t^k}\left(1-\beta+\beta e^{it}\right)\bigg|_{t=0} = \beta.$$

This result is consistent with Eq. (8.11). □

Exercise 8.1. The exponential distribution has a PDF given by

$$p(x) = \begin{cases} Ae^{-\lambda x}, & x \geq 0, \\ 0, & x < 0, \end{cases}$$

where the parameter $\lambda > 0$. Calculate:

(a) the normalization constant A,
(b) the *mean* and *variance* of the distribution,
(c) the characteristic function $G(t)$,
(d) the kth moment of the distribution (using $G(t)$).

8.2.5 *Cumulants*

The *cumulants* κ_k of a random variable X are defined via the characteristic function

$$\ln G(t) = \sum_{k=1}^{\infty} \frac{(it)^k}{k!} \kappa_k. \tag{8.15}$$

Using the properties of $G(t)$, the Taylor expansion in Eq. (8.15) starts from unity. By relating Eqs. (8.14) and (8.15), we obtain the following relations between the cumulants and the moments:

$$\kappa_1 = \mu_1,$$
$$\kappa_2 = \mu_2 - \mu_1^2 = \sigma^2.$$

The procedure naturally extends to higher-order moments and cumulants.

Moments and cumulants uniquely determine one another. If all the moments of two probability distributions are identical, their cumulants are also identical, and vice versa. In some cases, theoretical treatments using cumulants are simpler than those using moments.

Example 8.2.9. Find the characteristic function and the cumulants of the Poisson distribution (8.4).

Solution. The characteristic function for the Poisson distribution is

$$G(t) = \sum_{k=0}^{\infty} \frac{\lambda^k}{k!} e^{-\lambda} e^{itk} = e^{-\lambda} \sum_{k=0}^{\infty} \frac{(\lambda e^{it})^k}{k!} = \exp\left[\lambda(e^{it} - 1)\right].$$

Thus, the cumulant-generating function is

$$\ln G(t) = \lambda(e^{it} - 1),$$

from which we find that $\kappa_k = \lambda$ for all $k = 1, 2, \ldots$. □

Example 8.2.10 (The birthday problem). Assume that there are 366 days in a year. What is the probability, p_m, that m people in a room all have different birthdays?

Solution. Let $(b_1, b_2, \ldots, b_m)$ represent the birthdays of m people, where $b_i \in \{1, 2, \ldots, 366\}$. There are 366^m different birthday lists, all equally likely. To calculate p_m, we count the lists where no two people share the same birthday:

$$\prod_{i=1}^{m}(366 - i + 1).$$

Thus, the probability is

$$p_m = \prod_{i=1}^{m}\left(1 - \frac{i-1}{366}\right).$$

The probability that at least two people share the same birthday is $1 - p_m$. It turns out that $1 - p_{23} > 0.5$, and $1 - p_{22} < 0.5$. □

Exercise 8.2. Choose three points at random on a unit circle and interpret them as cuts that divide the circle into three arcs. Compute the expected length of the arc that contains the point $(1, 0)$.

8.3 Probabilistic Inequalities

In this section, we present several important probabilistic inequalities, which are useful for bounding probabilities in various contexts.

- **Markov inequality:** For any non-negative random variable X, we have

$$P(X \geq a) \leq \frac{\mathbb{E}[X]}{a}, \tag{8.16}$$

  where $a > 0$.
- **Chebyshev's inequality:** For any random variable X with mean μ and variance σ^2, we have

$$P(|X - \mu| \geq b) \leq \frac{\sigma^2}{b^2}, \tag{8.17}$$

  where $b > 0$.

- **Chernoff bound:** For any real-valued random variable X and $t \geq 0$, we have

$$P(X \geq a) = P(e^{tX} \geq e^{ta}) \leq \frac{\mathbb{E}[e^{tX}]}{e^{ta}}. \tag{8.18}$$

Example 8.3.1. Prove the Markov, Chebyshev and Chernoff inequalities.

Solution. We start by proving the Markov inequality (8.16). First, introduce the indicator function $\mathbb{1}(y)$, which is 1 if $y \geq 0$ and 0 otherwise. The probability $P(X \geq a)$ can be written as $\mathbb{E}[\mathbb{1}(X-a)]$. Since $\mathbb{1}(X-a) \leq X/a$ for all $X \geq 0$, taking expectations on both sides yields

$$\mathbb{E}[\mathbb{1}(X-a)] \leq \mathbb{E}\left[\frac{X}{a}\right],$$

which simplifies to the desired result $P(X \geq a) \leq \frac{\mathbb{E}[X]}{a}$. □

Note that equality in the Markov inequality holds if and only if $P(X \in \{0, a\}) = 1$.

Next, to prove Chebyshev's inequality (8.17), consider the auxiliary random variable $Y = (X - \mathbb{E}[X])^2 = (X-\mu)^2$, which is non-negative by definition. Applying the Markov inequality to Y with $a = b^2$ gives

$$P(Y \geq b^2) \leq \frac{\mathbb{E}[Y]}{b^2}.$$

Since $\mathbb{E}[Y] = \sigma^2$ and $P(Y \geq b^2) = P(|X-\mu| \geq b)$, we obtain Chebyshev's inequality:

$$P(|X-\mu| \geq b) \leq \frac{\sigma^2}{b^2}.$$ □

Finally, to prove the Chernoff bound (8.18), consider the random variable $Z = e^{tX}$, where $t \geq 0$ and X is real-valued. Since Z is non-negative, we can apply the Markov inequality to Z:

$$P(Z \geq e^{ta}) \leq \frac{\mathbb{E}[Z]}{e^{ta}}.$$

Noting that $P(X \geq a) = P(e^{tX} \geq e^{ta})$ and $\mathbb{E}[Z] = \mathbb{E}[e^{tX}]$, we obtain the Chernoff bound:

$$P(X \geq a) \leq \frac{\mathbb{E}[e^{tX}]}{e^{ta}}. \qquad \square$$

The Chernoff bound can be interpreted as the Markov inequality applied to the MGF. □

We will revisit these and other useful probabilistic inequalities discussing entropy in Section 8.5, where we will discuss how to compare probabilities effectively.

8.4 Random Variables: From One to Many

The transition from one to multiple random variables is a natural progression in probability theory. This section introduces key concepts that are fundamental to working with multivariate distributions. However, it's worth noting that we have already touched on aspects of multivariate probability distributions in the previous sections, particularly when we constructed more complex (but still univariate) distributions from simpler ones. For example, we generated a sequence of *independent* random variables $X_1, X_2, \ldots$ and then created a new variable by applying a function, such as a sum, to these original variables. Independence played a crucial role in these cases (e.g., when transitioning from the Bernoulli to Poisson distribution), but not all random variables are independent. In the following sections, we explore how to describe dependencies, or correlations, between variables in higher-dimensional statistics.

8.4.1 *Multivariate distributions, marginalization and conditional probability*

Consider an n-component random vector $\boldsymbol{X}$, with probability $P(\boldsymbol{x})$ representing the likelihood of observing a specific realization $\boldsymbol{x}$. The vector $\boldsymbol{X}$ satisfies the normalization condition $\sum_{\boldsymbol{x}} P(\boldsymbol{x}) = 1$. For example, if each component X_i takes values in $\Sigma = \{0, 1\}$, then $\boldsymbol{X}$ is a random vector of length n with binary entries (e.g., $\boldsymbol{x}$ could be $(1, 1, 0, \ldots, 1)$).

There are two fundamental questions we often ask about $P(\boldsymbol{x})$:

1. **Marginalization:** What is the probability of observing a state where one or more components take specific values?
2. **Conditional probability:** What is the probability of observing a state, given that one or more components are already known to take specific values?

Example 8.4.1. Let X_i represent the outcome of the ith toss of a fair coin, where $\Sigma = \{0, 1\}$, and $P(X_i = 0) = 1/2$ and $P(X_i = 1) = 1/2$. Let $\boldsymbol{X} = (X_1, X_2)$ be the random vector representing the outcome of two successive coin flips. The probabilities of each possible outcome are

$$\begin{aligned} P(\boldsymbol{X} = (0,0)) &= 1/4, \quad P(\boldsymbol{X} = (1,0)) = 1/4, \\ P(\boldsymbol{X} = (0,1)) &= 1/4, \quad P(\boldsymbol{X} = (1,1)) = 1/4. \end{aligned}$$

The following questions illustrate marginalization and conditional probability:

1. **Marginalization:** What is the probability that the first toss resulted in a "1"?
2. **Conditional probability:** Given that the first toss resulted in a "1," what are the possible outcomes of the second toss and their probabilities?

Solution.

1. To find the probability that the first toss resulted in a "1," sum the probabilities of the outcomes where $X_1 = 1$:

$$P(X_1 = 1) = P(\boldsymbol{X} = (1,0)) + P(\boldsymbol{X} = (1,1)) = 1/4 + 1/4 = 1/2.$$

2. Given that the first toss was "1," the remaining possible outcomes are $(1,0)$ and $(1,1)$, each with equal probability:

$$P(X_2 = 0 \,|\, X_1 = 1) = 1/2, \quad P(X_2 = 1 \,|\, X_1 = 1) = 1/2. \quad \square$$

Example 8.4.1 is straightforward because X_1 and X_2 are independent. More interesting scenarios arise when variables are dependent.

Consider the statistical Ising model, initially developed in physics to describe magnetization but also applicable to modeling phenomena like epidemics. The model defines a probability distribution over the 2^n-dimensional space Σ, given by

$$P(\boldsymbol{x}) = Z^{-1} \exp\left(\sum_{i=1}^{n-1} J x_i x_{i+1}\right),$$

$$Z = \sum_{\boldsymbol{x}} \exp\left(\sum_{i=1}^{n-1} J x_i x_{i+1}\right),$$

where J is the coupling constant between adjacent components and Z is the normalization constant (partition function) ensuring that the total probability is 1.

For $n = 2$, we obtain the bivariate distribution:

$$P(\boldsymbol{x}) = P(x_1, x_2) = \frac{\exp(J x_1 x_2)}{4\cosh(J)}.$$

The distribution $P(\boldsymbol{x})$ is called a **joint** or **multivariate** probability distribution, describing the probability of all components $(x_1, \ldots, x_n)$ together.

Conditional Probability: For the $n = 2$ case, the conditional probability of x_1 given x_2 is:

$$P(x_1|x_2) = \frac{P(x_1, x_2)}{\sum_{x_1} P(x_1, x_2)} = \frac{\exp(J x_1 x_2)}{2\cosh(J)}.$$

It satisfies the normalization condition: $\sum_{x_1} P(x_1|x_2) = 1$ for all x_2.

Marginalization: The joint distribution can also be marginalized over certain variables. For example, marginalizing over $x_2, \ldots, x_n$ gives the marginal probability:

$$P(x_1) = \sum_{\boldsymbol{x}\backslash x_1} P(\boldsymbol{x}) = \sum_{x_2,\ldots,x_n} P(x_1, \ldots, x_n).$$

Multivariate Gaussian (Normal) Distribution

Now, consider n zero-mean random variables $X_1, X_2, \ldots, X_n$ sampled i.i.d. from a generic Gaussian distribution:

$$p(x_1, \ldots, x_n) = \frac{1}{Z} \exp\left(-\frac{1}{2} \sum_{i,j=1}^{n} x_i A_{ij} x_j\right), \tag{8.19}$$

where A is a symmetric ($A = A^T$), positive definite ($A \succ 0$) matrix. If A is diagonal, the distribution factorizes, implying that each X_i is independent. The normalization constant Z, or partition function, is

$$Z = \frac{(2\pi)^{n/2}}{\sqrt{\det A}}.$$

The moments of this Gaussian distribution are

$$\forall i: \quad \mathbb{E}[X_i] = \mu_i,$$
$$\forall i,j: \quad \mathbb{E}[(X_i - \mu_i)(X_j - \mu_j)] = (A^{-1})_{ij} = \Sigma_{ij}.$$

Here, Σ_{ij} denotes the i, j component of the inverse of A and is called the *covariance matrix*, which is symmetric and positive definite. The multivariate Gaussian distribution with mean vector $\mu = (\mu_i | i = 1, \ldots, n)$ and covariance matrix Σ is denoted as $\mathcal{N}(\mu, \Sigma)$ or $\mathcal{N}_n(\mu, \Sigma)$.

The Gaussian distribution is remarkable for its *invariance* properties.

Theorem 8.4.2 (Invariance of the Gaussian Distribution Under Conditioning and Marginalization). *Let $X \sim \mathcal{N}_n(\mu, \Sigma)$ and decompose the n-dimensional vector X into two components: $X = (X_1, X_2)$, where X_1 is a p-dimensional subvector and X_2 is a q-dimensional subvector, with $p+q = n$. Assume that the mean vector μ and the covariance matrix Σ are partitioned as follows:*

$$\mu = (\mu_1, \mu_2), \quad \Sigma = \begin{pmatrix} \Sigma_{11} & \Sigma_{12} \\ \Sigma_{21} & \Sigma_{22} \end{pmatrix}, \tag{8.20}$$

where μ_1 and μ_2 are vectors and $\Sigma_{11}, \Sigma_{12}, \Sigma_{21}, \Sigma_{22}$ are submatrices of appropriate dimensions. Then:

- ***Marginalization:*** *The marginal distribution $p(x_1) := \int dx_2\, p(x_1, x_2)$ is Gaussian: $\mathcal{N}(\mu_1, \Sigma_{11})$.*
- ***Conditioning:*** *The conditional distribution $p(x_1|x_2) := \frac{p(x_1,x_2)}{p(x_2)}$ is Gaussian: $\mathcal{N}(\mu_{1|2}, \Sigma_{1|2})$, where*

$$\mu_{1|2} := \mu_1 + \Sigma_{12}\Sigma_{22}^{-1}(x_2 - \mu_2), \quad \Sigma_{1|2} := \Sigma_{11} - \Sigma_{12}^T\Sigma_{22}^{-1}\Sigma_{12}.$$

Proof. A useful technical exercise, involving basic linear algebra, requires deriving the explicit formula for the inverse of a positive definite matrix partitioned as in Eq. (8.20). □

8.4.2 Central limit theorem

Consider n i.i.d. random variables $X_1, X_2, \ldots, X_n$ drawn from a distribution with mean μ and variance $\sigma^2 > 0$. We are interested in the behavior of their sample mean:

$$Y_n = \frac{1}{n}\sum_{i=1}^{n} X_i.$$

What is the probability distribution of Y_n as n increases?

Theorem 8.4.3 (Weak Version of the Central Limit Theorem (CLT)). *The normalized sum $\sqrt{n}(Y_n - \mu)$ converges in distribution to a Gaussian with mean μ and variance σ^2, i.e.,*

$$n \to \infty : \quad \sqrt{n}\left(\frac{1}{n}\sum_{i=1}^{n} X_i - \mu\right) \sim \mathcal{N}(0, \sigma^2). \tag{8.21}$$

Let us outline the proof of the weak CLT (8.21) in the simple case where $\mu = 0$ and $\sigma^2 = 1$.

First, note that

$$\mu_1(Y_n\sqrt{n}) = 0.$$

Next, compute the second moment:

$$\mu_2(Y_n\sqrt{n}) = \mathbb{E}\left[\left(\frac{X_1 + \cdots + X_n}{\sqrt{n}}\right)^2\right] = \frac{\sum_i \mathbb{E}\left[X_i^2\right]}{n} + \frac{\sum_{i\neq j} \mathbb{E}\left[X_i X_j\right]}{n} = 1.$$

For the third moment, we find

$$\mu_3(Y_n\sqrt{n}) = \mathbb{E}\left[\left(\frac{X_1 + \cdots + X_n}{\sqrt{n}}\right)^3\right] = \frac{\sum_i \mathbb{E}[X_i^3]}{n^{3/2}} \to 0,$$

as $n \to \infty$, assuming that $\mathbb{E}[X_i^3] = O(1)$. The fourth moment satisfies

$$\mu_4(Y_n\sqrt{n}) = 3 = 3\mu_2(Y_n),$$

which is characteristic of a Gaussian distribution.

Example 8.4.4 (Sum of Gaussian Variables). Compute the probability density $p_n(y_n)$ of the random variable $Y_n = \frac{1}{n}\sum_{i=1}^{n} X_i$, where $X_1, X_2, \ldots, X_n$ are sampled i.i.d. from a normal distribution:

$$p(x) = \mathcal{N}(\mu, \sigma^2) = \frac{1}{\sqrt{2\pi}\sigma} \exp\left(-\frac{(x-\mu)^2}{2\sigma^2}\right).$$

Solution. First, recall that the characteristic function of a Gaussian distribution is also Gaussian:

$$G(t) = \int_{\mathbb{R}} e^{itx} p(x)\, dx = \exp\left(i\mu t - \frac{\sigma^2 t^2}{2}\right).$$

Now, evaluate the characteristic function for $p_n(y_n)$:

$$G_n(t) = \int_{\mathbb{R}^n} dx_1 \cdots dx_n \exp\left(i\frac{t}{n}\sum_{i=1}^{n} x_i\right) p(x_1)\cdots p(x_n) \Rightarrow (G(t/n))^n$$

$$= \exp\left(i\mu t - \frac{\sigma^2 t^2}{2n}\right).$$

Taking the inverse Fourier transform of $G_n(t)$ yields

$$p_n(y_n) = \int_{-\infty}^{\infty} \frac{dt}{2\pi} G_n(t) e^{-ity_n} = \frac{\sqrt{n}}{\sqrt{2\pi}\sigma} \exp\left(-\frac{n(y_n-\mu)^2}{2\sigma^2}\right).$$

□

Example 8.4.5 (Failure of the Central Limit Theorem). Consider $X_1, X_2, \ldots, X_n$ independently chosen from the Cauchy distribution:

$$p(x) = \frac{\gamma}{\pi}\frac{1}{x^2+\gamma^2}.$$

Calculate the probability density of the sample mean $Y_n = \frac{1}{n}\sum_{i=1}^{n} X_i$ and show that the CLT does not hold.

Solution. The characteristic function of the Cauchy distribution is

$$G(k) = \frac{\gamma}{\pi}\int_{-\infty}^{\infty} \frac{e^{ikx}\, dx}{x^2+\gamma^2} = e^{-\gamma|k|}.$$

For the sample mean Y_n, we have

$$G_n(k) = (G(k/n))^n = G(k).$$

Thus, for any n, the random variable Y_n follows the same Cauchy distribution as the individual samples, meaning that the CLT fails.

The reason is that the Cauchy distribution lacks a well-defined variance, which is a crucial requirement for the CLT to hold. (See Example 8.2.4.) □

Exercise 8.3. Assume you play a dice game 100 times. The payoffs for the game are as follows: \$0.00 for rolling a 1, 3, or 5; \$2.00 for rolling a 2 or 4; and \$26.00 for rolling a 6.

(1) What is the expected value of your winnings?
(2) What is the standard deviation of your winnings?
(3) Estimate the probability that you win at least \$400.

Exercise 8.4. Experiment with the CLT for different distributions mentioned in the lecture.

8.4.3 *Strong version of the central limit theorem*

The CLT holds for independent, but not necessarily identically distributed, variables. Furthermore, we may be interested in deviations beyond the standard deviation, i.e., large deviations. This leads us to the following powerful result.

Theorem 8.4.6 (Cramér's Theorem (Strong Version of the CLT)). *The normalized sum* $Y_n = \frac{1}{n}\sum_{i=1}^n X_i$ *of i.i.d. variables* $X_i \sim p_X(x)$ *satisfies*

$$\forall x > \mu : \quad \lim_{n\to\infty} \frac{1}{n} \log \mathrm{Prob}\,(Y_n \geq x) = -\Phi^*(x), \tag{8.22}$$

$$\Phi^*(x) := \sup_{t\in\mathbb{R}} (tx - \Phi(t)), \tag{8.23}$$

$$\Phi(t) := \log\left(\mathbb{E}[\exp(tX)]\right), \tag{8.24}$$

where $\Phi(t)$ *is the cumulant-generating function of* $p_X(x)$ *and* $\Phi^*(x)$ *is the Legendre–Fenchel transform of* $\Phi(t)$*, also called the Cramér function.*

Comments: (1) An informal version of Eq. (8.22) is

$$n \to \infty : \quad \mathrm{Prob}\,(Y_n) \propto \exp\left(-n\Phi^*(x)\right).$$

(2) The cumulant-generating function $\Phi(t)$ is related to the characteristic function $G(-it)$ of the random variable. (3) The weak version

of the CLT (8.21) is equivalent to approximating $\Phi^*(x)$ by a Gaussian distribution centered at its minimum.

Exercise 8.5. Prove the strong CLT (8.22, 8.23). [*Hint*: Use the saddle-point method to evaluate integrals.] Provide an example of an expectation where the details of $\Phi^*(x)$ beyond its minimum are significant for large n. Can you identify an example where the behavior is controlled by the tails of $\Phi^*(x)$?

Example 8.4.7. Compute the Cramér function for a Bernoulli process (unfair coin toss):

$$X = \begin{cases} 0, & \text{with probability } 1-\beta, \\ 1, & \text{with probability } \beta. \end{cases} \tag{8.25}$$

Solution. We compute

$$\Phi(t) = \log(\beta e^t + 1 - \beta),$$

and for $0 < x < 1$,

$$\Phi^*(x) = x \log \frac{x}{\beta} + (1-x) \log \frac{1-x}{1-\beta}. \qquad \square$$

The Cramér function for the Bernoulli process leads to the famous Stirling formula for the asymptotic behavior of factorials:

$$n! = \sqrt{2\pi n} \left(\frac{n}{e}\right)^n \left(1 + O\left(\frac{1}{n}\right)\right).$$

Theorem 8.4.8 (Chernoff Bound Version of the CLT, Adapted from Ref. [44]). *Let $X_1, \ldots, X_n$ be i.i.d. random variables with $\mathbb{E}[X_i] = \mu$ and a well-defined Cramér function $\Phi^*(x)$. Then,*

$$P\left(\sum_{i=1}^{n} X_i \geq nx\right) \leq \exp\left(-n\Phi^*(x)\right), \quad \forall x > \mu,$$

$$P\left(\sum_{i=1}^{n} X_i \leq nx\right) \leq \exp\left(-n\Phi^*(x)\right), \quad \forall x < \mu.$$

Proof. For $x > \mu$, we apply the Chernoff bound:

$$P\left(\sum_{i=1}^{n} X_i \geq nx\right) = P\left(e^{t\sum_{i=1}^{n} X_i} \geq e^{tnx}\right) \quad \text{for any } t > 0.$$

Using Markov's inequality,

$$P\left(e^{t\sum_{i=1}^{n} X_i} \geq e^{tnx}\right) \leq e^{-tnx}\mathbb{E}\left[e^{t\sum_{i=1}^{n} X_i}\right].$$

Since X_i are i.i.d.,

$$P\left(\sum_{i=1}^{n} X_i \geq nx\right) \leq \exp\left(-n\sup_{t>0}\left(tx - \Phi(t)\right)\right) = \exp\left(-n\Phi^*(x)\right).$$

For $x < \mu$, the proof is analogous, using $t < 0$. □

Exercise 8.6. Let $X = \sum_{i=1}^{n} X_i$, where the X_i are independent (but not necessarily identically distributed) Poisson random variables, $X_i \sim \text{Pois}(\lambda_i)$. Denote the characteristic function of X by $G_X(t)$ and the characteristic function of X_i by $G_{X_i}(t)$. Show that:

(1) $G_X(t) = \prod_{i=1}^{n} G_{X_i}(t)$.
(2) $X \sim \text{Pois}(\lambda)$, where $\lambda = \sum_{i=1}^{n} \lambda_i$.

8.4.4 *Bayes' theorem*

We have already seen how to derive both the conditional and marginal probability distributions from the joint probability distribution:

$$P(x|y) = \frac{P(x,y)}{P(y)}, \quad P(y|x) = \frac{P(x,y)}{P(x)}. \tag{8.26}$$

By combining these formulas to eliminate the joint probability distribution $P(x, y)$, we arrive at the well-known Bayes' theorem:

$$P(x|y)P(y) = P(y|x)P(x). \tag{8.27}$$

In Eqs. (8.26) and (8.27), both x and y may be multivariate. Rearranging Eq. (8.27) leads to the following form of Bayes' theorem:

$$P(x|y) = \frac{P(y|x)P(x)}{P(y)},$$

which is frequently used in the field of *Bayesian inference*. Here: – $P(x)$ is known as the *prior* distribution, representing the initial

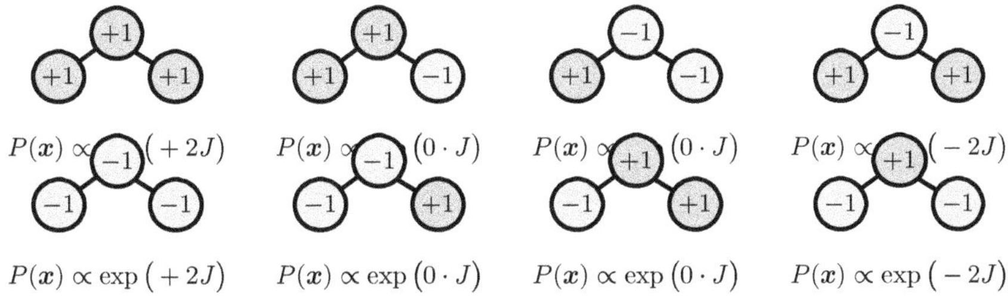

Figure 8.7. Set of all possible outcomes for the three-component Ising model and their relative probabilities. The normalization constant Z is found by summing the probabilities over all states.

degree of belief in x before observing y; – $P(x|y)$ is the *posterior* distribution, which quantifies the updated belief in x after observing y; – the term $\frac{P(y|x)}{P(y)}$ represents the amount of *evidence* or *support* that the observation y provides for x.

A helpful visual explanation of conditional probability can be found at http://setosa.io/ev/conditional-probability/.

Example 8.4.9. Consider the three-component Ising model. Compute the following:

(a) The normalization constant.
(b) The marginal probability $P(x_1)$.
(c) The conditional probability $P(x_3|x_1)$.

Solution. The set of all possible outcomes is shown in Fig. 8.7.

(a) The normalization constant Z is computed by summing the probabilities over all possible states:

$$Z = e^{2J} + e^0 + e^0 + e^{-2J} + e^{2J} + e^0 + e^0 + e^{-2J} = 4 + 4\cosh(2J).$$

Hence, the probabilities of the states are $P(1,1,1) = \frac{e^{2J}}{4+4\cosh(2J)}$, etc.

(b) The marginal probability $P(x_1)$ is obtained by summing over all possible values of x_2 and x_3:

$$\begin{aligned} P(x_1 = +1) &= P(1,1,1) + P(1,1,-1) + P(1,-1,-1) \\ &\quad + P(1,-1,1) \\ &= \frac{e^{2J} + e^0 + e^0 + e^{-2J}}{4 + 4\cosh(2J)} = \frac{1}{2}, \end{aligned}$$

$$\begin{aligned}P(x_1=-1) &= P(-1,-1,-1)+P(-1,-1,1)+P(-1,1,1)\\&\quad+P(-1,1,-1)\\&= \frac{e^{2J}+e^0+e^0+e^{-2J}}{4+4\cosh(2J)}=\frac{1}{2}.\end{aligned}$$

(c) The conditional probability $P(x_3|x_1)$ is given by

$$\begin{aligned}P(x_3=+1|x_1=+1) &= \frac{P(x_3=+1,x_1=+1)}{P(x_1=+1)}\\&= \frac{Z^{-1}(e^{2J}+e^{-2J})}{Z^{-1}(e^{2J}+1+1+e^{-2J})}=\frac{\cosh(2J)}{1+\cosh(2J)},\\P(x_3=+1|x_1=-1) &= \frac{P(x_3=+1,x_1=-1)}{P(x_1=-1)}\\&= \frac{Z^{-1}(e^0+e^0)}{Z^{-1}(e^{2J}+1+1+e^{-2J})}=\frac{1}{1+\cosh(2J)},\\P(x_3=-1|x_1=+1) &= \frac{P(x_3=-1,x_1=+1)}{P(x_1=+1)}\\&= \frac{Z^{-1}(e^0+e^0)}{Z^{-1}(e^{2J}+1+1+e^{-2J})}=\frac{1}{1+\cosh(2J)},\\P(x_3=-1|x_1=-1) &= \frac{P(x_3=-1,x_1=-1)}{P(x_1=-1)}\\&= \frac{Z^{-1}(e^{2J}+e^{-2J})}{Z^{-1}(e^{2J}+1+1+e^{-2J})}=\frac{\cosh(2J)}{1+\cosh(2J)}.\end{aligned}$$

□

Exercise 8.7. The joint probability density of two real random variables X_1 and X_2 is given by

$$\forall x_1,x_2\in\mathbb{R}:\quad p(x_1,x_2)=\frac{1}{Z}\exp(-x_1^2-x_1x_2-x_2^2).$$

(1) Calculate the normalization constant Z.
(2) Compute the marginal probability density $p(x_1)$.
(3) Find the conditional probability density $p(x_1|x_2)$.

8.5 Information-Theoretic View on Randomness

This section is accompanied by a Jupyter/Julia notebook, Entropy.ipynb, which is available on the author's living-book website https://sites.google.com/site/mchertkov/living-books/applied-math-book.

8.5.1 *Entropy*

Consider a random variable X that takes outcomes $x \in \mathcal{X}$. Our goal is to develop a systematic way to quantify the amount of information gained when a specific outcome occurs. We define the *information content* of an outcome x, denoted $h(x)$ (also referred to as *surprise*) and assume that it depends only on the probability of the outcome $P(x)$.

The question is: How should we quantify the information content? We begin by listing the properties we expect $h(x)$ to satisfy:

(1) Deterministic outcomes provide no new information. If an outcome is certain to occur, the information content must be zero:

$$h(x) = 0 \quad \text{if } P(x) = 1.$$

(2) Unlikely outcomes provide more information than likely outcomes. The information content should be a strictly decreasing function of the probability:

$$h(x_1) > h(x_2) \quad \text{for } P(x_1) < P(x_2).$$

(3) Independent events provide additive information. If two independent events occur, the total information should be the sum of the information for each individual event:

$$h(x, y) = h(x) + h(y) \quad \text{provided that } P(x, y) = P(x)P(y).$$

As illustrated in Fig. 8.8, with these criteria, we can show that only one family of continuous functions satisfies all the requirements. This leads to defining the information content as the negative logarithm of the probability:

$$h(x) = -\log\big(P(x)\big). \tag{8.28}$$

The base of the logarithm, or the scaling factor, can be chosen arbitrarily. In information theory, it is standard to use base 2, i.e., $\log \to \log_2$ in Eq. (8.28).

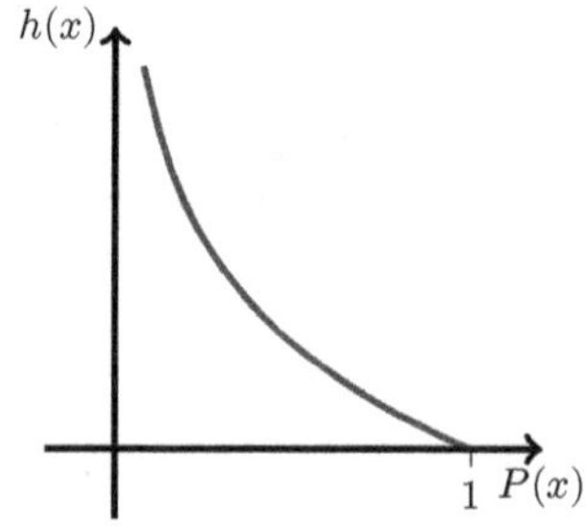

Figure 8.8. The information content $h(x)$ of an outcome x as a function of its probability. The family of functions $h(x) = -\log(P(x))$ is the only one that satisfies the desired properties of information content. The base of the logarithm determines the unit of measurement, with base 2 being common in information theory.

Terminology. In scientific contexts, the information gained from learning an outcome x is often called the *configurational entropy*, a term synonymous with the "surprise" of x.

We now define the *entropy* of a random variable X, which represents the expected amount of information (or average surprise) for all possible outcomes. This is the expectation of the configurational entropy:

$$\begin{aligned} H(X) &= -\mathbb{E}_{P(X)}\left[\log\big(P(X)\big)\right] = \sum_{x\in\mathcal{X}} P(x)h(x) \\ &= -\sum_{x\in\mathcal{X}} P(x)\log\big(P(x)\big), \end{aligned} \tag{8.29}$$

where x is drawn from the outcome space $\mathcal{X}$. Entropy can also be interpreted as a measure of uncertainty. For a deterministic process (i.e., when one outcome has probability 1), the entropy is zero since no new information is gained.

Terminology. Another term often used for entropy is the *measure of uncertainty*. In information theory, entropy is denoted by H, but in statistical mechanics, it is commonly represented by S.

To better understand entropy, consider the example of the Bernoulli process Bernoulli(β) (see Eq. (8.25)). In this case, there are two possible outcomes: $P(X = 1) = \beta$ and $P(X = 0) = 1 - \beta$.

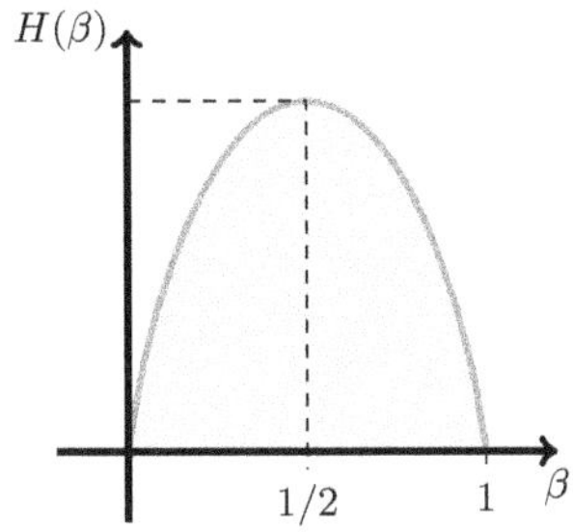

Figure 8.9. The entropy of a Bernoulli random variable as a function of β. The entropy is maximized at $\beta = 1/2$, corresponding to maximum uncertainty when the outcomes are equally likely. The entropy is zero at $\beta = 0$ and $\beta = 1$, where the outcomes are deterministic.

The entropy is given by

$$H = -\beta \log(\beta) - (1 - \beta) \log(1 - \beta).$$

This function is concave and reaches its maximum at $\beta = 1/2$, as shown in Fig. 8.9. This corresponds to the most uncertain scenario – when the outcomes are equally probable, as in the case of a fair coin flip. On the other hand, the entropy is zero when $\beta = 0$ or $\beta = 1$, which are the fully certain, deterministic cases.

The entropy formula (8.29) has the following important properties:

- $H \geq 0$, with equality if and only if the process is deterministic (i.e., $\exists x$ such that $P(x) = 1$).
- $H \leq \log(|\mathcal{X}|)$, where equality holds if x is uniformly distributed over $\mathcal{X}$.
- Entropy measures average uncertainty.
- Entropy is less than or equal to the average number of bits needed to describe the random variable (equality holds for uniform distributions).[a]
- Entropy is a lower bound on the average length of the shortest description of a random variable.

[a]For example, representing an integer smaller than n in binary requires $\log_2(n)$ bits. If all integers are equally probable, this is exactly the entropy of the distribution. For non-uniform distributions, the entropy is smaller.

Example 8.5.1. Zipf's law states that the frequency of the nth most frequent word in a randomly chosen English document can be approximated by

$$p_n = \begin{cases} \frac{0.1}{n}, & \text{for } n \in \{1, \ldots, 12367\}; \\ 0, & \text{for } n > 12367. \end{cases} \tag{8.30}$$

Assuming that English documents are generated by selecting words randomly according to Eq. (8.30), compute the entropy of "made-up English" per word.

Solution. Substituting the distribution (8.30) into the definition of entropy, we get

$$\begin{aligned} H &= -\sum_{n=1}^{12367} \frac{0.1}{n} \log_2 \frac{0.1}{n} \approx \frac{0.1}{\ln 2} \int_{10}^{123670} \frac{\ln x}{x} dx \\ &= \frac{1}{20 \ln 2} \left(\ln^2 123670 - \ln^2 10\right) \approx 9.9 \text{ bits.} \end{aligned}$$

□

It is well known from Shannon's work [45] that the entropy of the English alphabet per character is relatively low: around 1 bit. Therefore, the character-based entropy of typical English text is much smaller than its word-based entropy. This is intuitive: After the first few letters of a word, the rest of the word can often be guessed, whereas predicting the next word in a sentence is more difficult.

8.5.2 *Comparing probability distributions: Kullback–Leibler divergence*

The concepts of information content (or surprise) and entropy provide valuable tools for working with probability distributions. One of the most important tools is a method for comparing two probability distributions. Let X be a random variable taking values $x \in \mathcal{X}$. Suppose P_1 is the "true" probability distribution of X (considered as the ground truth) and P_2 is an approximation or model of P_1. The difference in information content between the two distributions for a particular outcome x is given by

$$\log\left(P_1(x)\right) - \log\left(P_2(x)\right) = \log\left(\frac{P_1(x)}{P_2(x)}\right).$$

The *Kullback–Leibler (KL) divergence* quantifies the expected difference in information content between the ground truth and its approximation. It is defined as the expectation of the log ratio of probabilities, taken with respect to the distribution P_1:

$$D(P_1\|P_2) := \sum_{x\in\mathcal{X}} P_1(x)\log\frac{P_1(x)}{P_2(x)}. \tag{8.31}$$

The KL divergence measures how much information is lost when approximating P_1 by P_2. Note that: KL divergence is not symmetric: $D(P_1\|P_2) \neq D(P_2\|P_1)$. It does not satisfy the triangle inequality, meaning it is not a true metric.

A metric $d(a,b)$ must satisfy three properties: (i) non-negativity: $d(a,b) \geq 0$ and $d(a,a) = 0$, (ii) symmetry: $d(a,b) = d(b,a)$ and (iii) the triangle inequality: $d(a,b) \leq d(a,c) + d(b,c)$. KL divergence fails the last two conditions. However, the second-order approximation (Hessian) of the KL divergence around its minimum (when $P_1 = P_2$) is known as the *Fisher information*, which satisfies all the properties of a metric.

Example 8.5.2. An illusionist uses a biased coin that comes up heads 70% of the time. Use KL divergence to quantify how much information would be lost if the biased coin were modeled as a fair coin.

Solution. Let the biased distribution be P_1 (the ground truth) and the fair distribution be P_2 (the approximation). The KL divergence between the two is

$$\begin{aligned} D(P_1\|P_2) = \mathbb{E}_{P_1}\left[\log\left(\frac{P_1}{P_2}\right)\right] &= \sum_x P_1(x)\log\left(\frac{P_1(x)}{P_2(x)}\right) \\ &= 0.3\log_2\left(\frac{0.3}{0.5}\right) + 0.7\log_2\left(\frac{0.7}{0.5}\right) \\ &= 0.3\log_2(0.6) + 0.7\log_2(1.4) \\ &\approx -0.221 + 0.329 = 0.118\,\text{bits}. \end{aligned}$$

Approximately 0.118 bits of information are lost by modeling the biased coin as a fair coin. □

Exercise 8.8. Let X be a random variable with known probability distribution $P_2(x)$, where $x \in \mathcal{X}$ and $\mathcal{X}$ is finite. Consider a vector $(P_1(x) \mid x \in \mathcal{X})$ that satisfies $P_1(x) \geq 0$ for all $x \in \mathcal{X}$ and $\sum_{x \in \mathcal{X}} P_1(x) = 1$. Show that $D(P_1 \| P_2)$ is non-negative and that its minimum is achieved when $P_1(x) = P_2(x)$ for all $x \in \mathcal{X}$, i.e.,

$$\arg \min_{(P_1(x) \mid x \in \mathcal{X})} D(P_1 \| P_2) \Bigg|_{\substack{\sum_{x \in \mathcal{X}} P_1(x) = 1 \\ \forall x \in \mathcal{X} : P_1(x) \geq 0}} = (P_2(x) \mid x \in \mathcal{X}). \tag{8.32}$$

8.5.3 *Joint and conditional entropy*

Entropy can be naturally extended to the case of multivariate statistics. If we have two discrete random variables, X and Y, with X taking values in $\mathcal{X}$ and Y in $\mathcal{Y}$, their *joint entropy* is defined as

$$H(X, Y) := -\mathbb{E}\left[\log\left(P(X, Y)\right)\right] = -\sum_{x \in \mathcal{X}, y \in \mathcal{Y}} P(x, y) \log\left(P(x, y)\right). \tag{8.33}$$

One might ask whether $H(X, Y) = H(X) + H(Y)$. This would imply that the total uncertainty in the system is the sum of the uncertainties in X and Y separately. To investigate, we consider the difference $H(X, Y) - H(X)$, which represents the remaining uncertainty in Y given that X is known:

$$\begin{aligned} H(X, Y) - H(X) &= -\sum_{x \in \mathcal{X}, y \in \mathcal{Y}} P(x, y) \log\left(P(x, y)\right) \\ &\quad + \sum_{x \in \mathcal{X}} P(x) \log\left(P(x)\right) \\ &= -\sum_{x \in \mathcal{X}, y \in \mathcal{Y}} P(x, y) \log\left(\frac{P(x, y)}{P(x)}\right). \end{aligned}$$

If X and Y are independent, $P(x, y) = P(x)P(y)$, and the result simplifies to $H(Y)$. However, if X and Y are dependent, we have

$P(x, y)/P(x) = P(y \mid x)$ by Bayes' theorem. This motivates the definition of *conditional entropy*:

$$H(Y \mid X) := H(X, Y) - H(X) = -\mathbb{E}\big[\log\big(P(Y \mid X)\big)\big]$$
$$= - \sum_{x \in \mathcal{X}, y \in \mathcal{Y}} P(x, y) \log\big(P(y \mid x)\big).$$

The relation between joint entropy and conditional entropy is given by the *chain rule*:

$$H(X, Y) = H(X) + H(Y \mid X), \tag{8.34}$$

as illustrated in Fig. 8.10.

This chain rule generalizes to the multivariate case, where for a collection of random variables $(X_1, \ldots, X_n) \sim P(x_1, \ldots, x_n)$, the entropy can be written as

$$H(X_1, \ldots, X_n) = \sum_{i=1}^{n} H(X_i \mid X_{i-1}, \ldots, X_1).$$

Note that the order in the chain is arbitrary.

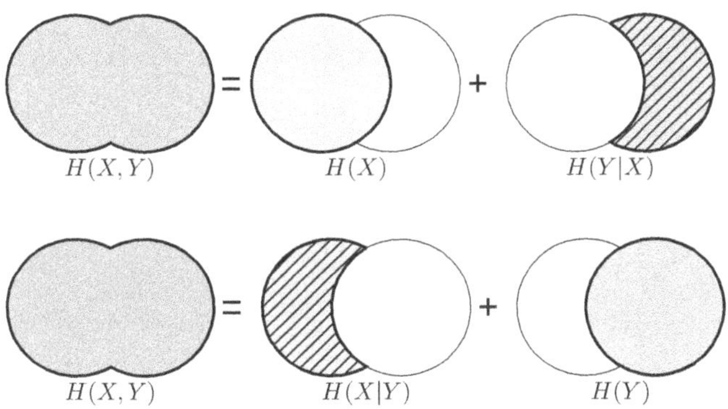

Figure 8.10. Venn diagrams illustrating the relationships between entropy, joint entropy and conditional entropy. Although these diagrams are useful for understanding information-theoretic concepts, note that they do not represent sets in the traditional sense. See Ref. [43, p. 141] for more details.

8.5.4 *Independence, dependence and mutual information*

When comparing two sources of information, say events x and y, one often considers the case where the two are independent, meaning that the joint probability satisfies $P(x, y) = P(x)P(y)$. In such a case, we also have $P(x|y) = P(x)$ and $P(y|x) = P(y)$. As a result, the *mutual information*, a measure of dependence between the two variables, is zero. Naturally, mutual information is defined as a quantitative measure of this dependence:

$$I(X;Y) = \mathbb{E}_{P(x,y)}\left[\log \frac{P(x,y)}{P(x)P(y)}\right] = \sum_{x\in\mathcal{X}}\sum_{y\in\mathcal{Y}} P(x,y)\log\frac{P(x,y)}{P(x)P(y)}. \tag{8.35}$$

Intuitively, mutual information measures how much knowledge of one random variable reduces uncertainty about the other. For instance, if X and Y are independent, knowing X provides no information about Y, and vice versa, leading to zero mutual information. On the other hand, if X is a deterministic function of Y, then all information conveyed by X is also conveyed by Y, and the mutual information equals the entropy of X (or equivalently, Y).

The mutual information is closely related to the respective entropies:

$$\begin{aligned} I(X;Y) &= H(X) - H(X|Y) = H(Y) - H(Y|X) \\ &= H(X) + H(Y) - H(X,Y), \end{aligned}$$

as illustrated in Fig. 8.11.

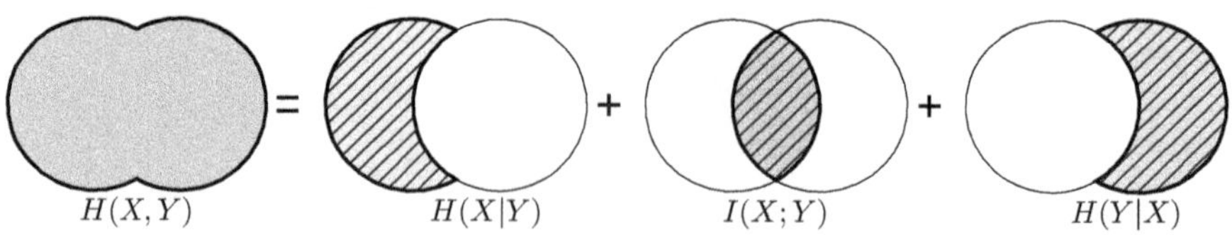

Figure 8.11. Venn diagram illustrating the relationship between mutual information and respective entropies for two random variables. It is customary in information theory to use Venn diagrams to depict entropies, conditional entropies and mutual information. However, note that the shapes in the diagram do not represent literal sets of objects. See, for example, Ref. [43, p. 141] for a detailed discussion with examples.

Mutual information also has the following properties:

$$I(X;Y) = I(Y;X) \quad \text{(symmetry)}.$$
$$I(X;X) = H(X) \quad \text{(self-information)}.$$

The conditional mutual information between two random variables X and Y, given another random variable Z, is defined as

$$I(X;Y|Z) := H(X|Z) - H(X|Y,Z) = \mathbb{E}_{P(x,y,z)}\left[\log \frac{P(x,y|z)}{P(x|z)P(y|z)}\right]. \tag{8.36}$$

Applying the entropy chain rule (8.34) to the mutual information of $(X_1,\ldots,X_n) \sim P(x_1,\ldots,x_n)$ results in

$$I(X_1,\ldots,X_n;Y) = \sum_{i=1}^{n} I(X_i;Y|X_{i-1},\ldots,X_1). \tag{8.37}$$

See Fig. 8.11 for a Venn diagram illustration of Eq. (8.37).

For extended discussions on entropy, mutual information and related concepts, we recommend consulting Ref. [43].

The notions of joint entropy, conditional entropy and mutual information can be illustrated through the following examples involving two random variables.

Example 8.5.3. Consider two Bernoulli random variables X and Y with the joint PMF $P(X,Y)$ given by the following:

	$y=0$	$y=1$
$x=0$	0	0.2
$x=1$	0.8	0

Compute the entropy of X, the joint entropy of X and Y and the conditional entropy of Y given X. Discuss the results.

Solution. The joint PMF indicates that the outcome of Y is completely determined by the outcome of X. Intuitively, we expect that all the information in the system is fully contained in X, and once X is known, no additional information can be gained from Y. Let's calculate to verify that intuition. The entropy of X is

$$H(X) = -\sum_{x \in \mathcal{X}} P(x) \log_2 \left(P(x)\right)$$
$$= -0.2 \log_2 (0.2) - 0.8 \log_2 (0.8) = 0.722.$$

The joint entropy of X and Y is

$$H(X,Y) = -\sum_{x,y \in \mathcal{X},\mathcal{Y}} P(x,y) \log_2 \left(P(x,y)\right)$$
$$= -0.8 \log_2(0.8) - 0.2 \log_2 (0.2) = 0.722.$$

For this situation, the expected information content of the entire system is the same as the information in X alone. No additional information is contributed by Y when X is known.

$$H(Y|X) = -\sum_{x,y \in \mathcal{X},\mathcal{Y}} P(x,y) \log_2 \left(P(y|x)\right)$$
$$= -0.8 \log_2(1) - 0.2 \log_2(1) = 0.$$

See Fig. 8.12 for an illustration. *Comment*: Similar calculations for $H(Y)$ and $H(X|Y)$ show that Y also contains all the information in the system, and once Y is known, X provides no additional information. □

Exercise 8.9. The joint probability distribution $P(x, y)$ for two random variables, X and Y, is given in Table 8.1. Compute the conditional probabilities $P(x|y)$ and $P(y|x)$, the marginal entropies $H(X)$ and $H(Y)$ as well as the mutual information $I(X;Y)$.

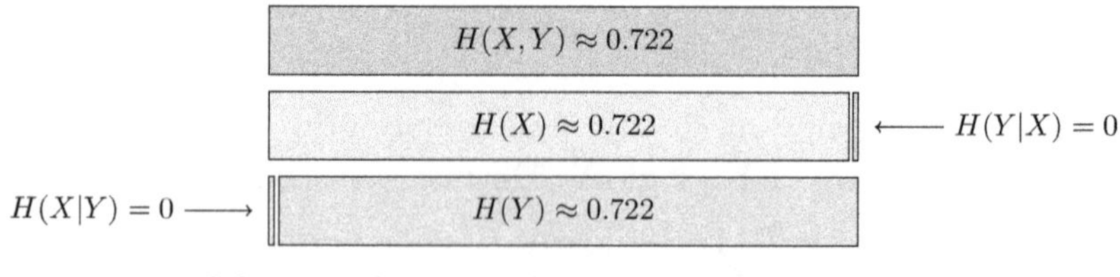

Figure 8.12. Schematic for Example 8.5.3. The entropy of the entire system (top) is the same as that of X (center). The conditional entropy of Y given X is zero, illustrated by the bar of "zero" width in the second row.

Table 8.1. Exemplary joint probability distribution $P(x, y)$ and marginal probability distributions $P(x)$ and $P(y)$ for random variables X and Y.

		X				
$P(x,y)$		x_1	x_2	x_3	x_4	$P(y)$
	y_1	1/8	1/16	1/32	1/32	1/4
Y	y_2	1/16	1/8	1/32	1/32	1/4
	y_3	1/16	1/16	1/16	1/16	1/4
	y_4	1/4	0	0	0	1/4
$P(x)$		1/2	1/4	1/8	1/8	

8.5.5 *Probabilistic inequalities for entropy and mutual information*

Consider the case where a one-dimensional random variable X is drawn from the real line, $x \in \mathbb{R}$, with probability density $p(x)$. When averaging a convex function of X, denoted $f(X)$, Jensen's inequality holds:

$$\mathbb{E}\left[f(X)\right] \geq f\left(\mathbb{E}\left[X\right]\right), \tag{8.38}$$

with equality if $p(x) = \delta(x)$. To illustrate, consider a Bernoulli-like distribution, $p(x) = \beta\delta(x - x_1) + (1 - \beta)\delta(x - x_0)$:

$$f\left(\mathbb{E}[X]\right) = f\left(x_1\beta + x_0(1 - \beta)\right) \leq \beta f(x_1) + (1 - \beta)f(x_0) = \mathbb{E}\left[f(X)\right],$$

where the inequality reflects the convexity of $f(x)$, as per its definition.

Figure 8.13 provides a graphical hint for the proof of Jensen's inequality.

Jensen's inequality applies to any space, and a notable consequence is that entropy, viewed as a function (or functional, in the continuous case) of probabilities at a specific state, is convex. This leads to several key results for entropy and mutual information:

- *Information inequality*:

$$D(p\|q) \geq 0, \quad \text{with equality iff } p = q.$$

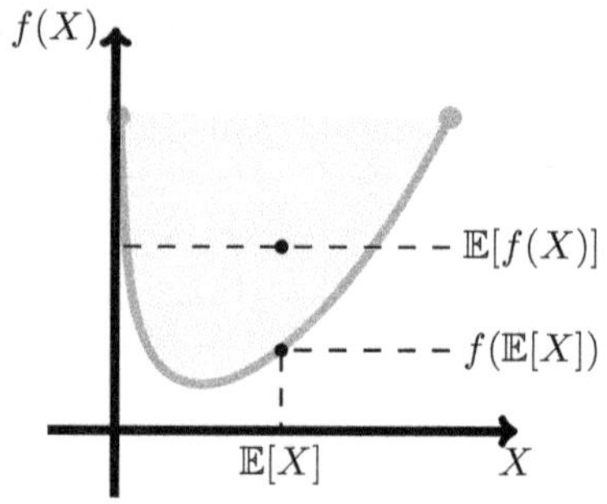

Figure 8.13. Graphical illustration of Jensen's inequality.

- *Conditioning reduces entropy*:

$$H(X|Y) \leq H(X) \quad \text{with equality iff } X \text{ and } Y \text{ are independent.}$$

- *Independence bound on entropy*:

$$H(X_1, \ldots, X_n) \leq \sum_{i=1}^{n} H(X_i) \quad \text{with equality iff } X_i \text{ are independent.}$$

Another useful result is the *log-sum inequality*:

$$\sum_{i=1}^{n} a_i \log \frac{a_i}{b_i} \geq \left(\sum_{i=1}^{n} a_i\right) \log \frac{\sum_{i=1}^{n} a_i}{\sum_{i=1}^{n} b_i},$$

with equality when a_i/b_i is constant. Convention: $0 \log 0 = 0$, $a \log(a/0) = \infty$ if $a > 0$, and $0 \log 0/0 = 0$.

Consequences of the log-sum inequality include the following:

- *Convexity of relative entropy*: $D(p\|q)$ is convex in p and q.
- *Concavity of entropy*: For $X \sim p(x)$, $H(P) := H_P(X)$ is concave in $P(x)$.
- *Concavity of mutual information in $P(x)$*: For $(X, Y) \sim P(x, y) = P(x)P(y|x)$, $I(X;Y)$ is concave in $P(x)$ for fixed $P(y|x)$.
- *Concavity of mutual information in $P(y|x)$*: For $(X, Y) \sim P(x, y) = P(x)P(y|x)$, $I(X;Y)$ is concave in $P(y|x)$ for fixed $P(x)$.

Example 8.5.4. Prove that $H(X) \leq \log_2 n$, where n is the number of possible values for the random variable $x \in X$.

Solution. Applying Jensen's inequality (8.38), we obtain

$$\mathbb{E}[\log_2 P(X)] \geq -\log_2 \mathbb{E}[1/P(X)],$$

where $\mathbb{E}[\log_2 P(X)] = -H(X)$ and $\mathbb{E}[1/P(X)] = n$. Therefore, $H(X) \leq \log_2 n$. □

Jensen's inequality leads to various other useful expressions for entropy, such as

$$H(X|Y) \leq H(X), \quad \text{with equality iff } X \text{ and } Y \text{ are independent,}$$

and, more generally,

$$H(X_1, \ldots, X_n) \leq \sum_{i=1}^{n} H(X_i), \quad \text{with equality iff all } X_i \text{ are independent.}$$

Chapter 9

Stochastic Processes

In this chapter, we delve into the key concepts of stochastic processes, an essential framework for understanding the probabilistic evolution of systems over time. These processes are widely used in various fields, including physics, biology, engineering and finance, to model random behavior in dynamic systems. A stochastic process can be thought of as a collection of random variables indexed by time or space, which often provides insights into real-world phenomena where deterministic models are insufficient.

We begin with the *Bernoulli process* in Section 9.1, a fundamental discrete-time, discrete-space model where each step results in a "success" or "failure" with a fixed probability. Next, in Section 9.2, we introduce the *Poisson process*, which extends the Bernoulli process to continuous time, serving as a model for random events happening at a constant rate. In Section 9.3, we explore the *Langevin equation*, an example of a *stochastic differential equation (SDE)*, which describes the evolution of continuous-time, continuous-space processes, typically found in physical systems subject to memoryless random forces such as Brownian motion. We then transition to *Markov processes* in Section 9.4, where the one-step memory simplifies the mathematical treatment of many real-world phenomena. This discussion paves the way for the *Markov decision processes (MDP)* in Section 9.5, which introduces controlled Markov processes for optimal decision-making, therefore connecting to the optimal control discussed in Chapter 7. Finally, we conclude with a brief discussion on *queuing theory* in Section 9.6, where we look at its application in modeling service systems such as networks and customer service frameworks.

For a more in-depth exploration of stochastic processes, readers are encouraged to consult the following texts. The classical reference by *Doob, Stochastic Processes* [46], offers rigorous mathematical foundations. For applications and further extensions into statistical physics, *Gardiner's Handbook of Stochastic Methods* [47] provides valuable insights, especially into SDEs. The book by *Grimmett and Stirzaker, Probability and Random Processes* [48], covers a wide range of topics, including Markov processes and random walks. *Oksendal's Stochastic Differential Equations: An Introduction with Applications* [49] is a valuable resource for readers interested in the theory and applications of SDEs.

In Chapter 8, we discussed random vectors and their associated probability distributions (for example, $\boldsymbol{X} = (X_1, \ldots, X_N)$ with $\boldsymbol{X} \sim P_{\boldsymbol{X}}$). In this chapter, we extend that discussion to stochastic processes, which are a natural generalization of random vectors. A *stochastic process* is a collection of random variables, often written as a path or sequence, $\{X_t \mid t = 1, \ldots, T\}$, or more compactly as $(X_t)_{t=1}^T$, where the components X_t take values from the same state space Σ. Typically, t represents time, and we consider both the case where T is finite, $T < \infty$, and the case where it is infinite, $T \to \infty$. Both the state space Σ and the index set t may be discrete or continuous.

We will explore three basic examples of stochastic processes: (a) the Bernoulli process, which is discrete in both space and time; (b) the Poisson process, which is discrete in space but continuous in time; and (c) a continuous-space and continuous-time process described by the Langevin equation, an example of an SDE. These are presented in Sections 9.1, 9.2 and 9.3, respectively.

These examples can also be classified by the amount of memory needed to generate them. In the first two cases (Bernoulli and Poisson processes), the random variables X_t are independent, or memoryless – meaning the outcome of each X_t does not affect, nor is it affected by, the outcomes of other X_t. However, this is not always the case. In the process described by the Langevin equation, the variables X_t are dependent, i.e., correlated over time.

Time correlations in a random path $\{X(t)\}$ described by an SDE can be complex and challenging to analyze. To simplify matters, one often considers a discrete-time approximation, known as a *Markov process*, discussed in Section 9.4. In a Markov process, the memory is limited to a single time step, meaning that X_t depends only on the outcome of the previous step, X_{t-1}.

Finally, in Section 9.5, we introduce the MDP, which is a controlled variation of the Markov process, conditioned on certain decisions. (As a bonus, we include a discussion on queuing theory in Section 9.6.)

9.1 Bernoulli Process (Discrete Space, Discrete Time)

A *Bernoulli process* is a sequence of independent Bernoulli random variables, often referred to as events or trials. In the case where each event can take one of two outcomes, typically labeled as "success" or "failure," a sample path of a Bernoulli process might look like $**S*S*S***S$, where S stands for "success," or equivalently as 00101010001, where 1 represents success and 0 represents failure.

In this chapter, we focus on stationary Bernoulli processes, meaning the probability of success remains constant for each X_t. Specifically, we define $P(\text{success}) = P(X_t = 1) = \beta$ and $P(\text{failure}) = P(X_t = 0) = 1 - \beta$ for every t.

Examples of processes modeled by a Bernoulli process include the number of "arrivals" observed at fixed intervals, such as the arrival of a monsoon on a given day during the Tucson summer, or any sequence of discrete updates, such as the random fluctuations of the stock market.

9.1.1 *Probability distribution of the total number of successes*

As discussed in Eq. (8.3), the number of successes k in n trials follows the binomial distribution:

$$\forall k = 0, \dots, n: \quad P(S = k \mid n, \beta) = \binom{n}{k} \beta^k (1-\beta)^{n-k},$$

The mean and variance can be found by computing $\mathbb{E}[S]$ and $\mathbb{E}[(S - \mathbb{E}[S])^2]$, respectively:

$$\begin{aligned} \text{mean}: \quad & \mathbb{E}[S] = n\beta, \\ \text{variance}: \quad & \mathrm{Var}(S) = \mathbb{E}[(S - \mathbb{E}[S])^2] = n\beta(1-\beta). \end{aligned}$$

9.1.2 *Probability distribution of the first success*

Let T_1 represent the number of trials until the first success (inclusive of the success event). The probability mass function (PMF) for the time of the first success is given by the product of the probabilities of $(t-1)$ failures followed by a success:

$$t = 1, 2, \cdots : \quad P(T_1 = t \mid \beta) = \beta(1-\beta)^{t-1} \quad \text{[Geometric PMF]}. \tag{9.1}$$

The distribution in Eq. (9.1) is called a geometric distribution. Verifying that this is a valid probability distribution involves summing a geometric series:

$$\sum_{t=1}^{\infty} \beta(1-\beta)^{t-1} = \beta\,(1-(1-\beta))^{-1} = 1.$$

The mean and variance of the geometric distribution are

$$\text{mean}: \qquad \mathbb{E}[T_1] = \frac{1}{\beta},$$

$$\text{variance}: \quad \mathrm{Var}(T_1) = \mathbb{E}[(T_1 - \mathbb{E}[T_1])^2] = \frac{1-\beta}{\beta^2}.$$

The Bernoulli process is memoryless, meaning that each outcome is independent of the past. If n trials have already taken place, the future sequence $X_{n+1}, X_{n+2}, \ldots$ is a Bernoulli process independent of the first n trials. Moreover, if no success has occurred in the first n trials, the PMF for the remaining trials is also geometric:

$$P(T - n = k \mid T > n, \beta) = \beta(1-\beta)^{k-1}.$$

9.1.3 *Probability distribution of the kth success*

Next, we consider the distribution of the kth success. Let T_k denote the number of trials until the kth success (inclusive). The PMF is given by

$$t = k, k+1, \ldots : \; P(T_k = t \mid \beta) = \binom{t-1}{k-1} \beta^k (1-\beta)^{t-k} \; \text{[Pascal PMF]},$$

The mean and variance are

$$\text{mean}: \qquad \mathbb{E}[T_k] = \frac{k}{\beta},$$

$$\text{variance}: \quad \operatorname{Var}(T_k) = \mathbb{E}[(T_k - \mathbb{E}[T_k])^2] = \frac{k(1-\beta)}{\beta^2}.$$

The combinatorial factor accounts for the number of ways k successes can be arranged in T_k trials.

Exercise 9.1. Define $\tau_k = T_k - T_{k-1}$ for $k = 2, 3, \ldots$, where τ_k is the inter-arrival time between the $(k-1)$st and kth successes. Derive the probability distribution function for the inter-arrival time τ_k.

The following sections introduce two continuous-time limits of Bernoulli processes: the Poisson process and Brownian motion.

9.2 Poisson Process (Discrete Space, Continuous Time)

Building on the material from the previous section, where we discussed the discrete-time, discrete-space Bernoulli process, we now extend our discussion to continuous time, leading to the *Poisson process*.

A Poisson process is formally defined as a sequence of random variables $\{N_t\}$ indexed by time, where N_t counts the number of independent arrivals over the interval $[0, t]$. Recall the Poisson distribution introduced in Section 8.1.1:

$$\forall k \in \{0, 1, 2, \ldots\}: \quad P(N = k; \tilde{\lambda}) = \frac{\tilde{\lambda}^k e^{-\tilde{\lambda}}}{k!}. \tag{9.2}$$

We now show that the outcomes of a Poisson process follow a Poisson distribution with parameter λt.

The distribution of the Poisson process can be derived by subdividing the interval $[0, t]$ into n subintervals of length $\Delta t := t/n$. For sufficiently small Δt, the probability of two or more arrivals in any subinterval is negligible, and the occurrences of arrivals in different subintervals are independent. Under these conditions, the probability of k arrivals in n subintervals can be modeled by a binomial distribution. The probability of an arrival, β, is proportional to the

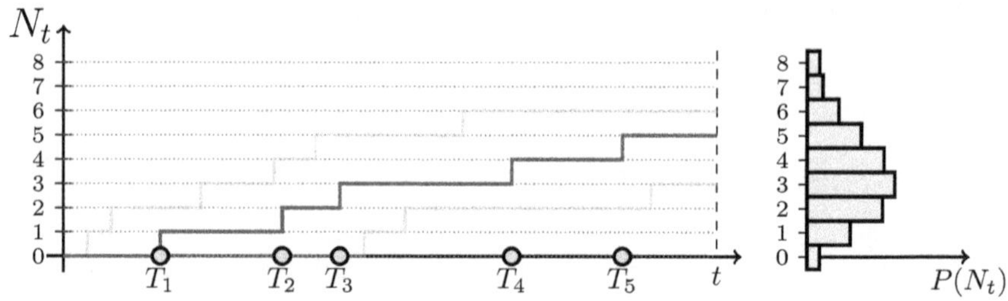

Figure 9.1. One realization of a Poisson process over the interval $[0, t]$ (middle curve) in which five "arrivals" occurred at random times $T_1, \dots, T_5$, making $N(t) = 5$. Two other realizations (top and low curves) show six and three arrivals on $[0, t]$. The random variable N_t follows a Poisson distribution (right) with parameter λt, where λ is a fixed rate.

subinterval length: $\beta \propto t/n$, or, equivalently, $\beta = \lambda t/n$, where λ is the constant of proportionality. Taking the limit as $n \to \infty$, we obtain

$$\begin{aligned} P(N_t = k; \lambda) &= \lim_{n\to\infty} \binom{n}{k} \beta^k (1-\beta)^{n-k} \\ &= \lim_{n\to\infty} \frac{n!}{k!(n-k)!} \left(\frac{\lambda t}{n}\right)^k \left(1 - \frac{\lambda t}{n}\right)^{n-k} \\ &= \lim_{n\to\infty} \frac{n^k + O(n^{k-1})}{k!} \left(\frac{\lambda t}{n}\right)^k \left(1 - \frac{\lambda t}{n}\right)^{n-k} \\ &= \frac{(\lambda t)^k e^{-\lambda t}}{k!}, \end{aligned} \tag{9.3}$$

where we have used the facts that $(1 - \frac{\lambda t}{n})^n \to e^{-\lambda t}$ and $(1 - \frac{\lambda t}{n})^k \to 1$. Note that the dimensionless parameter $\tilde{\lambda}$ in Eq. (9.2) is replaced by λt in Eq. (9.3), where λ has units of inverse time: $[\lambda] = [1/t]$. Relation between realizations of a Poisson process and a Poisson distribution is illustrated in Fig. 9.1.

Common examples of Poisson processes

- Email arrivals, assuming frequent checks.
- Collisions of high-energy beams in a particle accelerator (e.g., 10 MHz) with a small chance of actual collision.
- Radioactive decay of a nucleus, where the trial is to observe a decay within a small time interval.
- Spin flips in a magnetic field.

Properties of Poisson processes

A Poisson process has the following key properties:

- **Initialization:** No arrivals occur at time $t = 0$, i.e., $N(0) = 0$.
- **Independence:** The number of arrivals in two disjoint time intervals are independent.
- **Distribution:** The number of arrivals depends only on the length of the time interval, not on its location. Specifically, as $\Delta t \to 0$, we have $P(N(\Delta t) = 1) \to \lambda \Delta t$ and $P(N(\Delta t) \geq 2) = 0$.

These three properties are both necessary and sufficient to define a Poisson process with rate λt. A summary of the relationship between the Bernoulli and Poisson processes is given in Table 9.1.

Probability distributions of the first and kth arrival times

Let T_1 denote the (random) time of the first arrival. The probability density function (PDF) of T_1 can be derived by recalling that a PDF is the derivative of the cumulative distribution function (CDF) and recognizing that $P(T_1 < t)$ is equivalent to $P(N_t \geq 1)$, since the event that the first arrival occurs before time t is the same as having at least one arrival by time t:

$$\begin{aligned} P(T_1 = t) &= \lim_{\Delta t \to 0} \frac{1}{\Delta t}\big(P(T_1 < t + \Delta t) - P(T_1 < t)\big) = \frac{d}{dt} P(T_1 < t) \\ &= \frac{d}{dt} P(N_t \geq 1) = \frac{d}{dt}\big(1 - P(N_t = 0)\big) \\ &= \frac{d}{dt}\big(1 - e^{-\lambda t}\big) = \lambda e^{-\lambda t}. \end{aligned}$$

Table 9.1. Comparison between the Bernoulli and Poisson processes.

	Bernoulli	**Poisson**
Times of Arrival	Discrete	Continuous
Arrival Rate	p/trial	λ/unit time
PMF of Number of Arrivals	Binomial	Poisson
PMF of Inter-arrival Time	Geometric	Exponential
PMF of kth Arrival Time	Pascal	Erlang

Thus, the time of the first arrival follows an exponential distribution with parameter λ.

The process has two important properties:

- **Memorylessness:** If no arrival has occurred by time t, the density of the remaining time until the next arrival is still exponentially distributed.
- **Fresh starts:** The time of the next arrival is independent of the past and thus also follows an exponential distribution with parameter λ.

The probability that the first arrival occurs before time t can be computed by integrating the PDF

$$P(T_1 \leq t) = \int_0^t \lambda e^{-\lambda t'} dt' = 1 - \exp(-\lambda t).$$

By extension, the probability density of the kth arrival is given by the Erlang distribution:

$$p(T_k = t; \lambda) = \frac{\lambda^k t^{k-1} \exp(-\lambda t)}{(k-1)!}, \quad t > 0$$

(Erlang distribution of order k).

Merging and splitting Poisson processes

One of the key features of both Bernoulli and Poisson processes is their invariance under merging and splitting. We illustrate this with the Poisson process, though the same principles apply to the Bernoulli process.

Merging: Suppose $N_1(t)$ and $N_2(t)$ are two independent Poisson processes with rates λ_1 and λ_2, respectively. Define $N(t) = N_1(t) + N_2(t)$. This random process, as shown in Fig. 9.2, is a Poisson process with rate $\lambda_1 + \lambda_2$. The reasoning is as follows: Since $N_1(t)$ and $N_2(t)$ are independent with independent increments, their sum also

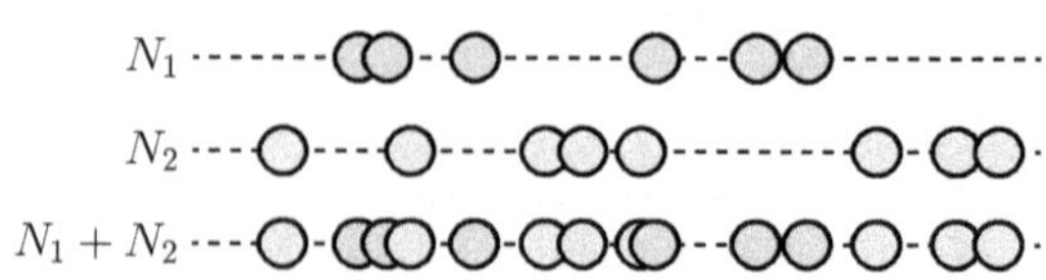

Figure 9.2. Merging and splitting Poisson processes.

has independent increments. The number of arrivals in an interval $(t, t+\tau]$ is Poisson$(\lambda_1\tau)$ and Poisson$(\lambda_2\tau)$, and their sum is Poisson $((\lambda_1+\lambda_2)\tau)$. This result can be extended to the sum of any number of independent Poisson processes.

Splitting: Let $N(t)$ be a Poisson process with rate λ. Now, split $N(t)$ into two processes, $N_1(t)$ and $N_2(t)$, based on a Bernoulli trial – whenever an arrival occurs, toss a coin to decide whether to assign the arrival to N_1 or N_2 with probabilities β and $1-\beta$, respectively. The resulting processes satisfy the following:

- $N_1(t)$ is a Poisson process with rate $\lambda\beta$.
- $N_2(t)$ is a Poisson process with rate $\lambda(1-\beta)$.
- $N_1(t)$ and $N_2(t)$ are independent.

Example 9.2.1. Astronomers estimate that large meteors hit Earth on average once every 1000 years, and the number of hits follows a Poisson distribution.

(a) What is the probability of observing at least one meteor next year?
(b) What is the probability of observing no meteor hits within the next 1000 years?
(c) Calculate the PDF of T_k, where T_k represents the appearance time of the kth meteor.

Solution. The probability of observing k meteors in a time interval $[0, t]$ is given by

$$P(k \mid t) = \frac{(\lambda t)^k}{k!} e^{-\lambda t},$$

where $\lambda = 0.001$ (events per year) is the average rate. Simplifying the notation, $P(k \mid t, \lambda) \to P(k \mid t)$.

(a) $P(k > 0 \text{ meteors next year}) = 1 - P(0 \mid 1) = 1 - e^{-0.001} \approx 0.001$.
(b) $P(k = 0 \text{ meteors in the next 1000 years}) = P(0 \mid 1000) = e^{-1} \approx 0.37$.
(c) The PDF of T_k is given by

$$p(T_k) = \frac{\lambda^k T_k^{k-1}}{(k-1)!} e^{-\lambda T_k}.$$

□

Exercise 9.2. Customers arrive at a store according to a Poisson process with a rate of 5 per hour, and 40% of arrivals are men.

(a) Compute the probability that at least 10 customers enter between 10 and 11 a.m.
(b) Compute the probability that exactly 5 women enter between 10 and 11 a.m.
(c) Compute the expected inter-arrival time of men.
(d) Compute the probability that no men arrive between 2 and 4 p.m.

9.3 Stochastic Processes that are Continuous in Space and Time

The stochastic processes discussed so far have been memoryless. In this section, we explore the dynamics of continuous variables governed by the *Langevin equation*. We derive the associated Fokker–Planck equation, which describes the temporal evolution of the probability distribution for a system's state. As an illustrative example, we consider stochastic dynamics in free space (without drift), where the Fokker–Planck equation simplifies to the diffusion equation, describing Brownian motion.

9.3.1 *Random walks on the integers*

For a binomial process $\boldsymbol{Y} = (Y_1, \ldots, Y_n)$ with outcomes ± 1 occurring with equal probability, the stochastic process $\boldsymbol{X}$ defined as $X_j := \sum_{i=1}^{j} Y_i$ is known as a *random walk* on the integers. The PMF of this random walk is binomial, and as n increases, it converges to a Gaussian distribution with mean zero and variance $\sqrt{n}$, in accordance with the central limit theorem.

Example 9.3.1. A grasshopper is dropped on a number line and proceeds to take a random walk, making a unit jump to the right with probability β or to the left with probability $1 - \beta$. The starting location of the grasshopper is random: $X_0 = 0$ with probability 0.7 and $X_0 = 5$ with probability 0.3. Find the probability distribution of the grasshopper's position at time t.

Solution. First, consider the case where $X_0 = 0$. After one jump, the grasshopper can be at $X_1 = -1$ (left) or $X_1 = +1$ (right). After two jumps, the possible positions are $X_2 = -2$, $X_2 = 0$ or $X_2 = +2$. Let $F(n,k)$ denote the number of ways to make n left and right jumps that result in ending at position $X_n = k$. Specifically, $F(n,k) = \binom{n}{(n+k)/2}$ if $(n+k)/2$ is an integer between 0 and n, and $F(n,k) = 0$ otherwise. The probability of reaching position k in n steps from $X_0 = 0$ is $F(n,k)\beta^k(1-\beta)^{n-k}$. Repeating this process for $X_0 = 5$ and applying Bayes' theorem gives

$$\begin{aligned} P(X_n = k) &= P(X_n = k \mid X_0 = 0)P(X_0 = 0) \\ &\quad + P(X_n = k \mid X_0 = 5)P(X_0 = 5) \\ &= 0.7F(n,k)\beta^k(1-\beta)^{n-k} + 0.3F(n,k-5)\beta^k(1-\beta)^{n-k}. \end{aligned}$$

□

Exercise 9.3. Modify Example 9.3.1 to account for the random starting location $P(X_0 = x) = 2^{-|x|}/3$.

9.3.2 *From random walks to Brownian motion*

Example 9.3.2. Consider a random walk on the line with n jumps at times $t = \Delta, 2\Delta, \ldots, n\Delta$, where $\Delta t = 1/n$. Find the PMF of the random walk, where each jump is $\pm\sqrt{\Delta}$ with equal probability: $P(X_{j+1} = X_j + \sqrt{\Delta}) = P(X_{j+1} = X_j - \sqrt{\Delta}) = 1/2$. Show that the stochastic process has zero mean and unit variance when $t = 1$.

Solution. The probability $P(X_n = (n-2k)\sqrt{\Delta})$ is given by $\binom{n}{k}\left(\frac{1}{2}\right)^n$. Recognizing that $X_n = \sqrt{\Delta}(2Y - n)$, where $Y \sim \mathrm{B}(n, 1/2)$, we calculate the mean and variance:

$$E[X_n] = \sqrt{\Delta}(2E[Y] - n) = 0, \quad \mathrm{Var}(X_n) = 4\Delta\mathrm{Var}(Y) = 1. \quad \square$$

Example 9.3.2 can be extended using the central limit theorem, which shows that the CDF of the random walk converges to the CDF of a standard normal distribution as $n \to \infty$. This result is key to approximating high-dimensional discrete-time stochastic processes by continuous-time processes and vice versa. Solutions to continuous-time processes are often easier to analyze, while discrete processes are easier to compute numerically.

The central limit theorem further implies that the steps of a random walk need not be Bernoulli trials. Any random walk with i.i.d. steps converges to the same limit as long as the variance of the step sizes scales as $1/\sqrt{n}$ as $n \to \infty$. This limiting process is known as Brownian motion.

9.3.3 *The Langevin equation in continuous and discrete time*

Many 1D stochastic processes can be described by the Langevin equation in both continuous and discrete time:

$$\begin{aligned} \dot{x} &= v(x) + \sqrt{2D}\xi(t), \quad \langle \xi(t) \rangle = 0, \\ &\langle \xi(t_1)\xi(t_2) \rangle = \delta(t_1 - t_2), \end{aligned} \tag{9.4}$$

$$\begin{aligned} x_{n+1} - x_n &= \Delta v(x_n) + \sqrt{2D\Delta}\eta(t_n), \quad \langle \eta(t_n) \rangle = 0, \\ &\langle \eta(t_n)\eta(t_k) \rangle = \delta_{kn}, \end{aligned} \tag{9.5}$$

where $\eta/\sqrt{\Delta} \to \xi$ as $\Delta \to 0$ to ensure the limit is well-defined.

In Eq. (9.4), the first term describes deterministic drift, and the second term, known as the *Langevin term*, introduces random noise with mean zero and variance controlled by D. The discrete-time form in Eq. (9.5) represents the same process, where η is independent noise at each time step. These equations describe the evolution of a particle's position $x \in \mathbb{R}$ under both deterministic and random forces. The random term models environmental uncertainty or random interactions with other "invisible" particles.

To understand the origin of the $\sqrt{\Delta}$ factor in Eq. (9.5), consider the case of zero drift, $v(x) = 0$, which corresponds to Brownian motion. In this case, the Langevin equation simplifies to

$$x(t) = \sqrt{2D} \int_0^t \xi(t')\, dt', \tag{9.6}$$

$$\langle x^2(t) \rangle = 2D \int_0^t dt_1 = 2Dt, \tag{9.7}$$

assuming $x(0) = 0$. In the discrete version, the infinitesimal form of Eq. (9.7) becomes

$$\langle (x_{n+1} - x_n)^2 \rangle = 2D\Delta,$$

which is derived directly from the Brownian motion version of Eq. (9.5).

9.3.4 *The Wiener process: A rigorous definition of Brownian motion*

The notations introduced for the SDE in its continuous form (9.4) and discrete form (9.5) are commonly used in physics and parts of applied mathematics. Although intuitive and simple, they rely on the concept of the δ-function, a generalized function, making them formally ambiguous. This is analogous to the ambiguity of using the δ-function as a source term in a linear ODE for a Green function. To resolve this, we typically regularize the δ-function or use it under an integral.

In theoretical mathematics, statistics and engineering, we more often encounter the SDE restated in differential form:

$$dx(t) = v(x(t))\,dt + \sqrt{2D}\,dW(t), \tag{9.8}$$

where $W(t)$ denotes the *standard Brownian motion*, also called the Wiener process. Some mathematical literature uses $dB(t)$ instead of $dW(t)$. Formally, the Wiener process is defined as follows.

Definition 9.3.3 (Wiener Process). The *Wiener process*, $\{W(t)\}_{t\geq 0^+}$, is a continuous-time stochastic process in $\mathbb{R}$ characterized by:

1. $W(0) = 0$;
2. $W(t)$ is almost surely continuous (i.e., with probability 1, $W(t)$ is continuous in t);
3. $W(t)$ has stationary independent increments that are normally distributed with mean zero and variance proportional to the length of the increment: $W_t - W_{t'} \sim \mathcal{N}(0, t - t')$ for $0 \leq t' \leq t$.

The differential form (9.8) of the Langevin equation is advantageous because it leads naturally to the following integral form, resolving the aforementioned ambiguity:

$$\begin{aligned} x(t+\Delta) - x(t) &= \int_t^{t+\Delta} dx(t') \\ &= \int_t^{t+\Delta} v(x(t'))dt' + \sqrt{2D}\int_t^{t+\Delta} dW(t'), \end{aligned}$$

Here, the first term on the right-hand side is a standard (Lebesgue) integral, and the second term is the so-called *Itô integral.* From Eqs. (9.6) and (9.7), and consistent with the Wiener process definition, the Itô integral's heuristic interpretation is that, as $\Delta \to 0$, the increment $x(t+\Delta) - x(t)$ becomes Gaussian with mean zero and variance $2D\Delta$.

9.3.5 *From the Langevin equation to the path integral*

The Langevin equation, in both continuous (9.4) and discrete (9.5) forms, relates the dynamic increment of $x(t)$ to stochastic dynamics of the δ-correlated source $\xi(t_n)/\sqrt{\Delta} = \phi_n$, characterized by the PDF

$$p(\eta_1, \ldots, \eta_N) = (2\pi)^{-N/2} \exp\left(-\frac{1}{2}\sum_{n=1}^{N} \eta_n^2\right). \tag{9.9}$$

Equations (9.4), (9.5) and (9.9) form the starting point for further derivations and can be used for simulating the Langevin equation on a computer, generating many paths simultaneously. Other simulation methods, such as the telegraph process, also exist.

To express η_n in terms of x_n from Eq. (9.5) and substitute it into Eq. (9.9), we obtain

$$\begin{aligned} &p(\eta_0, \ldots, \eta_{N-1}) \to p(x_1, \ldots, x_N \mid x_0) \\ &\qquad = \frac{1}{(4\pi D)^{N/2}} \exp\left(-\frac{1}{4D\Delta}\sum_{n=0}^{N-1} (x_{n+1} - x_n - \Delta v(x))^2\right). \end{aligned}$$

This is the explicit expression for the measure over a path in discretized form. In continuous form (as a notational shortcut), it becomes

$$p\{x(t)\} \propto \exp\left(-\frac{1}{4D}\int_0^T dt\,(\dot{x} - v(x))^2\right).$$

This object is called the *path integral* (in both physics and mathematics), or the Feynman–Kac integral.

9.3.6 *From the path integral to the Fokker–Planck equation*

The PDF of a path is a useful general concept, but we may also wish to marginalize it to extract the PDF for the position x_N at step N, given an initial position x_0 at t_0. This involves marginalizing the joint PDF of the path:

$$\begin{aligned} p_N(x_N) &= \int dx_0 \cdots dx_{N-1}\, p(x_0, \ldots, x_N) \\ &= \int dx_1 \cdots dx_{N-1}\, p(x_1, \ldots, x_N \mid x_0) p_0(x_0). \end{aligned} \tag{9.10}$$

It is convenient to derive the relationship between $p_N(\cdot)$ and $p_0(\cdot)$ in steps, integrating over $dx_0, \ldots, dx_{N-1}$ sequentially. For Brownian motion (where $v = 0$), the first step of the induction is

$$\begin{aligned} p_1(x_1) &= (4\pi D)^{-1/2} \int dx_0 \exp\left(-\frac{(x_1 - x_0)^2}{4D\Delta}\right) p_0(x_0) \\ &= (4\pi D)^{-1/2} \int d\epsilon \exp\left(-\frac{\epsilon^2}{4D\Delta}\right) p_0(x_1 + \epsilon) \qquad (9.11) \\ &\approx (4\pi D)^{-1/2} \int d\epsilon \exp\left(-\frac{\epsilon^2}{4D\Delta}\right) \\ &\quad \times \left(p_0(x_1) + \epsilon \partial_{x_1} p_0(x_1) + \frac{\epsilon^2}{2} \partial_{x_1}^2 p_0(x_1)\right) \\ &= p_0(x_1) + \Delta D \partial_{x_1}^2 p_0(x_1), \qquad (9.12) \end{aligned}$$

where the transition from Eq. (9.11) to Eq. (9.12) involves a Taylor expansion in ϵ, assuming $\epsilon \sim \sqrt{\Delta}$ and keeping only the leading terms. The resulting Gaussian integrals are straightforward, leading to the discretized version of the diffusion equation:

$$\partial_t p(x \mid t) = D \partial_x^2 p(x \mid t), \tag{9.13}$$

where $p(x \mid t)$ represents the probability of being at position x at time t. Naturally, for Brownian motion, this reduces to the diffusion equation for the marginal PDF.

By restoring the deterministic drift term $v(x)$, we generalize the diffusion equation to the Fokker–Planck equation:

$$\partial_t p(x \mid t) + \partial_x \left(v(x) p(x \mid t)\right) = D \partial_x^2 p(x \mid t). \tag{9.14}$$

9.3.7 *Analysis of the Fokker–Planck equation: General features and examples*

The Fokker–Planck equation (9.14) is a linear, deterministic partial differential equation (PDE) that governs the evolution of the probability density in both phase space x and time t. Although derived for a particle moving in one dimension ($\mathbb{R}$), the approach extends naturally to higher dimensions ($\mathbb{R}^d$, $d = 1, 2, \dots$) and to compact continuous spaces, such as a circle, sphere or torus.

The equation describes how probabilities evolve, driven by two terms: *diffusion*, which originates from the stochastic source, and *advection*, which results from the deterministic force. The linearity of the Fokker–Planck equation does not necessarily simplify the problem; it represents a transition from nonlinear, stochastic ODEs to linear PDEs, which is common in mathematical physics.

The Fokker–Planck equation can also be written in flux form:

$$\partial_t p(x \mid t) + \partial_x J(t; x) = 0,$$

where $J(t; x)$ represents the flux of probability at position x and time t. This formulation emphasizes the conservation of probability, as integrating over the entire space yields $\partial_t \int dx\, p(x \mid t) = 0$.

In steady-state conditions, where $\partial_t p(x \mid t) = 0$, the flux J is constant, leading to the special case of zero-flux equilibrium, typical in statistical mechanics.

The simplest case, Brownian motion, results in the diffusion equation. A key example in equilibrium statistical physics arises when the drift velocity is the gradient of a potential $U(x)$: $v(x) = -\partial_x U(x)$. For an overdamped particle connected to the origin by a spring, this gives the stationary solution of the Fokker–Planck equation, known as the *Gibbs distribution*:

$$p(x \mid t)\Big|_{t \to \infty} \to p_{\mathrm{st}}(x) = Z^{-1} \exp\left(-\frac{U(x)}{D}\right),$$

where Z is a normalization constant.

9.3.8 *Examples and exercises*

Example 9.3.4. Consider the motion of a Brownian particle in a parabolic potential, $U(x) = \gamma x^2/2$. This situation typically arises when a particle is located near the minimum or maximum of a potential. The Langevin equation (9.4) in this case becomes

$$\frac{dx}{dt} + \gamma x = \sqrt{2D}\xi(t), \quad \langle \xi(t) \rangle = 0, \quad \langle \xi(t_1)\xi(t_2) \rangle = \delta(t_1 - t_2). \tag{9.15}$$

Write a formal solution of Eq. (9.15) for $x(t)$ as a functional of $\xi(t)$. Compute $\langle x^2(t) \rangle$ as a function of t, and interpret the results. Write the Kolmogorov–Fokker–Planck (KFP) equation for $p(x|t)$ and solve it for the initial condition $p(x|0) = \delta(x)$.

Solution. Multiplying Eq. (9.15) by the integrating factor $e^{\gamma t}$, we obtain

$$\frac{d}{dt}\left(x(t)e^{\gamma t}\right) = \sqrt{2D}\xi(t)e^{\gamma t},$$

which, after integrating both sides, gives the formal solution

$$x(t)e^{\gamma t} = x(0) + \sqrt{2D}\int_0^t \xi(t')e^{\gamma t'}\, dt'.$$

Simplifying, we get

$$x(t) = x(0)e^{-\gamma t} + \sqrt{2D}\int_0^t \xi(t')e^{-\gamma(t-t')}\, dt'.$$

To compute $\langle x^2(t) \rangle$, we first calculate the mean and variance:

$$\begin{aligned}
\langle x(t) \rangle &= \left\langle x(0)e^{-\gamma t} + \sqrt{2D}\int_0^t \xi(t')e^{-\gamma(t-t')}\, dt' \right\rangle \\
&= \langle x(0) \rangle e^{-\gamma t} + \sqrt{2D}\int_0^t \langle \xi(t') \rangle e^{-\gamma(t-t')}\, dt' \\
&= x(0)e^{-\gamma t},
\end{aligned}$$

where we used $\langle \xi(t) \rangle = 0$.

Next, we compute $\langle x^2(t)\rangle$:

$$\begin{aligned}\langle x^2(t)\rangle &= \left\langle \left(x(0)e^{-\gamma t} + \sqrt{2D}\int_0^t \xi(t')e^{-\gamma(t-t')}\,dt' \right)^2 \right\rangle \\ &= \langle x(0)^2\rangle e^{-2\gamma t} + 2D\int_0^t\int_0^t \langle \xi(t')\xi(t'')\rangle e^{-\gamma(t-t')}e^{-\gamma(t-t'')}\,dt'\,dt'' \\ &= x(0)^2e^{-2\gamma t} + 2De^{-2\gamma t}\int_0^t e^{2\gamma t'}\,dt' \\ &= x(0)^2e^{-2\gamma t} + \frac{D}{\gamma}\left(1-e^{-2\gamma t}\right).\end{aligned}$$

Interpretation: (i) The contribution to $\langle x^2(t)\rangle$ from the initial condition decays as $e^{-2\gamma t}$. (ii) At small times, $t \ll 1/\gamma$, the particle does not feel the potential, and we recover the usual diffusion: $\langle x^2(t)\rangle \approx 2Dt$. (iii) At larger times, $t \gg 1/\gamma$, the dispersion saturates at $\langle x^2(t)\rangle \approx D/\gamma$.

The KFP equation for this system is

$$\partial_t p(x|t) = (\gamma\partial_x x + D\partial_x^2)p(x|t).$$

For the initial condition $p(x|0) = \delta(x)$, the solution (Green function) is

$$p(x|t) = \frac{1}{\sqrt{2\pi\langle x^2(t)\rangle}}\exp\left(-\frac{x^2}{2\langle x^2(t)\rangle}\right).$$

This solution describes a time-dependent Gaussian distribution. □

Example 9.3.5. Prove that the moments $\langle x^{2k}(t)\rangle$ for Brownian motion in $\mathbb{R}$ satisfy the recurrence relation:

$$\partial_t\langle x^{2k}\rangle = 2k(2k-1)D\langle x^{2(k-1)}\rangle.$$

Solve this equation for a particle starting from $x = 0$ at $t = 0$.

Solution. From the definition of the kth moment and the diffusion equation (9.13), we have

$$\begin{aligned}\partial_t \langle x^{2k} \rangle &= \partial_t \int_{-\infty}^{+\infty} x^{2k} p(x|t)\, dx \\ &= D \int_{-\infty}^{+\infty} x^{2k} \partial_{xx} p(x|t)\, dx \\ &= D \int_{-\infty}^{+\infty} 2k(2k-1) x^{2k-2} p(x|t)\, dx \\ &= 2k(2k-1) D \langle x^{2(k-1)} \rangle.\end{aligned}$$

The solution for $\langle x^{2k}(t) \rangle$ follows from this recurrence relation with the initial condition $x = 0$ at $t = 0$. □

Example 9.3.6 (Brownian Motion in a Parabolic Potential). The conditional probability distribution for a Brownian particle in a parabolic potential, $U(x) = \alpha x^2$, is governed by the advection-diffusion equation:

$$D\partial_x^2 p + \alpha \partial_x (xp) = \partial_t p.$$

Write the stochastic ODE for the underlying process $x(t)$, and compute the respective statistical moments $\langle x^k(t) \rangle$.

Solution. The corresponding Langevin equation is

$$\dot{x} = -\alpha x + \sqrt{2D}\xi(t).$$

We verify this by following steps similar to Example 9.3.4.

Let $\mu_k(t) := \langle x^k(t) \rangle$. From the KFP equation, we derive the following recurrence relation:

$$\partial_t \mu_k(t) = -k\alpha \mu_k(t) + k(k-1)\mu_{k-2}(t),$$

where the boundary terms vanish at $\pm\infty$.

For $p(x|0) = \delta(x)$, we have $\mu_0(t) = 1$ and $\mu_1(t) = 0$. Solving this recurrence relation gives $\mu_{2k+1}(t) = 0$ for all odd moments. For even moments, the second moment can be computed as

$$\partial_t \mu_2(t) = -2\alpha \mu_2(t) + 2D,$$

yielding $\mu_2(t) = \frac{D}{\alpha}\left(1 - e^{-2\alpha t}\right)$. □

Example 9.3.7. Consider the following expectation over the Langevin term, $\xi(t)$:

$$\Psi(t;x) = \left\langle \exp\left(\int_0^t d\tau Q(x(\tau))\right)\right\rangle_{x(t)=x,\ x(0)=0} \tag{9.16}$$

$$\xrightarrow{N\to\infty} \tag{9.17}$$

$$\begin{aligned}\Psi_N(x) &= \int dx_0 \cdots dx_{N-1}\, p\,(x, x_{N-1}, \ldots, x_1 | x_0)\, \delta\,(x_0)\\ &\quad \times \exp\left(\Delta\left(Q(x_{N-1}) + \cdots + Q(x_0)\right)\right)\\ &= \int \frac{dx_1 \cdots dx_{N-1}}{(4\pi D\Delta)^{(N-1)/2}} \exp\left(-\frac{(x - x_{N-1})^2 + \cdots + x_1^2}{4D\Delta}\right)\\ &\quad \times \exp\left(\Delta\left(Q(x_{N-1}) + \cdots + Q(0)\right)\right),\end{aligned}$$

where $x(t)$ is Brownian motion, which satisfies $\dot{x}(t) = \xi(t)$ with initial condition $x(0) = 0$. The function $Q(x(t))$ is a bounded function of x over $\mathbb{R}$. Equations (9.16) and (9.17) provide continuous-time and discrete-time versions, respectively, of the same expectation.

- (a) Derive the partial differential equation governing the evolution of $\Psi(t;x)$ in terms of t and x.
- (b) Propose a method for computing the first and second moments of $\int_0^t d\tau Q(x(\tau))$,

$$\phi^{(1)}(t) := \left\langle \int_0^t d\tau Q(x(\tau))\right\rangle_{x(0)=0},$$

$$\phi^{(2)}(t) := \left\langle \left(\int_0^t d\tau Q(x(\tau))\right)^2\right\rangle_{x(0)=0},$$

explicitly without solving the PDE derived in (a), which lacks an explicit solution for a general $Q(x(t))$.

Solution. (a) Start by noting that $\Psi(0;x) = \delta(x)$. When $Q(x) = 0$, the corresponding PDE reduces to the diffusion equation (Eq. (9.14)) with $p(t;x)$ replaced by $\Psi(t;x)$.

To rewrite Eq. (9.17) as a recurrence, we express it as follows:

$$\forall k = 1, \dots, N: \quad \Psi_k(x) = \int \frac{dx_{k-1}}{\sqrt{4\pi D\Delta}} \times \exp\left(-\frac{(x - x_{k-1})^2}{4D\Delta} + \Delta Q(x_{k-1})\right) \Psi_{k-1}(x_{k-1}),$$

where $\Psi_0(x) = \delta(x)$. Changing the integration variable, $x_{k-1} \to \epsilon = x_{k-1} - x$, we keep the Gaussian term in the integrand intact, expanding all other terms in a Taylor series around ϵ. Then, performing the resulting Gaussian integrals yields the differential version:

$$\forall k = 1, \dots, N: \quad \Psi_k(x) = \left(1 + \Delta Q(x) + \Delta D\partial_x^2 + O(\Delta^2)\right) \Psi_{k-1}(x).$$

Passing to the continuous-time limit gives the PDE governing $\Psi(t;x)$ as

$$\partial_t \Psi(t;x) = Q(x)\Psi(t;x) + D\partial_x^2 \Psi(t;x), \quad \Psi(0;x) = \delta(x). \tag{9.18}$$

(b) To compute the moments $\phi^{(1)}(t)$ and $\phi^{(2)}(t)$ without directly solving the PDE, let us substitute $Q(x)$ in Eq. (9.18) by $\delta * Q(x)$ and expand the PDE in a Taylor series in δ. Evaluating the series for the zero, first and second powers in δ, we derive a series of diffusion equations:

$$\partial_t \psi^{(0)}(t;x) = D\partial_x^2 \psi^{(0)}(t;x), \quad \psi^{(0)}(0;x) = \delta(x),$$
$$\partial_t \psi^{(1)}(t;x) = Q(x)\psi^{(0)}(t;x) + D\partial_x^2 \psi^{(1)}(t;x), \quad \psi^{(1)}(0;x) = 0,$$
$$\partial_t \psi^{(2)}(t;x) = Q(x)\psi^{(1)}(t;x) + D\partial_x^2 \psi^{(2)}(t;x), \quad \psi^{(2)}(0;x) = 0,$$

where

$$n = 0, 1, \dots: \quad \psi^{(n)}(t;x) := \left\langle \left(\int_0^t d\tau Q(x(\tau)) \right)^n \right\rangle_{x(t)=x,\ x(0)=0}.$$

Each of these equations can be solved explicitly:

$$\psi^{(0)}(t;x) = \frac{1}{\sqrt{4\pi Dt}} \exp\left(-\frac{x^2}{4Dt}\right),$$

$$\psi^{(1)}(t;x) = \int_0^t \int_{-\infty}^{\infty} d\xi\, d\tau \frac{1}{\sqrt{4\pi D(t-\tau)}} \exp\left(-\frac{(x-\xi)^2}{4D(t-\tau)}\right) \times Q(\xi)\psi^{(0)}(\tau;\xi),$$

$$\psi^{(2)}(t;x) = \int_0^t \int_{-\infty}^{\infty} d\xi\, d\tau \frac{1}{\sqrt{4\pi D(t-\tau)}} \exp\left(-\frac{(x-\xi)^2}{4D(t-\tau)}\right) \times Q(\xi)\psi^{(1)}(\tau;\xi).$$

Finally, we obtain $\phi^{(k)}(t)$ by integrating over x:

$$\phi^{(k)}(t) = \int dx\, \psi^{(k)}(t;x), \quad k = 0, 1, 2. \qquad \square$$

Exercise 9.4 (Self-Propelled Particle). A self-propelled particle moves in the xy-plane with fixed speed v_0. Its velocity components, in polar coordinates, are

$$\dot{x} = v_0 \cos\varphi, \quad \dot{y} = v_0 \sin\varphi,$$

where φ evolves according to the stochastic equation:

$$\frac{d\varphi}{dt} = \sqrt{2D}\xi(t),$$

with $\xi(t)$ representing Gaussian white noise, $\langle \xi(t_1)\xi(t_2)\rangle = \delta(t_1 - t_2)$. The initial conditions are $x(0) = y(0) = 0$ and $\varphi(0) = 0$.

(a) Calculate $\langle x(t)\rangle$ and $\langle y(t)\rangle$.
(b) Calculate $\langle r^2(t)\rangle = \langle x^2(t)\rangle + \langle y^2(t)\rangle$.

Hint: Derive the probability density of φ and use it to calculate the moments. Consider using derivatives with respect to t.

9.4 Markov Process (Discrete Space, Discrete Time)

This section is accompanied by a Jupyter/Julia notebook, MarkovChain.ipynb, which is available on the author's living-book website https://sites.google.com/site/mchertkov/living-books/applied-math-book.

When studying stochastic processes, it may be tempting to assume that random events are independent and identically distributed (i.i.d.). However, in many real-world systems, the process evolves in such a way that the future states are influenced by the past. Often, this influence only extends to the most recent state, rather than the entire history. This property defines a *Markov process* (MP), also known as a *Markov chain* (MC), characterized by the Markov property:

$$P(X_{t+1} \mid X_t, X_{t-1}, X_{t-2}, \dots) = P(X_{t+1} \mid X_t).$$

In other words, the future state depends only on the present state and not on any of the previous states.

9.4.1 *Transition probabilities*

The Markov property allows us to represent an MC as a random walk on a directed graph. The vertices of the graph represent the possible states, and the edges represent the transitions between the states, each associated with a probability of transitioning.

Formally, we define a directed graph as $\mathcal{G} = (\mathcal{V}, \mathcal{E})$, where $\mathcal{V}$ is the set of vertices (states) and $\mathcal{E}$ is the set of directed edges representing the possible transitions between the states. Transitions are governed by probabilities $P(X_{t+1} = j \mid X_t = i)$, denoted p_{ji} or $p_{j\leftarrow i}$. These transition probabilities must satisfy the following conditions:

$$\forall (j \leftarrow i) \in \mathcal{E} : \quad p_{ji} \geq 0$$

and

$$\forall i : \quad \sum_{j:(j\leftarrow i)\in\mathcal{E}} p_{ji} = 1.$$

Thus, the combination of the graph $\mathcal{G}$ and the transition probabilities $p = \{p_{ji}\}$ fully defines an MC.

We will primarily focus on *stationary* MCs, where the transition probabilities p_{ji} do not change over time. However, many of the results and methods apply to time-dependent processes as well.

9.4.2 *Sample trajectories and simulation*

One effective way to analyze an MC is by generating sample trajectories, which simulate the evolution of the system. A sample trajectory starts at a specific initial state and evolves by randomly transitioning between states according to the transition probabilities.

Generating a sample trajectory involves the following steps: (1) Initialize the system in a given state. (2) At each time step, select the next state based on the transition probabilities from the current state. (3) Repeat until the desired length of the trajectory is reached.

This process can also be reversed, using observed data to verify whether a sequence of events follows the rules of an MC and, if so, to estimate the underlying transition probabilities.

Example 9.4.1. Describe how to generate a sample trajectory for the MC illustrated in Fig. 9.3. Assume the system begins in state A at time 0.

Solution. Since the system starts in state A at time 0, set $X_0 = A$. At time 1, the system can either remain in state A with probability 0.7 or transition to state B with probability 0.3. Formally, $P(X_1 = A \mid X_0 = A) = 0.7$ and $P(X_1 = B \mid X_0 = A) = 0.3$. To simulate this, draw a random number in the interval $[0, 1)$. If the random number is in $[0, 0.7)$, set $X_1 = A$; otherwise, set $X_1 = B$.

Repeat the process for each subsequent time step: Generate a random number and then transition from the current state to the next state according to the appropriate transition probabilities. A sample trajectory could look something like $AABABBAA\cdots$. □

MCs describe a discrete-time stochastic process where jumps between states occur at specific, discrete time steps. The system evolves over time as a random sequence of states, where the probability of moving to the next state is determined solely by the current state.

A single trajectory (or sample path) might look like

$$i_1(0), i_2(1), \ldots, i_k(t_k),$$

where $i_1, \ldots, i_k \in \mathcal{V}$ are states at respective time steps. Similarly, multiple trajectories can be generated as

$$n = 1, \ldots, N: \quad i_1^{(n)}(0), i_2^{(n)}(1), \ldots, i_k^{(n)}(t_k),$$

where N is the number of trajectories sampled.

9.4.3 *Trajectory statistics*

We can compute various statistics from these trajectories, such as: 1. the proportion of time spent in a particular state; 2. the probability that the system takes more than k steps to return to a given state after leaving it.

While it might seem that statistics gathered from a single trajectory are representative of the entire MC, this is not always true for all MCs. The key property that guarantees that individual trajectories represent the overall process is called *ergodicity*, which will be discussed in the following section.

Basic properties of Markov chains

Definition 9.4.2 (Irreducible). An MC is said to be *irreducible* if it is possible to reach any state from any other state. Formally,

$$\forall i, j \in \mathcal{V}: \quad \exists n > 0, \quad \text{such that } P(X_n = j \mid X_0 = i) > 0. \tag{9.19}$$

The MC shown in Fig. 9.3 is irreducible because every state is accessible from any other state. However, if we set $p_{BA} = 0$ and $p_{AA} = 1$, the chain becomes *reducible* since state B can no longer be accessed from state A.

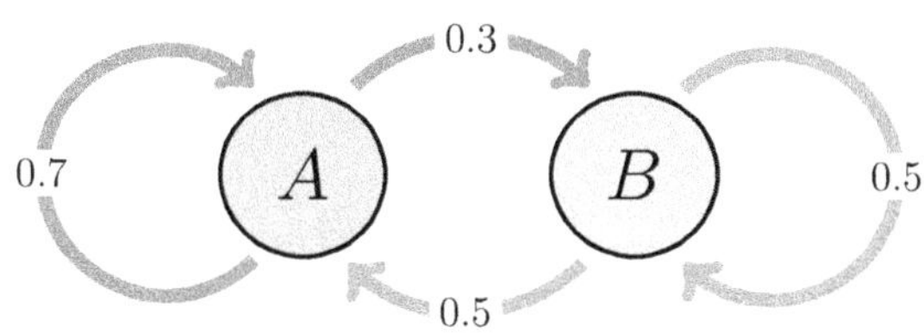

Figure 9.3. An example of a two-state Markov chain.

Definition 9.4.3 (Aperiodicity). A state i has *period* k if returns to state i only occur at multiples of k. The period is defined as

$$k = \text{greatest common divisor}\,\{n > 0 : P(X_n = i \mid X_0 = i) > 0\}\,.$$

If $k = 1$, the state is said to be *aperiodic*. An MC is called *aperiodic* if all its states are aperiodic.

For an irreducible MC, if at least one state is aperiodic, all states are aperiodic. Any irreducible MC with at least one self-loop (i.e., $p_{ii} > 0$ for some i) is aperiodic. In Fig. 9.3, the MC is aperiodic. However, if the self-loops at states A and B are removed, the chain becomes periodic with period two.

Example 9.4.4. Consider the MC in Fig. 9.4. Is this MC irreducible or reducible? Periodic or aperiodic?

Solution. The MC in Fig. 9.4 is *irreducible* because each state is accessible from every other state. However, the chain is *periodic* with period 3 because, starting from state C, the system returns to C only at steps that are multiples of 3 (i.e., 3, 6, 9, etc.). To make the chain aperiodic, we could add a self-loop to any of the states. □

Example 9.4.5. Consider the MC in Fig. 9.5. Is this MC irreducible or reducible? Is it periodic or aperiodic?

Solution. The MC in Fig. 9.5 is *reducible* because states A and B cannot be accessed from state C. Once the system enters state C, it remains there permanently. The chain is also *periodic*, as state A has period 2. □

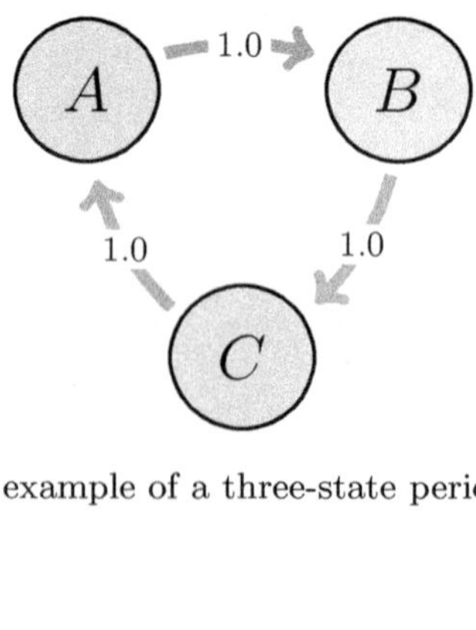

Figure 9.4. An example of a three-state periodic Markov chain.

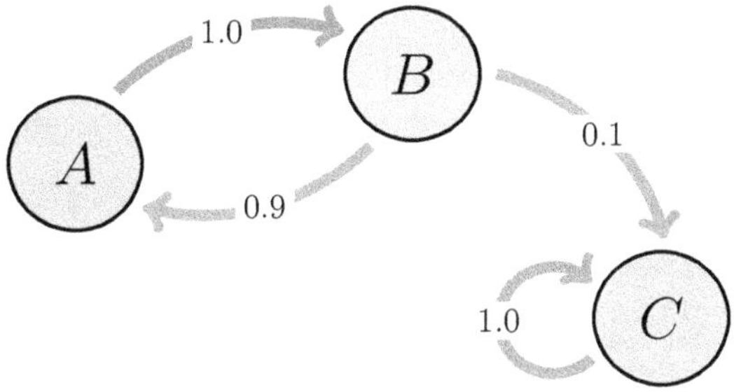

Figure 9.5. An example of a three-state reducible Markov chain.

Recurrence and return times

One key question in MC analysis is whether the system will return to a given state after leaving it, and if so, how long this return is expected to take. The first return time to a state is defined as

$$\tau_1 = \inf\{n \geq 1 : X_n = X_0\}.$$

The expected return time to state i is given by

$$\mathbb{E}[\tau_1 \mid X_0 = i] = \sum_{n=1}^{\infty} nP(\tau_1 = n).$$

Example 9.4.6. Consider the MC illustrated in Fig. 9.3. (a) Compute the probability that the first return to state A occurs in exactly n steps. (b) Compute the expected return time to state A.

Solution. (a) If the first return occurs in exactly one step, the state must transition directly from A back to A (self-loop). If the first return occurs in exactly n steps, with $n \geq 2$, the state must transition from A to B in step 1, remain in state B for $n-2$ steps and then return from B to A on the nth step. Therefore,

$$P(\tau_1 = n \mid X_0 = A) = \begin{cases} 0.7, & \text{if } n = 1, \\ 0.3 \cdot (0.5)^{n-2} \cdot 0.5, & \text{if } n \geq 2. \end{cases}$$

(b) The expected return time to state A is

$$\begin{aligned} \mathbb{E}[\tau_1 \mid X_0 = A] &= 1 \cdot P(\tau_1 = 1) + 2 \cdot P(\tau_1 = 2) + 3 \cdot P(\tau_1 = 3) + \cdots \\ &= 1 \cdot 0.7 + 2 \cdot 0.3 \cdot 0.5 + 3 \cdot 0.3 \cdot (0.5)^2 + \cdots = 1.6. \end{aligned}$$

□

Example 9.4.7. Consider the MC illustrated in Fig. 9.4. (a) Compute the probability that the first return to state A occurs in exactly n steps. (b) Compute the expected return time to state A.

Solution. (a) When the chain transitions out of state A, it must transition to B, then from B to C, and finally from C back to A. Thus, the first return to A occurs in exactly three steps:

$$P(\tau_1 = n | X_0 = A) = \begin{cases} 1, & \text{if } n = 3, \\ 0, & \text{otherwise.} \end{cases}$$

(b) The expected return time to state A is given by

$$\begin{aligned} \mathbb{E}[\tau_1 | X_0 = A] &= 1 \cdot P(\tau_1 = 1) + 2 \cdot P(\tau_1 = 2) + 3 \cdot P(\tau_1 = 3) + \cdots \\ &= 1 \cdot (0) + 2 \cdot (0) + 3 \cdot (1) + 4 \cdot (0) + \cdots = 3. \end{aligned}$$ □

Example 9.4.8. Consider the MC illustrated in Fig. 9.6.

(a) Compute the probability that the first return to state A_1 occurs in exactly n steps.
(b) Compute the expected return time to state A_1.

Solution. (a) Note that if the first return time is n, then the MC must transition from A_1 to A_2, then to A_3 and continue sequentially through states $A_4, \ldots, A_{n-1}$, before finally returning to A_1 on the nth step. Since each transition has probability $\frac{1}{2}$, we have

$$P(\tau_1 = n | X_0 = A_1) = \frac{1}{2} \cdot \frac{1}{2} \cdots \frac{1}{2} = \frac{1}{2^n}.$$

(b) The expected return time to state A_1 is given by summing over the probabilities of each possible return time:

$$\begin{aligned} \mathbb{E}[\tau_1 | X_0 = A_1] &= 1 \cdot P(\tau_1 = 1) + 2 \cdot P(\tau_1 = 2) + 3 \cdot P(\tau_1 = 3) + \cdots \\ &= 1 \cdot \frac{1}{2} + 2 \cdot \frac{1}{4} + 3 \cdot \frac{1}{8} + 4 \cdot \frac{1}{16} + \cdots = 2. \end{aligned}$$ □

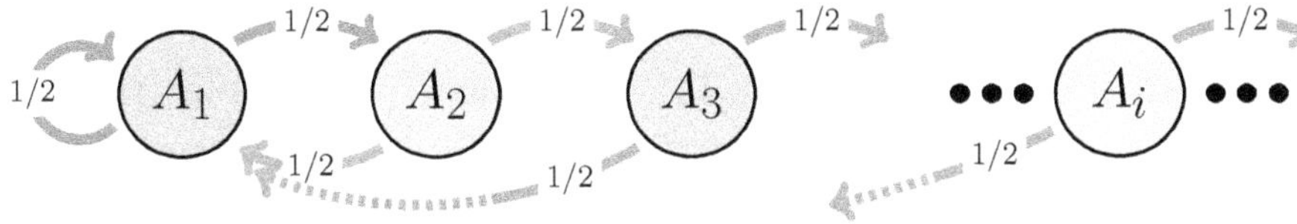

Figure 9.6. An example of a Markov chain with countably infinite states. In this chain, the transition probabilities are $P(A_{n+1} \leftarrow A_n) = 1/2$ and $P(A_1 \leftarrow A_n) = 1/2$ for all n. The chain is *positive recurrent*, meaning that the system is expected to return to any given state in a finite number of steps. This example is analyzed in detail in Example 9.4.8.

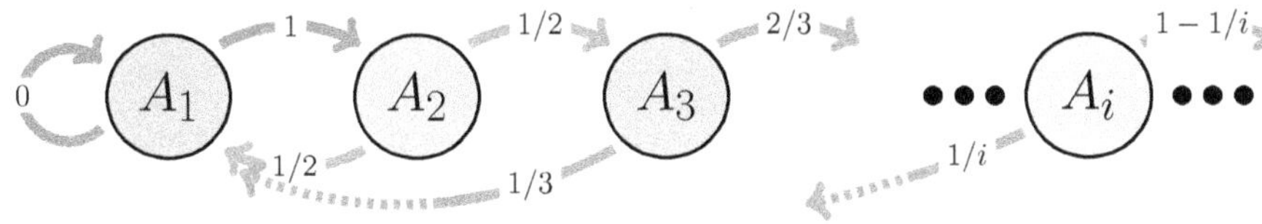

Figure 9.7. An example of a Markov chain with countably many states. In this example, the transition probabilities are given by $P(A_{n+1} \leftarrow A_n) = 1 - \frac{1}{n}$ and $P(A_1 \leftarrow A_n) = \frac{1}{n}$. This Markov chain is recurrent but not positive recurrent. See Example 9.4.9.

Example 9.4.9. Consider the MC illustrated in Fig. 9.7. (a) Compute the probability that the first return to state A_1 occurs in exactly n steps. (b) Compute the expected return time to the state A_1.

Solution.

(a) Note that for the first return to state A_1 to occur in exactly n steps, the chain must sequentially transition through states $A_1 \to A_2 \to A_3 \cdots \to A_{n-1}$ and finally return to A_1 at step n. Thus,

$$P(\tau_1 = n | X_0 = A_1) = \frac{1}{1} \cdot \frac{1}{2} \cdot \frac{2}{3} \cdot \frac{3}{4} \cdots \frac{n-2}{n-1} \cdot \frac{1}{n} = \frac{1}{(n-1)n}.$$

(b) The expected return time to state A_1 is given by summing over all possible return times:

$$\mathbb{E}[\tau_1 | X_0 = A_1] = 1 \cdot P(\tau_1 = 1) + 2 \cdot P(\tau_1 = 2) + 3 \cdot P(\tau_1 = 3) + \cdots$$
$$= 1 \cdot 0 + 2 \cdot \frac{1}{2} + 3 \cdot \frac{1}{6} + 4 \cdot \frac{1}{12} + \cdots \to \infty.$$

Thus, the expected return time to state A_1 is infinite. □

Recurrence, transience and positive recurrence

Definition 9.4.10 (Transient, Recurrent, Positive Recurrent). A state i is said to be *transient* if, starting from state i, there is a nonzero probability that the system will never return to i. A state i is *recurrent* if the system is guaranteed to eventually return to i. A state is *positive recurrent* if the expected return time to state i is finite.

In systems with a countable number of states, the distinction between recurrence and positive recurrence becomes important for long-term behavior.

Ergodicity

Definition 9.4.11 (Ergodic). A state is said to be *ergodic* if it is both aperiodic and positive recurrent. An MC is *ergodic* if it is irreducible and all states are ergodic.

Ergodicity ensures that, in the long run, the system behaves the same regardless of its initial state. An ergodic MC guarantees that time-averaged properties of individual trajectories will converge to the same values as ensemble averages.

Ergodicity and sampling

MCs are commonly used to generate samples from a desired distribution. Imagine a particle traveling on a graph where transitions between states (vertices) occur based on the edge weights (probabilities). If the MC is *ergodic*, the particle's probability distribution converges to a *stationary distribution* after sufficient time – this is when the chain is said to be *mixed*. Once the chain is mixed, the particle's trajectory can be treated as representative of the distribution.

This method allows us to gather useful information about the distribution, such as moments or expectation values of functions, by analyzing the particle's trajectory.

Consider the task of generating a random string of n bits. There are 2^n possible configurations, which can be represented on a hypercube graph with 2^n vertices. Each vertex corresponds to one of the bit strings, and each vertex has n neighbors, representing strings that differ by a single bit (see Fig. 9.8). The MC moves along the edges

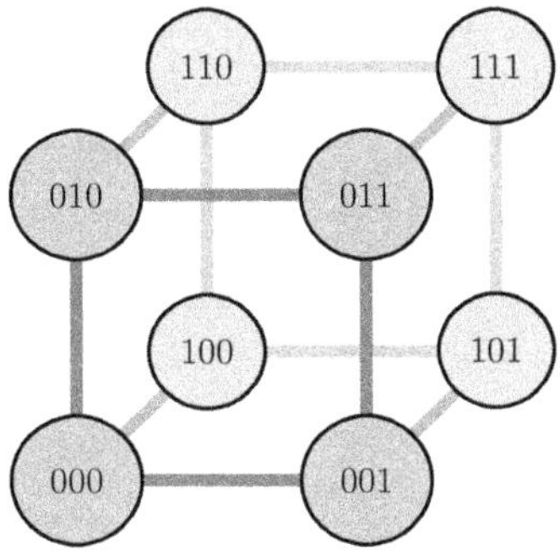

Figure 9.8. Sampling on a hypercube: Each vertex corresponds to a binary string configuration, and edges represent transitions where one bit is flipped at a time.

of this hypercube, flipping one bit at a time. After running the chain long enough, the trajectory will approximate a random sample of bit strings from the uniform distribution.

A crucial question is, how long must we wait before the MC becomes mixed (i.e., forgets its initial state)? To answer this, we need to examine the MC from a mathematical perspective.

9.4.4 *Evolution of the probability state vector*

We initially defined an MC in terms of a weighted, directed graph $(\mathcal{V}, \mathcal{E}, p)$ and used sample trajectories for analysis. A more rigorous approach involves the *probability state vector*, which gives the probabilities of being in each state at time t. Formally, the probability state vector is

$$\pi(t) := (\pi_i(t))_{i \in \mathcal{V}} \quad \text{where } \pi_i(t) := P(X_t = i).$$

Thus, $\pi_i(t) \geq 0$ and $\sum_{i \in \mathcal{V}} \pi_i(t) = 1$.

The probability state vector evolves over time according to the equation

$$\pi_i(t+1) = \sum_{j:(i \leftarrow j) \in \mathcal{E}} p_{ij} \pi_j(t), \quad \forall i \in \mathcal{V}, \quad t = 0, 1, 2, \ldots.$$

In matrix form, this can be written as

$$\pi(t+1) = P\pi(t),$$

where $P := \{p_{ij}\}$ is the *transition probability matrix*. The (i, j) element of P represents the probability of transitioning from state j to state i. Since P is a matrix of probabilities, it is a *stochastic matrix*, defined as follows.

Definition 9.4.12. A matrix is called *stochastic* if all its components are non-negative and the entries in each column sum to 1.

To analyze how the system evolves over k steps, we apply the transition matrix repeatedly:

$$\pi(t+k) = P^k \pi(t).$$

Thus, understanding the long-term behavior of the MC boils down to analyzing the properties of P^k as $k \to \infty$.

Example 9.4.13. Find the stochastic matrix associated with the MC in Fig. 9.3. Is the chain reducible?

Solution. For the MC in Fig. 9.3, the transition matrix is

$$P = \begin{pmatrix} 0.7 & 0.5 \\ 0.3 & 0.5 \end{pmatrix}. \tag{9.20}$$

To determine whether the chain is irreducible, we compute powers of P for large k:

$$P^2 = \begin{pmatrix} 0.64 & 0.60 \\ 0.36 & 0.40 \end{pmatrix}, \quad P^{10} \approx P^{100} \approx \begin{pmatrix} 0.625 & 0.625 \\ 0.375 & 0.375 \end{pmatrix}. \tag{9.21}$$

Since both rows converge to the same values and all entries become positive, we conclude that every state can be reached from any other state. Therefore, the MC is *irreducible*. □

Example 9.4.14. Find the stochastic matrix associated with the MC in Fig. 9.5. Is the chain reducible?

Solution. For the MC in Fig. 9.5, the transition matrix is

$$P = \begin{pmatrix} 0.8 & 0.9 & 0.0 \\ 0.2 & 0.0 & 0.0 \\ 0.0 & 0.1 & 1.0 \end{pmatrix}.$$

Calculating subsequent powers of P,

$$P^2 = \begin{pmatrix} 0.82 & 0.72 & 0.00 \\ 0.16 & 0.18 & 0.00 \\ 0.02 & 0.10 & 1.00 \end{pmatrix}, \quad P^{10} \approx \begin{pmatrix} 0.71 & 0.65 & 0.00 \\ 0.14 & 0.133 & 0.00 \\ 0.14 & 0.21 & 1.00 \end{pmatrix}.$$

Since the third state is inaccessible from the first two states (the third column contains zero entries), the MC is *reducible*. □

Steady-state analysis

Definition 9.4.15 (Stationary Distribution). The probability state vector π^* (if it exists) that satisfies

$$\pi^* = p\pi^*, \tag{9.22}$$

is called the *stationary distribution* or *invariant measure*. (Recall that to be a state vector, each component must be non-negative and the components must sum to unity.)

Theorem 9.4.16 (Existence of a Stationary Distribution). *An MC has a stationary distribution if and only if it is ergodic. Equivalently, an MC has a stationary distribution if and only if it is aperiodic and all of its states are positive recurrent.*

Solving Eq. (9.22) for the example in Eq. (9.20), we find the stationary distribution to be

$$\pi^* = \begin{pmatrix} 0.625 \\ 0.375 \end{pmatrix},$$

which is consistent with Eq. (9.21).

In general, the stochastic matrix for an ergodic MC has a unique eigenvalue, $\lambda^* = 1$. The stationary distribution π^* is the ℓ^1-normalized eigenvector associated with this unit eigenvalue:

$$\pi^* = \frac{e}{\sum_i e_i},$$

where e is the corresponding eigenvector.

An important practical consequence of ergodicity is that the steady state is both unique and universal. Universality means that the steady state does not depend on the initial conditions. But why do we care about uniqueness, invariance with respect to the initial condition and ergodicity? The answer lies in the fact that these properties enable powerful techniques to explore complex phase spaces. One such technique is Markov chain Monte Carlo (MCMC), which we will discuss in Chapter 10. For now, it is crucial to understand that different properties of MCs (and later MCMC algorithms) allow us to efficiently solve complex inference and machine learning problems in data science and related disciplines.

When we shift from analyzing a specific MC to designing one, we aim for properties such as uniqueness, invariance and ergodicity. Ergodicity guarantees convergence to a unique and desirable probability distribution. Moreover, MCMC enables us to generate samples drawn from *any* desired probability distribution. While generating independent samples is generally challenging, MCs provide an efficient way to bypass the need for independence by generating dependent samples, which can eventually become independent through multiple iterations, ensuring that the initial state is forgotten after sufficiently many steps.

Spectrum of the transition matrix and the speed of convergence to the stationary distribution

Assume that the transition matrix p is diagonalizable (i.e., it has $n = |p|$ linearly independent eigenvectors). Then, p can be decomposed via eigen-decomposition as

$$p = U\Sigma U^{-1},$$

where $\Sigma = \text{diag}(\lambda_1, \ldots, \lambda_n)$, with $1 = \lambda_1 \geq |\lambda_2| \geq \cdots \geq |\lambda_n|$, and U is the matrix of eigenvectors, each normalized to have an ℓ^2 norm equal to 1. The evolution of an initial stochastic vector $\pi(0)$ over discrete time $t = 1, 2, \ldots$ is then given by

$$\pi(t) = p^t \pi(0) = (U\Sigma U^{-1})^t \pi(0) = U\Sigma^t U^{-1} \pi(0).$$

We can express the initial vector $\pi(0)$ as a linear combination of the normalized eigenvectors u_i of p:

$$\pi(0) = \sum_{i=1}^{n} a_i u_i.$$

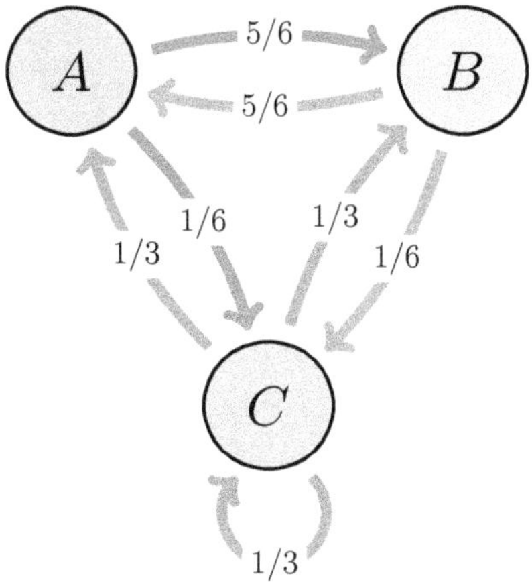

Figure 9.9. Illustration of the detailed balance (DB) condition.

Using the orthonormality of the eigenvectors, we obtain

$$\pi(t) = \lambda_1 \left(a_1 u_1 + a_2 \left(\frac{\lambda_2}{\lambda_1} \right)^t u_2 + \cdots + a_n \left(\frac{\lambda_n}{\lambda_1} \right)^t u_n \right). \tag{9.23}$$

Since $\lim_{t\to\infty} \pi(t) = \pi^* = u_1$, we find that $a_1 = 1$. The remaining terms on the right-hand side of Eq. (9.23) describe the rate of convergence of $\pi(t)$ to the steady state as $t \to \infty$. The convergence is exponential in t, with a rate determined by $\log(\lambda_1/\lambda_2)$.

Example 9.4.17. Find the eigenvalues of the MC shown in Fig. (9.9) with the transition matrix

$$p = \begin{pmatrix} 0 & 5/6 & 1/3 \\ 5/6 & 0 & 1/3 \\ 1/6 & 1/6 & 1/3 \end{pmatrix}. \tag{9.24}$$

What determines the speed of convergence of this MC to a steady state?

Solution. We begin by noting that p is a stochastic matrix. If the initial probability distribution is $\pi(0)$, then after t steps, the distribution is

$$\pi(t) = p^t \pi(0).$$

As t increases, $\pi(t)$ approaches the stationary distribution π^* (since the MC is ergodic, as can be verified). The stationary distribution satisfies

$$p\pi^* = \pi^*.$$

Thus, π^* is an eigenvector of p with eigenvalue 1. For the matrix in Eq. (9.24), the eigenvalues are $\lambda_1 = 1$, $\lambda_2 = 1/6$ and $\lambda_3 = -5/6$, with the corresponding eigenvectors as follows:

$$\pi^* = \left(\frac{2}{5}, \frac{2}{5}, \frac{1}{5}\right)^T, \quad u_2 = \left(-\frac{1}{2}, -\frac{1}{2}, 1\right)^T, \quad u_3 = (-1, 1, 0)^T.$$

If we start in state "A," i.e., $\pi(0) = (1, 0, 0)^T$, we can write the initial state as a linear combination of the eigenvectors:

$$\pi(0) = \pi^* - \frac{u_2}{5} - \frac{u_3}{2}.$$

The state at time t is then

$$\pi(t) = p^t \pi(0) = \pi^* - \frac{\lambda_2^t}{5} u_2 - \frac{\lambda_3^t}{2} u_3.$$

Since $|\lambda_2| < 1$ and $|\lambda_3| < 1$, in the limit of $t \to \infty$, $\pi(t)$ converges to π^*. The speed of convergence is determined by the eigenvalue with the greatest absolute value, other than 1. □

It is worth noting that this example illustrates a general principle summarized by the following theorem (see Ref. [40] for details).

Theorem 9.4.18 (Perron–Frobenius Theorem). *An ergodic Markov chain with transition matrix p has a unique stationary eigenvector π^* with eigenvalue 1, and all other eigenvectors correspond to eigenvalues with absolute values strictly less than 1.*

Some additional properties of Markov chains

Definition 9.4.19 (Reversible). An MC is called *reversible* if there exists a stationary distribution π^* such that

$$\forall \{i, j\} \in \mathcal{E}: \quad p_{ji}\pi_i^* = p_{ij}\pi_j^*, \tag{9.25}$$

where $\{i, j\}$ denotes the undirected edge between states i and j, assuming that both directed edges $(i \leftarrow j)$ and $(j \leftarrow i)$ are elements of the set $\mathcal{E}$.

In physics, this property is also known as *detailed balance* (DB). If we introduce the so-called ergodicity matrix

$$Q := (Q_{ji} = p_{ji}\pi_i^* \mid (j \leftarrow i) \in \mathcal{E}),$$

then DB implies that Q is symmetric, i.e., $Q = Q^T$. An MC where this property does not hold is called *irreversible*, and the asymmetry in $Q - Q^T$ arises from nonzero probability flows (or currents). For instance, for the system illustrated in Fig. (9.3),

$$Q = \begin{pmatrix} 0.7 \times 0.625 & 0.5 \times 0.375 \\ 0.3 \times 0.625 & 0.5 \times 0.375 \end{pmatrix} = \begin{pmatrix} 0.4375 & 0.1875 \\ 0.1875 & 0.1875 \end{pmatrix},$$

Q is symmetric, which shows that even though $p_{12} \neq p_{21}$, there is no net probability flow between states 1 and 2, as the population distribution π_1^* and π_2^* are different. In fact, for any two-state MC, the steady state always satisfies DB.

It is worth noting that if a stationary distribution π^* satisfies the DB condition (9.25) for an MC $(\mathcal{V}, \mathcal{E}, p)$, it will also be the steady state of another MC $(\mathcal{V}, \mathcal{E}, \tilde{p})$ that satisfies the more general *balance* condition (or *global balance*) given by

$$\sum_{j:(j \leftarrow i) \in \mathcal{E}} \tilde{p}_{ji}\pi_i^* = \sum_{j:(i \leftarrow j) \in \mathcal{E}} \tilde{p}_{ij}\pi_j^*. \tag{9.26}$$

This suggests that multiple different MCs (i.e., dynamics) can lead to the same steady state. DB is simply a specific case of the general balance condition (9.26).

The distinction between DB and general balance can be interpreted in terms of flows in the state space, such as water currents. In the hydrodynamic analogy, a reversible MC corresponds to an irrotational flow, while an irreversible chain allows for rotational flow (e.g., vortices). More formally, in the irreversible case, the skew-symmetric part of the ergodic flow matrix, $Q = (\tilde{p}_{ij}\pi_j^* \mid (i \leftarrow j))$, is nonzero. This allows for the following cycle decomposition:

$$Q_{ij} - Q_{ji} = \sum_\alpha J_\alpha \left(C_{ij}^\alpha - C_{ji}^\alpha\right),$$

where α indexes the cycles on the graph of states with adjacency matrices C^α and J_α represents the magnitude of the probability flux in cycle α.

The cycle decomposition can be used to modify an MC without changing its steady-state distribution. However, care must be taken when adding cycles to ensure that all transition probabilities in the resulting matrix $\tilde{p}$ remain positive (the stochasticity of the matrix is guaranteed by construction). This method, along with techniques such as *lifting* or *replication*, can improve the mixing rate, thereby accelerating convergence to the steady state – a highly desirable property for efficient sampling of π^*.

Example 9.4.20. Given the stationary distribution $\pi^* = (\pi_1^*, \pi_2^*, \pi_3^*)$, construct a (3×3) transition matrix p that (a) satisfies the general balance condition and (b) satisfies DB. Are the constructions unique? Find the spectrum of the transition matrix in case (b) and verify that the Perron–Frobenius theorem (9.4.18) holds. In case (b), formulate and solve an example of the fastest mixing MC. Can this solution be generalized to find the fastest mixing MC of size n, given $\pi^* = (\pi_1^*, \ldots, \pi_n^*)$? Return to the three-state MC and impose the constraint that all diagonal elements of the transition probability matrix (corresponding to self-loops) are zero, i.e., $p(1,1) = p(2,2) = p(3,3) = 0$. Is the MC unique in this case? Is it ergodic?

Example 9.4.21. Let $\Sigma = \{x_0, x_1, \ldots, x_{K-1}\}$ be K equidistant points on a circle, i.e., $x_k = e^{2\pi i k/K}$. Let $\alpha, \beta \in (0,1)$ be constants satisfying $\alpha + \beta + \gamma = 1$. Consider the random walk (X_t) defined by

$$\begin{aligned} P(X_{t+1} = x_{k+1} \mid X_t = x_k) &= \alpha, \\ P(X_{t+1} = x_{k-1} \mid X_t = x_k) &= \beta, \\ P(X_{t+1} = x_k \mid X_t = x_k) &= \gamma. \end{aligned}$$

Let π be the unique stationary distribution.

(a) For what values of α, β and γ (and K) is the MC ergodic?
(b) What is the stationary distribution? (Provide intuitive arguments.)
(c) For what values of α and β does the MC satisfy DB?
(d) Let p denote the transition matrix. Find exact expressions for the eigenvalues of p.
Hint: The linear transformation represented by p is a convolution operator and can be diagonalized by the discrete Fourier transform.

(e) The *spectral gap* of p is $1 - |\lambda'|$, where λ' is the second-largest eigenvalue (in absolute value) of p. The size of the spectral gap determines the rate of convergence to the stationary distribution. Suppose $\gamma = 0.98$ and $\alpha = \beta$. Find the spectral gap of p to leading order in $1/K$ as $K \to \infty$.

(f) Are there initial distributions that converge to the stationary distribution faster than the rate determined by the second-largest eigenvalue? If so, provide an example. If not, explain why.

Solution. (a) The MC is irreducible if $\gamma < 1$ and aperiodic if $\gamma > 0$. If K is odd, then $\alpha, \beta > 0$ also ensures aperiodicity.

(b) The stationary distribution satisfies

$$\beta\pi(x_{k+1}) + \alpha\pi(x_{k-1}) + \gamma\pi(x_k) = \pi(x_k).$$

The uniform distribution $\pi(x_k) = 1/K$ solves this equation. Since the chain is irreducible, the uniform distribution is the unique stationary distribution.

(c) DB holds if and only if $\beta = \alpha$, i.e., $\beta = \alpha = (1-\gamma)/2$.

(d) The transition matrix is circulant, meaning it acts by convolution on K-vectors and can be diagonalized by the discrete Fourier transform. The eigenvalues are given by

$$\lambda_\ell = \gamma + (\alpha+\beta)\cos\left(\frac{2\pi\ell}{K}\right) + i(\alpha-\beta)\sin\left(\frac{2\pi\ell}{K}\right).$$

(e) For $\alpha = \beta = (1-\gamma)/2$, the second-largest eigenvalue to leading order in $1/K$ as $K \to \infty$ is $\lambda_{\pm 1} \approx 1 - 2\pi^2/K^2$, yielding a spectral gap of $2\pi^2/K^2$.

(f) No, the convergence rate is determined by the second-largest eigenvalue. Any initial distribution will converge at a rate determined by this eigenvalue. □

Exercise 9.5 (Hardy–Weinberg Law). Consider a gene inheritance experiment involving rabbits, where each rabbit has a pair of genes: GG (dominant), Gg (hybrid) or gg (recessive). The offspring inherits a gene from each parent with equal probability. Starting with a rabbit of a given type, the offspring is always mated with a hybrid.

(a) Write down the transition matrix P for the MC. Is the chain irreducible and aperiodic?

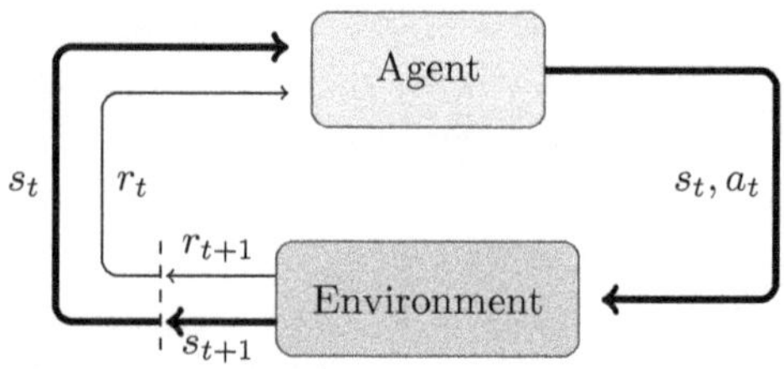

Figure 9.10. Schematic of a Markov decision process: Starting from a state s_t and receiving a reward r_t, an agent selects an action. The environment, influenced by uncertainty, transitions to a new state, s_{t+1}, based on the transition probability $P(s_{t+1}|s_t, a_t)$. The next reward r_{t+1} is determined by $r(s_t, a_t, s_{t+1})$. Figure adapted from Ref. [50].

(b) Starting with a hybrid rabbit, compute $\pi(1), \pi(2), \pi(3)$, where $\pi(n)$ denotes the probability distribution of gene types in generation n. Is there a discernible pattern?
(c) Calculate P^n for general n and discuss the behavior of $\pi(n)$.
(d) Compute the stationary distribution $\pi^* = (\pi^*_{GG}, \pi^*_{Gg}, \pi^*_{gg})$. Does DB hold?

9.5 Stochastic Optimal Control: Markov Decision Process

In the preceding sections, we have discussed MPs, and two chapters ago, we explored optimal control. Now, we combine these two concepts to introduce the *Markov decision process* (MDP).

An MDP is a stochastic, discrete-state, discrete-time formulation of the stochastic optimal control problem, described as follows [50], see also Fig. 9.10:

- **Given:**
  - A set of states, S (e.g., nodes in a graph, or squares in a 4×4 grid as discussed in later examples).
  - A set of actions, A (corresponding to arrows connecting the nodes/squares).
  - A transition probability function, $\mathcal{P} : S \times A \times S \to [0, 1]$, which defines the probability of transitioning from one state to another, dependent on the chosen action, i.e., $\mathcal{P}(s, a, s') = P(s_{t+1} = s' \mid s_t = s, a_t = a)$.

- A reward function, $r : S \times A \times S \to \mathbb{R}$, which assigns rewards $r(s, a, s')$ for transitioning from state s to state s' after taking action a.
- The problem is considered over an infinite time horizon.
- A discount factor γ^t, representing the diminishing value of rewards over time.

- **Goal:**
  - Maximize the expected sum of rewards over a policy, $\pi : S \to A$, which maps each state s_t to an action a_t:

$$\pi^* = \arg\max_{\pi(\cdot)} \mathbb{E}\left[\sum_{t=0}^{\infty} \gamma^t r(s_t, \pi(s_t), s_{t+1})\right], \tag{9.27}$$

  where the expectation is taken over the random transitions governed by $\mathcal{P}(s, a, s')$.

Note that in this formulation, both the reward function $r(\cdot,\cdot,\cdot)$ and the transition probabilities $\mathcal{P}(\cdot,\cdot,\cdot)$ are independent of time. Generalizations to time-dependent cases are possible but will not be covered here.

9.5.1 *Bellman equation and dynamic programming*

The expectation in Eq. (9.27) is referred to as the global reward, which is the reward accumulated over the entire (infinite) time horizon under a given (but not necessarily optimal) policy, π. However, it is also useful to discuss the expected reward over a finite time horizon, τ, known as the value function:

$$\begin{aligned}
\forall \tau \in [1, \ldots, \infty],\ \forall s_0 :\ V_\tau^\pi(s_0) &:= \mathbb{E}_{s_1, s_2, \ldots}\left[\sum_{t=0}^{\tau-1} \gamma^t r(s_t, \pi(s_t), s_{t+1})\right] \\
&= \sum_{s_1, \ldots, s_\tau} \left(\prod_{t'=0}^{\tau-1} \mathcal{P}(s_{t'}, \pi(s_{t'}), s_{t'+1})\right) \\
&\quad \times \sum_{t=0}^{\tau-1} \gamma^t r(s_t, \pi(s_t), s_{t+1}),
\end{aligned}$$

which depends on the initial state s_0 and the policy $\pi(\cdot)$.

This expression can be rewritten recursively in terms of the value function at step $\tau - 1$:

$$V_\tau^\pi(s_0) = \sum_{s_1,\ldots,s_\tau} \left(\prod_{t'=0}^{\tau-1} \mathcal{P}(s_{t'}, \pi(s_{t'}), s_{t'+1}) \right) \Big(r(s_0, \pi(s_0), s_1)$$

$$+\gamma \sum_{t=0}^{\tau-2} \gamma^t r(s_{t+1}, \pi(s_{t+1}), s_{t+2}) \Big) \tag{9.28}$$

$$= \mathbb{E}_{s_1} \left[r(s_0, \pi(s_0), s_1) + \gamma V_{\tau-1}^\pi(s_1) \right]. \tag{9.29}$$

Next, we define the optimal value function, $V_\tau^*(s)$, which maximizes the reward over all policies:

$$\forall \tau,\ \forall s:\ V_\tau^*(s) := \max_{\pi(\cdot)} V_\tau^\pi(s).$$

By optimizing both sides of Eq. (9.29) over the policy and replacing s_1 by s', we obtain the following Bellman recursion:

$$\forall \tau,\ \forall s:\ V_\tau^*(s) = \max_a \mathbb{E}_{s'} \left[r(s, a, s') + \gamma V_{\tau-1}^*(s') \right]. \tag{9.30}$$

This recursion, introduced by Bellman in 1948, allows us to solve the optimal control problem in Eq. (9.27) by iteratively solving Eq. (9.30). Once the optimal value is found, we can also determine the optimal policy:

$$\forall \tau,\ \forall s:\ \pi_\tau^*(s) = \arg\max_a \sum_{s'} \mathcal{P}(s, a, s') \left[r(s, a, s') + \gamma V_{\tau-1}^*(s') \right],$$

which defines the policy as a mapping from states S to actions A. These equations, known as Bellman equations, represent a core example of *dynamic programming.*

9.5.2 *Value-iteration and policy-iteration algorithms*

A few important remarks are in order:

- The recursive Bellman equations (9.30) naturally lead to the **value-iteration** algorithm, which is detailed in Algorithm 5 and demonstrated in the Grid World example discussed in the following section.

- In addition to the state-based value function, we can define the value of taking action a_0 in state s_0 under policy π, known as the *action-value* function or the Q-function:

$$\forall\tau, s_0, a_0 : \; Q^{\pi}_{\tau}(s_0, a_0) = \mathbb{E}_{s_1, s_2, \ldots}\left[r(s_0, a_0, s_1) + \sum_{t=1}^{\tau-1} \gamma^t r(s_t, \pi(s_t), s_{t+1})\right]$$
$$= \mathbb{E}_{s'}\left[r(s_0, a_0, s') + \gamma V^{\pi}_{\tau-1}(s')\right]. \tag{9.31}$$

Thus, instead of working with $V^{\pi}_{\tau}(s_0)$, the entire dynamic programming approach can be restated in terms of the Q-function. The action-value version of Eq. (9.29) becomes

$$\forall\tau, s_0, a_0 : \; Q^{\pi}_{\tau}(s_0, a_0) = \mathbb{E}_{s'}\left[r(s_0, a_0, s') + \gamma \max_{a'} Q^{\pi}_{\tau-1}(s', a')\right]. \tag{9.32}$$

- Another approach to solving the optimization problem (9.27) is the **policy-iteration** algorithm. This method alternates between two steps until convergence: (a) **policy evaluation**, where the value iteration algorithm solves Eqs. (9.29) for a fixed policy, either for a set number of steps or until a tolerance level is reached; and (b) **policy improvement**, where the policy is updated as follows:

$$\forall\tau, \; \forall s_0 : \; \pi_{\tau}(s) \leftarrow \arg\max_{a} \sum_{s'} \mathcal{P}(s, a, s') \left[r(s, a, s') + \gamma V^{\pi}_{\tau-1}(s')\right].$$

This policy-iteration approach can be advantageous when the policy converges faster than the value function.

9.5.3 *MDP: Grid World example*

An MDP can be seen as an interactive probabilistic game, where the objective is to define optimal transition rates between states in order to achieve a certain goal. Once these transition rates are set (whether optimal or sub-optimal), the system behaves like an MP.

Let us explore the Grid World game with the rules shown in Fig. 9.11. The agent moves on a 3×4 grid, with certain walls blocking its path. The agent's actions do not always succeed as intended: For example, if the agent chooses to move north, it will succeed 80% of the time (if no wall is present), but 10% of the time the agent will

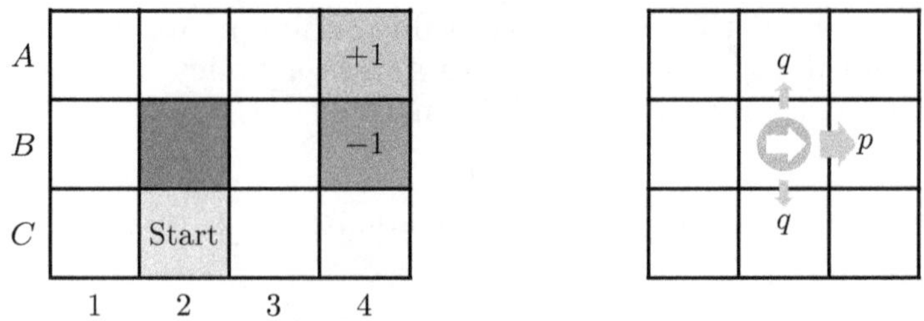

Figure 9.11. Left: A canonical example of an MDP from the "Grid World" game. Right: When the agent chooses to move to the right, this happens with probability p, while moving up or down occurs with probability q.

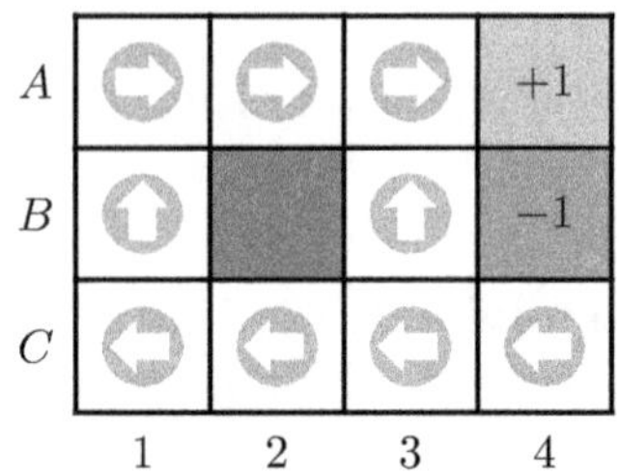

Figure 9.12. Optimal set of actions (arrows) for each state in Grid World.

accidentally move west and 10% of the time east. If a wall blocks the intended movement, the agent stays in place. The game ends when the agent reaches one of two terminal states, receiving a reward of $+1$ or a penalty of -1, respectively. No rewards are given for visiting any other state.

We can translate these rules into mathematical terms as follows:

$$\begin{aligned}
&s \in \{(C,1),(C,2),(C,3),(C,4),(B,1),(B,2),(B,4),\\
&\qquad (A,1),(A,2),(A,3),(A,4)\},\\
&a \in \{\uparrow,\downarrow,\leftarrow,\rightarrow\},\\
&P\left((B,1)|(C,1),\uparrow\right)=0.8,\ \ P\left((C,1)|(C,1),\uparrow\right)\\
&\qquad =0.1,\ \ P\left((C,2)|(C,1),\uparrow\right)=0.1,\\
&P\left((C,2)|(C,1),\rightarrow\right)=0.8,\ \ P\left((C,1)|(C,1),\rightarrow\right)\\
&\qquad =0.1,\ \ P\left((B,1)|(C,1),\rightarrow\right)=0.1,\ldots,
\end{aligned}$$

$$r(s,a,s') = \begin{cases} +1, & \text{if } s = (A,4), \\ -1, & \text{if } s = (B,4), \\ 0, & \text{if } s \neq (A,4),(B,4). \end{cases}$$

Here, $V_\tau^*(s)$ represents the expected sum of rewards when starting from state s and acting optimally for a horizon of τ steps.

To find the optimal actions (policy), we use the value iteration algorithm – with a pseudo-code shown in Algorithm 5 – which solves the value iteration equations (9.30).

This algorithm is also illustrated in Figs. 9.12 and 9.13, which show the rules and the value iteration process for the Grid World example. Some values for the optimal value function at the first three iterations are calculated as follows:

$$\begin{aligned} V_1^*((A,3)) &= P((A,4) \mid (A,3), \rightarrow) * V_0^*((A,4)) * \gamma \\ &= 0.8 * 1 * 0.9 \approx 0.72, \end{aligned}$$

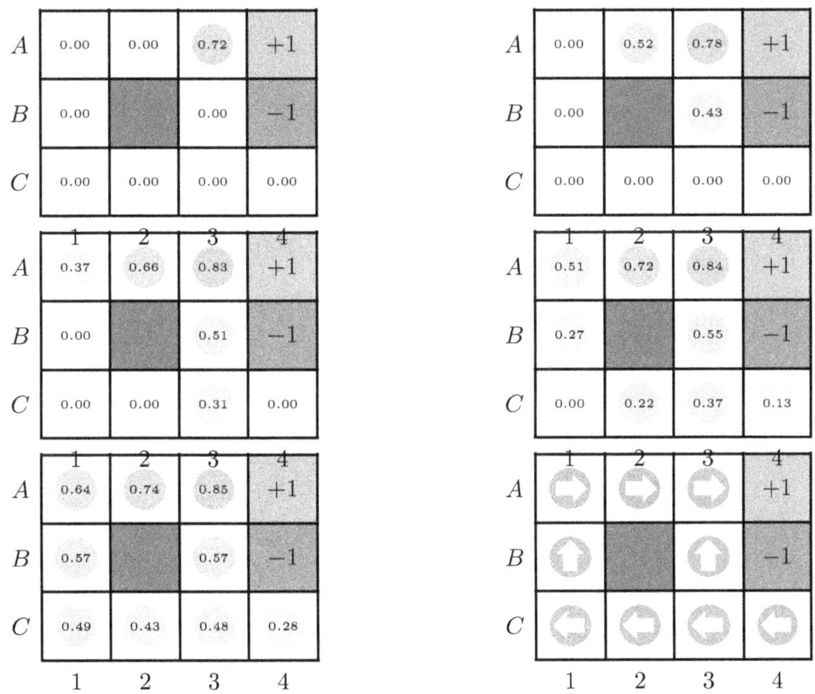

Figure 9.13. Value iteration in Grid World.

Algorithm 5 MDP – Value Iteration (finite horizon)

Input: Set of states S; set of actions A; transition probabilities $P(s' \mid s, a)$; rewards $r(s, a, s')$; discount factor γ

Initialize: $\forall s: \quad V_0^*(s) = 0$
for $\tau = 0, \ldots, H - 1$ **do**
$\forall s: \quad V_{\tau+1}^*(s) \leftarrow \max_a \sum_{s'} P(s' \mid s, a)\left[r(s, a, s') + \gamma V_\tau^*(s')\right]$
3pt – [Bellman update] – the expected sum of rewards when starting from state s and acting optimally for $\tau + 1$ steps.
end for

$$\begin{aligned}
V_2^*((A,3)) &= (P((A,4) \mid (A,3), \rightarrow) * V_1^*((A,4)) \\
&\quad + P((A,3) \mid (A,3), \rightarrow) * V_1^*((A,3))) * \gamma \\
&\approx (0.8 * 1 + 0.1 * 0.72) \approx 0.78, \\
V_3^*((2,3)) &= (P((A,3) \mid (2,3), \uparrow) * V_2^*((A,3)) \\
&\quad + P((B,4) \mid (2,3), \uparrow) * V_2^*((B,4))) * \gamma \\
&\approx (0.8 * 1 * 0.72 + 0.1 * (-1)) * 0.9 \approx 0.43.
\end{aligned}$$

Empirically, we observe that as $\tau \rightarrow \infty$, the optimal value functions converge to a well-defined limit, indicating that the expected sum of rewards becomes stationary when acting optimally.

Exercise 9.6. Consider the following modification of the Grid World example: The discount factor is $\gamma = 1$, and the terminal penalty at state $(B, 4)$ is increased to -2:

$$r(s, a, s') = \begin{cases} +1, & \text{if } s = (A, 4), \\ -2, & \text{if } s = (B, 4), \\ 0, & \text{if } s \neq (A, 4), (B, 4). \end{cases}$$

(a) Compute the optimal values of the expected sum of rewards for the first three iterations of Algorithm 5, i.e., find $\forall s$: $V_1^*(s), V_2^*(s), V_3^*(s)$.
(b) Find the optimal policy for the first three iterations, i.e., $\forall s$: $a_1^* = \pi_1^*(s), a_2^* = \pi_2^*(s), a_3^* = \pi_3^*(s)$.

(c) Reformulate the original MDP problem (9.27) as a **linear programming** problem.
Hint: Use the linearity of the value iteration procedure, as described in Eq. (9.30).

(d) So far, we have considered a deterministic policy, where π is a map from S to A. It can be advantageous to consider a **stochastic policy**, where different actions can be taken in a state with specific probabilities. In this case, $\pi(s, a)$ becomes a function of both state and action, describing the probability of taking action a in state s. Suggest a stochastic policy modification of Eq. (9.27).

We will revisit the MDP (and the Grid World example) in Section 10.5, where we will discuss reinforcement learning in more detail.

9.6 Queuing Networks (*)

9.6.1 *Queuing: A bit of history and applications*

*A number of books have been written on queuing theory. We recommend the book by Frank Kelly and Elena Yudovina [51].

The foundations of queuing theory were laid by Agner Krarup Erlang, a Danish engineer who worked for the Copenhagen Telephone Exchange. In 1909, he published the first paper on what is now called queuing theory. Erlang modeled the number of telephone calls arriving at an exchange using a Poisson process. He solved the $M/D/1/\infty$ queuing model in 1917 and extended this to the $M/D/k/\infty$ model in 1920.

The notation used in queuing theory has become standardized, making it easier to describe various queuing systems. This field is traditionally considered a part of operations research, with strong connections to stochastic processes. For example, in $M/D/k/\infty$:

- **M** stands for Markov or memoryless, meaning that arrivals occur according to a Poisson process. Arrivals may also be deterministic, denoted by D.

*This is an auxiliary section which can be skipped during the first reading. Material from this section will not be included in the midterm or final exams.

- **D** indicates deterministic, implying that jobs arriving at the queue require a fixed (deterministic) amount of service. Service times can also be stochastic, where Markovian service times are denoted by M and general (non-Markovian) service times by G (generic).
- k represents the number of servers at the queuing node. If there are more jobs than servers, the extra jobs must queue and wait for service.
- ∞ refers to the allowed size of the queue (waiting room), meaning there is no limit on the number of jobs in the queue.

In this course, we focus on systems with an infinite waiting room capacity, thereby omitting the last argument.

The $M/M/1$ queue is a fundamental model where a single server handles jobs arriving according to a Poisson process, with exponentially distributed service times. In contrast, the $M/G/1$ queue generalizes the service time distribution to any arbitrary distribution (G for general).

Many mathematicians and engineers have contributed to the development of queuing theory since the 1930s, including Pollaczek, Khinchin, Kendall, Kingman, Jackson and Kelly.

Applications of queuing theory span a wide range of fields: call centers, logistics at various scales, manufacturing systems, supermarket checkouts, electric vehicle charging stations and any system where arrivals and processing fit this framework. The primary objectives in such systems are to:

- manage the queue (control its size),
- keep processing units fully utilized,
- keep waiting times under control.

9.6.2 *Single open queue: Birth–death process and Markov chain representation*

Let us explore the $M/M/1$ model in detail. We begin by using the Java Modeling Tool (JMT), which can be downloaded from http://jmt.sourceforge.net/Download.html.

This process is often referred to as a birth–death process, a name that comes from the MC representation shown in Fig. (9.14). The MC has infinitely many states, each representing the number of customers in the system (including the waiting room). The arrival of customers follows a Poisson process with rate λ, and all customers are assumed

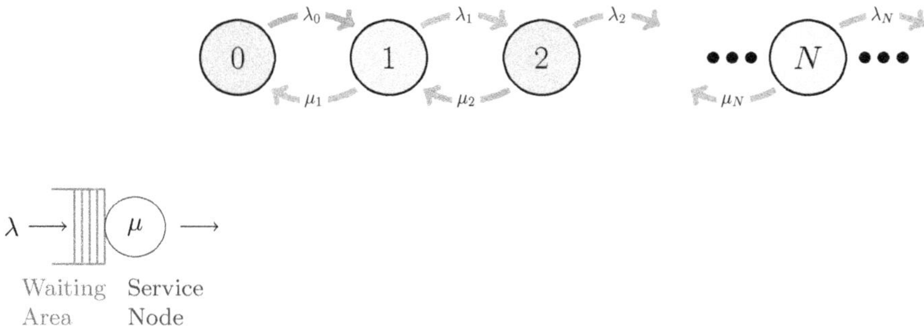

Figure 9.14. Left: Markov chain representation of the $M/M/1$ queue, where for all i, $\lambda_i = \lambda$ and $\mu_i = \mu$. Right: Simplified graphical representation of a single queue.

to be identical. The service rate is denoted by μ, where customers are taken from the waiting room when a server becomes available.

Since both arrivals and services follow Poisson processes, the system is entirely governed by Poisson statistics (recall that both splitting and merging of Poisson processes result in another Poisson process).

Steady-state analysis: We begin by analyzing the steady state of the MC. Let P_i denote the probability that the system is in state i, i.e., that there are i customers in the system.

The balance equations are:

$$\begin{aligned}
&\#\ 0 \text{ customers:} \quad \underbrace{\mu P_1}_{\text{service completion}} = \underbrace{\lambda P_0}_{\text{arrival}},\\
&\#\ 1 \text{ customer:} \quad \lambda P_0 + \mu P_2 = (\lambda+\mu)P_1,\\
&\#\ \text{n customers:} \quad \lambda P_{n-1} + \mu P_{n+1} = (\lambda+\mu)P_n,
\end{aligned} \tag{9.33}$$

By solving these equations iteratively and using the normalization condition $\sum_{i=0}^{\infty} P_i = 1$, we find

$$P_n = \left(\prod_{i=0}^{n-1}\frac{\lambda}{\mu}\right)P_0 = \left(\frac{\lambda}{\mu}\right)^n P_0 = \rho^n P_0,$$

$$1 = \sum_{n=0}^{\infty} P_n = P_0 \sum_{n=0}^{\infty}\rho^n = \frac{P_0}{1-\rho},$$

$$P_n = (1-\rho)\rho^n,$$

where $\rho := \lambda/\mu$ is called the traffic intensity.

The average number of customers in the queue is given by

$$\mathbb{E}[\text{Queue Length}] = \sum_{n=0}^{\infty} nP_n = (1-\rho)\sum_{n=0}^{\infty} n\rho^n = \frac{\rho}{1-\rho}.$$

We observe that as $\rho \to 1$, the average queue length tends to infinity, indicating that a steady state exists only if $\rho < 1$. This condition is known as the stability criterion.

Exercise 9.7. Consider a single $M/M/m$ queue, i.e., a system with m servers. Derive the steady-state solution and determine the modified stability criterion. Can a single queue system with $m = 2$ servers become unstable?

Transient dynamics: In this simple queuing system, we can also study the transient dynamics. The steady-state equations (9.33) are modified to

$$\forall n : \quad \frac{d}{dt}P_n = \underbrace{\lambda P_{n-1} + \mu P_{n+1}}_{\text{arrivals}} - \underbrace{(\lambda+\mu)P_n}_{\text{departures}}. \tag{9.34}$$

The solution to this system can be found in analytic form:

$$P_k(t) = e^{-(\lambda+\mu)t}\Bigg(\rho^{(k-i)/2} I_{k-i}(at) + \rho^{(k-i-1)/2} I_{k+i+1}(at) + (1-\rho)\rho^k \sum_{j=k+i+2} \rho^{-j/2} I_j(at)\Bigg), \tag{9.35}$$

where $a = 2\sqrt{\lambda\mu}$ and $I_k(x)$ is the modified Bessel function of the first kind. This solution assumes that the system starts in state i at $t = 0$.

Exercise 9.8. Derive Eq. (9.35) from Eq. (9.34). Compute the distribution of the busy period of the server, and assuming a first-come-first-served policy, compute the distribution of waiting time and total time in the system.

9.6.3 *Generalization to Jackson networks: Product solution for the steady state*

The single-queue analysis can be extended to networks, such as the one shown in Fig. (9.15). Consider the case of an $M/M/\infty$ network, where there are infinitely many servers and jobs are processed immediately upon arrival (no waiting). The arrival and processing rates are indexed by the nodes and edges of the network. We study $P(n_1, \ldots, n_N; t)$, the probability distribution over the entire network, and write the balance (master) equation for any state at any time:

$$\begin{aligned}\frac{\partial}{\partial t}P(\boldsymbol{n};t) = \sum_{(i,j)\in\mathcal{E}} \lambda_{ij}\Bigg(&\underbrace{(n_i+1)P(\ldots,n_i+1,\ldots,n_j-1,\ldots;t)}_{\text{jobs leaving node } i \text{ for node } j} \\ &- \underbrace{n_i P(\ldots,n_i,\ldots,n_j,\ldots;t)}_{\text{jobs staying at node } i}\Bigg) \\ &+ \sum_{i\in\mathcal{V}} \lambda_{0i}\left(P(\ldots,n_i-1,\ldots;t) - P(\ldots,n_i,\ldots;t)\right) \\ &+ \sum_{i\in\mathcal{V}} \lambda_{i0}\left((n_i+1)P(\ldots,n_i+1,\ldots;t) - n_i P(\ldots,n_i,\ldots;t)\right),\end{aligned} \tag{9.36}$$

where $\mathcal{V}$ and $\mathcal{E}$ represent the set of nodes and directed edges of the network.

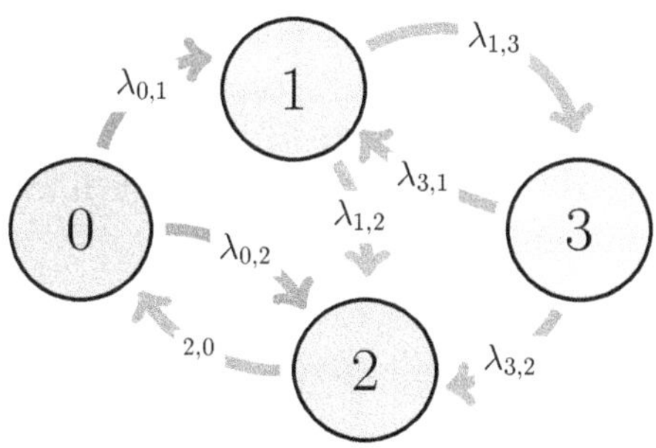

Figure 9.15. Example of a queuing network.

Exercise 9.9. Derive the $M/M/m$ version of Eq. (9.36).

Remarkably, the complex equation (9.36) admits an explicit steady-state solution for any graph:

$$P(\boldsymbol{n}) = Z^{-1} \prod_{i\in\mathcal{V}} \frac{h_i^{n_i}}{n_i!},$$

$$\forall i \in \mathcal{V}: \quad h_i \sum_{j\neq 0}^{(i,j)\in\mathcal{E}} \lambda_{ij} + \sum_{j\neq 0}^{(j,i)\in\mathcal{E}} \lambda_{ji} h_j + \lambda_{0i} - \lambda_{i0} h_i = 0, \tag{9.37}$$

known as the product-form solution. The equation (9.37) is a linear inhomogeneous equation for the vector $\boldsymbol{h} = (h_i \mid i \in \mathcal{V})$, which can be written as $\hat{\Lambda}\boldsymbol{h} = \boldsymbol{\lambda}_{\text{in}}$, where

$$\boldsymbol{\lambda}_{\text{in}} := (-\lambda_{0i} \mid i \in \mathcal{V}), \quad \hat{\Lambda} := (\Lambda_{ij} \mid (i,j) \in \mathcal{E}),$$

and

$$\Lambda_{ij} = \begin{cases} -\lambda_{i0} - \sum_k^{(i,k)\in\mathcal{E}} \lambda_{ik}, & i = j, \\ \lambda_{ji}, & i \neq j. \end{cases}$$

A few important points:

- This product-form solution is **not** a Gibbs (equilibrium) distribution.
- The existence of a steady-state solution (stability) requires that (a) $\hat{\Lambda}$ is non-singular and (b) all components of $\boldsymbol{h}$ are positive.
- Each component of $\boldsymbol{h}$ represents the effective ratio of the steady-state arrival rate to the service rate at each node, generalizing the single-server traffic intensity ρ.

Exercise 9.10. Generalize the steady-state formula and reformulate the stability condition for the general $M/M/m$ case.

Exercise 9.11. Derive the skewed Detailed Balance (DB) relation and show how analyzing the skewed DB leads to the product-form solution for the steady state.

For further reading, see Refs. [51] and [52].

9.6.4 *Heavy traffic limit*

This discussion is mainly based on the material from http://www.columbia.edu/~ww2040/A1a.html.

The heavy traffic limit applies in two cases (or combinations thereof, which we will not discuss here):

- The number of servers is fixed, but traffic intensity λ/μ approaches unity from below. In this case, the queue length behaves like *reflected Brownian motion.*
- Traffic intensity is fixed, but both the number of servers and the arrival rate increase to infinity. Here, the queue length converges to a normal distribution.

Exercise 9.12. Assume that $1-\rho \ll 1$ (where $\rho = \lambda/\mu$). Estimate the following:

- How much time does a typical customer spend in the system?
- How long does it take for the system to transition from a typical filling level to being empty?

Hint: The system's changes occur on a much longer timescale than the service time for a single customer.

We can analyze a queuing system (Q-system) with many customers by separating it into two time scales: *fluid* and *diffusive*. Let $X(t)$ represent some process related to the Q-system. The fluid rescaling of $X(t)$ by a factor of n is given by $\bar{X}_n(t) = nX(nt)$, where time is measured in units of n, and the state (the number of customers) is also measured in units of n. As $n \to \infty$, we seek a limit where $n^{-1}X(nt) \to \bar{X}(t)$, with $\bar{X}(t)$ being the *fluid limit.*

At this scale, when $n \to \infty$, both the arrival and service processes converge to their fluid limits, λt and μt, respectively. This means that they behave deterministically. Since queuing behavior arises from variability, on the fluid scale, where both input and output are deterministic, **the queue effectively disappears**, and there is no meaningful queuing behavior. In this regime, the queue length may increase linearly and indefinitely if $\rho > 1$, decrease to zero and remain there if $\rho < 1$ or remain constant if $\rho = 1$. For queuing networks, we observe piecewise linear changes in queue lengths, reflecting the dynamics on the fluid scale. On this scale, changes in the queue occur over a time span of order n, and stochastic fluctuations are scaled down to zero, rendering them insignificant.

To capture the system's stochastic fluctuations around the fluid limit, we use *diffusion scaling*, which focuses on the difference between the process and its fluid limit. In diffusion scaling, time is measured in units of n, but the state (number of customers) is measured in units of $\sqrt{n}$. The diffusion rescaling of $A(t)$ by n is defined as

$$\hat{A}_n(t) = \sqrt{n}\left(\bar{A}_n(t) - \bar{A}(t)\right).$$

As $n \to \infty$, we seek a limit (analogous to the central limit theorem) where $A_n(t)$ converges in distribution to $\hat{A}(t)$, which describes the *diffusion limit* – a diffusion process such as Brownian motion or reflected Brownian motion. The diffusion limit captures the random fluctuations of the system around the deterministic fluid limit.

Chapter 10

Elements of Inference and Learning

In this chapter, we explore the fundamental concepts of statistical inference and learning, both of which are crucial to data analysis and decision-making under uncertainty. We begin in Section 10.1 with *statistical inference*, covering key tasks such as sampling from probability distributions, computing marginal probabilities and identifying the most likely configurations. Special attention is given to Monte Carlo (MC) methods, including direct sampling (DS) and Markov chain Monte Carlo (MCMC) techniques. In Section 10.2, we extend these ideas into variational approximations for computing marginals and normalizations (so-called partition functions), linking the discussion to variational calculus over *graphical models* (GMs) – expressing factorizations of multivariate probability distribution functions via graphs. We also explore special cases where the graphs are trees, where dynamic programming methods, as discussed in Chapter 7, become efficient, scaling linearly with the number of state space components. Section 10.3 introduces the theory of learning, focusing on sufficient statistics and maximum likelihood (ML) estimation, while Section 10.4 addresses the role of neural networks in approximating functions – an essential technique in modern machine learning and artificial intelligence. Finally, in Section 10.5, we delve into reinforcement learning, covering important algorithms such as temporal difference learning and Q-learning.

This chapter is accompanied by three Jupyter/Julia notebooks, sampling.ipynb, MCMC.ipynb and RL.ipynb, which are available on the author's living-book website: https://sites.google.com/site/mchertkov/living-books/applied-math-book.

For a deeper understanding of the topics covered in this chapter, several advanced references are recommended. For a comprehensive treatment of stochastic simulation techniques, including rare-event simulation and MCMC, see *Stochastic Simulation* by Asmussen and Glynn [53]. *Pattern Recognition and Machine Learning* by Bishop [54] offers a solid introduction to inference and machine learning. For a more technical exploration of graphical models and variational methods, refer to *Graphical Models, Exponential Families, and Variational Inference* by Wainwright and Jordan [55] and to the author's living book, *INFERLO: Inference, Learning and Optimization with Graphical Models* [56]. Additionally, *Reinforcement Learning: An Introduction* by Sutton and Barto [50] is an essential resource for reinforcement learning techniques, covering both policy-based and value-based methods.

Statistical inference refers to a set of tasks and operations performed over a statistical model of a given phenomenon. These tasks, assuming knowledge of the statistical model, typically include: (a) sampling from the probability distribution, (b) computing marginal probabilities and (c) identifying the most likely configuration or state. However, the statistical model may not always be known, in which case it must first be learned before solving the inference problem.

In this chapter, we first introduce statistical inference and then shift our focus to learning the statistical models that we aim to infer.

10.1 Statistical Inference: Sampling and Stochastic Algorithms

10.1.1 *Monte Carlo algorithms: General concepts and direct sampling*

This section is best understood in conjunction with the accompanying Jupyter notebook, sampling.ipynb, available on the author's webpage at https://sites.google.com/site/mchertkov/living-books/applied-math-book.

MC methods refer to a broad class of algorithms that rely on repeated random sampling to obtain numerical results. These methods are named after Monte Carlo, the famous city known for gambling, where randomness plays a central role. MC algorithms can be used for tasks such as numerical integration (e.g., computing weighted sums, expectations or marginal probabilities) as well as optimization.

Sampling, in the context of inference, refers to selecting a subset of individuals or configurations from a statistical population to estimate characteristics of the entire population.

There are two main types of sampling in MC methods:

- **Direct sampling (DS):** This method focuses on drawing independent samples from a given distribution.
- **Markov chain Monte Carlo (MCMC):** In contrast to DS, MCMC draws correlated samples according to an underlying Markov chain (MC), and the goal is to reach a target distribution over time.

We illustrate both types using the simple example of the "pebble game," where we estimate the value of π by sampling within the interior of a circle.

Direct sampling vs. MCMC for the "pebble game"

In this example, we want to sample uniformly from within a circle using another uniform distribution over a square that contains the circle. In **DS**, we generate random samples uniformly from the square (using two independent random variables) and reject any samples that fall outside the circle.

In **MCMC**, we start from an initial sample (a point in the square) and generate subsequent samples by adding random independent shifts to both coordinates of the point. If the sample moves outside the square, it reappears on the opposite side (periodic boundary conditions). Although the sample "walks" within the square, only those that fall inside the circle are counted to estimate the area of the circle.

Direct sampling by mapping

Direct sampling by mapping involves applying a deterministic transformation to samples drawn from a known distribution to generate samples from a new target distribution. This method is exact and produces independent random samples from the new distribution. (We discuss the formal criteria for independence in the following.)

For example, suppose we want to generate samples from the exponential distribution $y_i \sim p(y) = \exp(-y)$, where $y \in [0, \infty)$, but we only have access to a uniform distribution over $[0, 1]$. The transformation $y_i = -\log(x_i)$, where x_i are uniform random variables, generates the desired exponentially distributed samples.

Another classic example is the **Box–Muller algorithm**, which transforms two uniform random variables into two normally distributed random variables. The mapping is given by

$$x = \sqrt{-2\log\psi}\cos(2\pi\theta), \quad y = \sqrt{-2\log\psi}\sin(2\pi\theta),$$

where ψ, θ are uniform random variables in $[0, 1]$. For further details and numerical examples, refer to the sampling.ipynb notebook, available on the author's webpage at https://sites.google.com/site/mchertkov/living-books/applied-math-book.

Direct sampling by rejection

Let us now demonstrate how to generate samples from a positive half-Gaussian distribution using rejection sampling. We start by sampling from an exponential distribution:

$$x \sim p_0(x) = \begin{cases} e^{-x}, & x > 0, \\ 0, & \text{otherwise}. \end{cases}$$

Next, to obtain a sample from the positive half of the Gaussian distribution

$$x \sim p(x) = \begin{cases} \sqrt{\frac{2}{\pi}} e^{-x^2/2}, & x > 0, \\ 0, & \text{otherwise}, \end{cases}$$

we accept the sample with the probability

$$p(x) = \frac{1}{M}\sqrt{\frac{2}{\pi}} e^{x - x^2/2},$$

where M is a constant chosen to ensure that $p(x) \leq 1$ for all $x > 0$ (in this case, $M \approx 1.32$).

Rejection sampling is particularly useful when the probability densities are known only up to a multiplicative constant. This situation frequently arises in statistical models with complex normalization terms (also known as partition functions). For further details and illustration, refer to the accompanying Jupyter notebook, sampling.ipynb, available on the author's webpage at https://sites.google.com/site/mchertkov/living-books/applied-math-book.

We also recommend the following resources for additional reading on DS and MC methods:

- *Introduction to Direct Sampling*, a chapter from the "Monte Carlo Lecture Notes" by J. Goodman, NYU.
- *Lecture on Monte Carlo Sampling*, part of the UC Berkeley course on "Bayesian Modeling and Inference" by M. Jordan.

Importance sampling

MC methods are often used to compute sums, integrals and expectations. Suppose we need to compute the expectation of a function, $f(x)$, over a distribution, $p(x)$, i.e., $\mathbb{E}_p[f(x)] = \int dx\, p(x) f(x)$, but $f(x)$ and $p(x)$ are concentrated around very different regions of x. In this case, many MC samples drawn from $p(x)$ may be "wasted."

Importance sampling addresses this problem by adjusting the sampling distribution from $p(x)$ to a new distribution $\tilde{p}(x)$. We then compute

$$\mathbb{E}_p[f(x)] = \int dx\, \tilde{p}(x) \frac{f(x)p(x)}{\tilde{p}(x)} = \mathbb{E}_{\tilde{p}}\left[\frac{f(x)p(x)}{\tilde{p}(x)}\right].$$

For example, consider the case where $p(x) = \frac{1}{\sqrt{2\pi}} e^{-x^2/2}$ and $f(x) = e^{-(x-4)^2/2}$. In this case, we can improve the sampling efficiency by using $\tilde{p}(x) = \frac{1}{\sqrt{\pi}} e^{-(x-2)^2}$.

While importance sampling is a powerful tool, in high-dimensional problems, it can be challenging to guess the correct proposal distribution $\tilde{p}(x)$. One solution is to search for a good $\tilde{p}(x)$ adaptively.

For a comprehensive review of importance sampling, we recommend the lecture notes by A. Owen, which can be found at this link, and/or check the adaptive importance sampling package. We also recommend Ref. [53] for advanced reading.

Brute-force sampling

This method requires access to a uniform sampling algorithm, `rand()`, which generates random numbers in $[0, 1]$. The interval is divided into subintervals proportional to the weights of all possible states, and `rand()` is used to select a state. While this approach provides independent samples, it is often impractical due to memory constraints, as it requires storing all possible configurations.

Direct sampling from a multivariate distribution with a partition function oracle

Suppose we have an oracle capable of computing the partition function (i.e., the normalization constant) of a multivariate probability distribution, as well as any marginal probabilities. Can this oracle generate independent samples?

The answer is yes, through the following **decimation algorithm**, which generates independent samples $x \sim P(x)$, where $x = (x_i \mid i = 1, \ldots, N)$.

The correctness of the algorithm follows from the chain rule for probability distributions:

$$P(x_1, \ldots, x_n) = P(x_1)P(x_2 \mid x_1)P(x_3 \mid x_1, x_2) \cdots P(x_n \mid x_1, \ldots, x_{n-1}).$$

This algorithm can be understood as a hierarchical splitting of the interval $[0, 1]$, first according to $P(x_1)$, then subdividing for $P(x_2 \mid x_1)$ and so on.

Although the partition function oracle is often exponentially expensive in the size of the problem, in some cases it can be computed efficiently.

Exercise 10.1. Consider the Ising model, where the probability of a binary vector (spin configuration) x is given by

$$p(x) = \frac{\exp(-\beta E(x))}{Z}, \quad E(x) = -\frac{1}{2} \sum_{\{i,j\} \in \mathcal{E}} x_i J_{ij} x_j + \sum_{i \in \mathcal{V}} h_i x_i, \tag{10.1}$$

$$\text{where } Z = \sum_{x} \exp(-\beta E(x)). \tag{10.2}$$

For the case $h = 0$ and uniform interaction $J_{ij}\beta = -1$ over an $n \times n$ grid graph with nearest-neighbor interactions, implement the decimation algorithm (Algorithm 6). Test its performance for $n = 2, 3, 4, 5$. Analyze how the time required to generate i.i.d. samples depends on n and explain your results.

Algorithm 6 Decimation Algorithm

Input: Probability distribution $P(x)$ and a partition function oracle.

1: Initialize: $x^{(d)} = \varnothing$, $I = \varnothing$
2: **while** $|I| < N$ **do**
3: Randomly pick i from $\{1, \dots, N\} \setminus I$
4: Let $x^{(I)} = (x_j \mid j \in I)$
5: Compute $P(x_i \mid x^{(d)}) = \sum_{x \setminus x_i; x^{(I)} = x^{(d)}} P(x)$ using the oracle.
6: Generate a random sample $x_i \sim P(x_i \mid x^{(d)})$
7: Update $I \leftarrow I \cup \{i\}$
8: Update $x^{(d)} \leftarrow x^{(d)} \cup \{x_i\}$
9: **end while**

Output: $x^{(\text{dec})}$ is an independent sample from $P(x)$.

10.1.2 *Inference via Markov chain Monte Carlo*

This section is best understood in conjunction with the accompanying Jupyter notebook, MCMC.ipynb, available on the author's webpage at https://sites.google.com/site/mchertkov/living-books/applied-math-book.

MCMC methods are a class of algorithms for sampling from a probability distribution by constructing an MC that converges to the desired target distribution. MCMC is particularly useful when direct sampling is infeasible due to unknown or complex normalization factors in the distribution.

There are various flavors of MCMC methods, such as heat bath, Glauber dynamics, Gibbs sampling, Metropolis–Hastings (MH), cluster algorithms and warm-start algorithms. Although these methods may differ in approach, they all rely on constructing transition probabilities between states. The actual stationary distribution of the system may not be known exactly but is often known up to a normalization constant, known as the *partition function*. In the following, we explore two key MCMC methods in detail: Gibbs sampling and MH.

Gibbs sampling

Consider a multivariate probability distribution, $\pi(x)$, that is known up to an unknown normalization constant. Specifically, we have access to $\tilde{\pi}(x)$, which is proportional to $\pi(x)$, such that $\pi(x) = \tilde{\pi}(x)/Z$, where $Z = \sum_x \tilde{\pi}(x)$ is the partition function.

In this scenario, direct sampling methods, such as the decimation algorithm, are not feasible due to the unknown partition function. Instead, we resort to correlated sampling methods such as Gibbs sampling, which progressively generates samples that eventually approximate independent samples as the algorithm proceeds.

Gibbs sampling is an MCMC method where we generate samples by updating one variable at a time, conditioning on the current values of the other variables. The key feature of Gibbs sampling is that sampling from the *conditional distribution* of a single variable, given all others, is often computationally efficient, even if sampling from the full joint distribution is not.

We describe the Gibbs sampling process in Algorithm 7. The algorithm starts with an initial sample $x^{(t)}$, selects a component at random, computes the conditional distribution for this component and draws a sample from it. The process is repeated iteratively until the system reaches convergence, which can be identified empirically (e.g., by monitoring the stability of the histograms or observables).

Algorithm 7 Gibbs Sampling

Input: Conditional distributions $p(x_i \mid x_{\sim i})$ for all $i \in \{1, \ldots, N\}$.
Initialize with sample $x^{(t)}$.

loop Repeat until convergence
  Draw i uniformly from $\{1, \ldots, N\}$.
  Sample $x_i \sim p(x_i \mid x_{\sim i}^{(t)})$.
  Update $x_i^{(t+1)} = x_i$.
  For all $j \neq i$, set $x_j^{(t+1)} = x_j^{(t)}$.
  Output $x^{(t+1)}$ as the next sample.
end loop

Example 10.1.1. Illustrate Gibbs sampling using the Ising model. Construct the corresponding MC and demonstrate that the algorithm satisfies the detailed balance (DB) condition.

Solution. Starting with a state $x^{(t)}$, we randomly select a spin x_i and consider two possible configurations: $x_i = +1$ and $x_i = -1$. The conditional probabilities for each configuration, given the current state of all other spins $x_{\sim i}^{(t)}$, are calculated as

$$p_+ = p(x_i = +1 \mid x_{\sim i}^{(t)}), \quad p_- = p(x_i = -1 \mid x_{\sim i}^{(t)}),$$

where $p_+ + p_- = 1$ and

$$\frac{p_+}{p_-} = e^{-\beta \Delta E}, \quad \Delta E = E(x_i = +1, x_{\sim i}^{(t)}) - E(x_i = -1, x_{\sim i}^{(t)}).$$

The algorithm updates x_i to $+1$ with probability p_+ and to -1 with probability p_-.

The MC is defined on the 2^N-dimensional hypercube, where N is the number of spins, and the states are the spin configurations. The DB condition is satisfied because the probability fluxes between the states $x_i = +1$ and $x_i = -1$ are equal:

$$Q_{-+} = \frac{1}{Z} e^{-\beta E(x_i = -1, x_{\sim i}^{(t)})} p_+, \quad Q_{+-} = \frac{1}{Z} e^{-\beta E(x_i = +1, x_{\sim i}^{(t)})} p_-. \qquad \square$$

Metropolis–Hastings sampling

The MH algorithm is a general MCMC method that constructs a MC with a desired stationary distribution $\tilde{\pi}(x)$, which is known up to a normalization constant. The MH algorithm introduces a proposal distribution $q(x' \mid x)$, which generates candidate samples based on the current state. Starting from an initial state $x^{(t)}$, the algorithm proceeds by drawing a candidate sample x' from the proposal distribution $q(x' \mid x^{(t)})$ and accepting or rejecting the proposal based on the acceptance probability:

$$\alpha = \min \left\{ 1, \frac{q(x_t \mid x')\tilde{\pi}(x')}{q(x' \mid x_t)\tilde{\pi}(x_t)} \right\}. \tag{10.3}$$

If the proposal is accepted, $x^{(t+1)} = x'$; otherwise, the current state is retained. By construction, the MH algorithm satisfies DB:

$$\forall x, x' : \quad \text{MH trans. prob.}(x' \leftarrow x)\tilde{\pi}(x) = \text{MH trans. prob.}(x \leftarrow x')\tilde{\pi}(x'),$$

ensuring that the chain converges to the desired distribution.

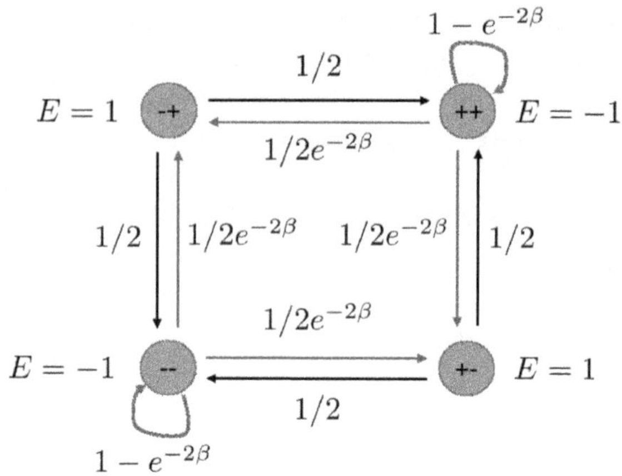

Figure 10.1. Example of a Markov chain induced by the Metropolis–Hastings algorithm for a two-spin Ising model.

Example 10.1.2. Consider the MC shown in Fig. 10.1. Demonstrate that this MC is ergodic and can be interpreted as a specific instance of the MH algorithm (Algorithm 8) for a two-spin Ising model. Determine the proposal distribution in this case, and identify the stationary distribution. Does the MC obey DB? Additionally,

---

Algorithm 8 Metropolis–Hastings Sampling

Input: Given target distribution $\tilde{\pi}(x)$ and proposal distribution $q(x' \mid x)$. Start with sample x_t.

loop Repeat until convergence
  Draw $x' \sim q(x' \mid x^{(t)})$.
  Compute acceptance probability $\alpha = \frac{q(x_t|x')\tilde{\pi}(x')}{q(x'|x^{(t)})\tilde{\pi}(x^{(t)})}$.
  Draw $\beta \sim U([0,1])$ (uniform random number).
  if $\beta < \alpha$ **then**
    $x^{(t)} \leftarrow x'$ [accept].
  else
    x' is rejected.
  end if
  Output $x^{(t)}$ as the next sample.
end loop

---

explore how the steady-state distribution would change if rejections were removed from the consideration.

Solution. The state space has cardinality $2^2 = 4$, represented by states $x = (x_1 = \pm 1, x_2 = \pm 1)$. By inspecting the MC, we observe that it is aperiodic and positive recurrent, which implies ergodicity. Notably, some transition probabilities are exactly $1/2$, naturally corresponding to cases where Eq. (10.3) yields a unity acceptance probability.

The self-loops in the MC represent rejections, or cases where a proposed state is not accepted. These observations suggest that this MC is indeed an example of the MH Algorithm 8 with

$$\tilde{\pi}(x_1, x_2) = \exp(-\beta E(x)), \quad E(x = (x_1, x_2)) = -\frac{x_1 x_2}{2},$$
$$q(x'|x) = \exp\left(-\beta\left(E(x') - E(x)\right)\right).$$

Thus, the stationary distribution corresponds to a ferromagnetic Ising model with unit pairwise strength and no external magnetic field. The proposal distribution $q(x'|x)$ allows only single-spin flips, prohibiting simultaneous flips of both spins. A proposal is accepted only if it leads to a positive energy gain, i.e., when the algorithm proposes a transition from a misaligned to an aligned state.

If rejections are ignored, this is equivalent to setting $\beta = 0$, corresponding to an infinite temperature. In this case, the stationary distribution becomes uniform, leading to a paramagnetic phase. □

The choice of the MH proposal distribution significantly affects the mixing time of the resulting algorithm. Ideally, we seek a proposal that promotes rapid mixing. Although evaluating the mixing time analytically is challenging, it can be estimated empirically with heuristics. If the largest distance between states (measured by the number of elementary steps in the algorithm) is L, then the MH algorithm, performing a random walk, covers this distance in approximately $T \sim L^2$ steps. This provides a lower bound on the mixing time. However, the actual mixing time – i.e., the time required to reach a sample that is nearly independent of the initial sample – may be considerably slower, especially if rejections occur frequently. This slow, diffusive exploration of the state space by the MH algorithm is directly linked to DB.

The specific form of the MH proposal used in Example 10.1.2 for the two-spin Ising model generalizes to the so-called Glauber dynamics, which we illustrate in Algorithm 9. (Refer to MCMC.ipynb available on the author's living-book website https://sites.google.com/site/mchertkov/living-books/applied-math-book for a snippet of the Glauber algorithm on a 128×128 square lattice for more insight.)

Algorithm 9 Glauber Sampling

Input: Ising model on a graph (refer to Eq. (10.1)). Initialize with a sample x.

```
loopUntil convergence
    Randomly select a node i.
    Set x_i ← −x_i (flip spin).
    Compute α = exp( x_i ( Σ_{j∈V:{i,j}∈E} J_ij x_j − 2h_i ) ).
    Draw β ~ U([0,1]), a uniform i.i.d. random variable from [0,1].
    if α < β < 1 then
        Set x_i ← −x_i (reject the proposed flip).
    end if
    Output: x as a sample.
end loop
```

Exercise 10.2 (Spanning Trees). Let G be an undirected complete graph. The following MCMC algorithm yields a uniform stationary distribution over all spanning trees of G: Start with an initial spanning tree; add a random edge from G (forming a cycle); remove a random edge from this cycle; repeat.

Suppose now that G is a positively weighted graph, where each edge e has a cost $c_e > 0$. Design an MCMC algorithm that samples from the set of spanning trees of G, with a stationary distribution proportional to the total weight of the spanning tree under the following conditions: (i) The weight of any spanning tree of G is the sum of its edge costs. (ii) The weight of any spanning tree of G is the product of its edge costs.

Additionally: (iii) Estimate the average weight of a spanning tree using uniform sampling. (iv) Implement these algorithms on a (4×4)

square lattice with randomly assigned weights. Verify that the algorithm converges to the correct distribution.

For additional resources on sampling and computations for the Ising model, see https://www.physik.uni-leipzig.de/~janke/Paper/lnp739_079_2008.pdf.

Exactness and convergence

An MCMC algorithm is said to be "exact" if it can be demonstrated that the generated samples converge to the desired stationary distribution. However, "convergence" can be understood in different ways, depending on the strength of the result.

The strongest form of convergence, which we refer to as the **exact independence test** (note that this is a custom term), states that at each step, the algorithm generates an independent sample from the target distribution. To prove this form of convergence, one would need to show that, in the limit of an infinite number of samples, the empirical correlation between consecutive samples vanishes. Specifically, for arbitrary functions $f(x)$ and $g(x)$, we require

$$\lim_{N\to\infty} \frac{1}{N} \sum_{n=0}^{N} f(x_n) g(x_{n-1}) \to \mathbb{E}[f(x)]\mathbb{E}[g(x)], \tag{10.4}$$

where the expectations on the right-hand side are well-defined.

A weaker form of convergence, which we call **asymptotic convergence**, implies that as $N \to \infty$, the algorithm approximates the target distribution and can reproduce its moments. Formally, for any well-behaved function $f(x)$,

$$\lim_{N\to\infty} \frac{1}{N} \sum_{n=0}^{N} f(x_n) \to \mathbb{E}[f(x)], \tag{10.5}$$

where the expectation on the right-hand side is defined with respect to the target distribution.

The weakest form of convergence, which we term **parametric convergence**, applies when the target estimate is only reached in a specific limit with respect to a certain parameter. This form of convergence is common in statistical physics and computer science, where one studies the so-called thermodynamic limit, in which the number of degrees of freedom (e.g., the number of spins in the Ising model) tends to infinity. In this case, we require

$$\lim_{s \to s_*} \lim_{N \to \infty} \frac{1}{N} \sum_{n=0}^{N} f_s(x_n) \to \mathbb{E}[f_{s_*}(x)]. \tag{10.6}$$

For further reading on MCMC (and MC methods in general) and their convergence properties, we recommend Persi Diaconis' article "The Mathematics of Mixing Things Up" and also Ref. [40] for more details.

Exact Monte Carlo sampling (Has it converged yet?)

This discussion follows Chapter 32 of MacKay's book [43], and for modern references, discussions and codes, see the website [57] on perfectly random sampling with MCs.

One major challenge with MCMC methods is determining how long one needs to run the MC before the generated samples approximate independent samples from the target distribution. If the chain is stopped too early, the resulting empirical distribution (e.g., a histogram) will deviate from the true target distribution. A key question here is, how long should the MC be run to guarantee convergence? This is a difficult problem, and in many cases, it is not possible to provide a rigorous answer.

However, there is a clever method called the Propp–Wilson exact sampling algorithm, also known as **coupling from the past**, which allows us to test for *exact convergence* on the fly for certain cases. The Propp–Wilson method is based on three main ideas:

- The first key idea involves the concept of **trajectory coalescence**. If we run MCMC chains starting from different initial conditions but ensure that they share the same random number generator, their trajectories in phase space can eventually coalesce (i.e., they meet at the same point). Once coalescence occurs, the chains will remain coalesced, indicating that the initial conditions have been "forgotten."

  However, simply running all possible initial conditions forward in time until coalescence occurs does not provide an exact sample because the coalescence point might still be biased by the initial conditions.

- The second idea is to **sample from a time T_0 in the past**. Instead of starting at the present, we run the simulation backward from time T_0 in the past, gradually approaching the present. If coalescence occurs during this process, the sample at the present time is an unbiased, exact sample. If coalescence does not occur, we move further into the past, reusing the same random numbers, and repeat the process until coalescence is achieved before the present time. The resulting sample is then exact.
- The final idea addresses the challenge of testing for all possible initial conditions, which are often too numerous to track explicitly. For a certain class of models – the so-called **attractive models** – it is possible to **reduce the number of necessary trials**. These are often referred to as **ferromagnetic** models in physics, where the preferred configuration is one in which the variables are aligned (i.e., both have the same value). In such models, monotonicity (submodularity) ensures that paths do not cross. Therefore, it suffices to track the limiting trajectories, which allows us to deduce the behavior of all other trajectories from these extremal cases.

10.2 Statistical Inference: General Relations, Calculus of Variations and Trees

This section follows material from the author's living book, *INFERLO: Inference, Learning & Optimization with Graphical Models*, available at https://sites.google.com/site/mchertkov/research/living-books.

10.2.1 *From the Ising model to (factor) graphical models*

Let's briefly revisit what we've learned so far about the Ising model. It is fully described by Eqs. (10.1) and (10.2). The weight of a "spin" configuration is given by Eq. (10.1). For now, let's disregard the normalization factor Z and observe that the weight naturally factorizes. Specifically, it is a product of pairwise terms, each describing an "interaction" between spins. This factorization can be represented through a graph, where spins correspond to graph nodes, and the interactions (or pairwise factors) are represented as edges.

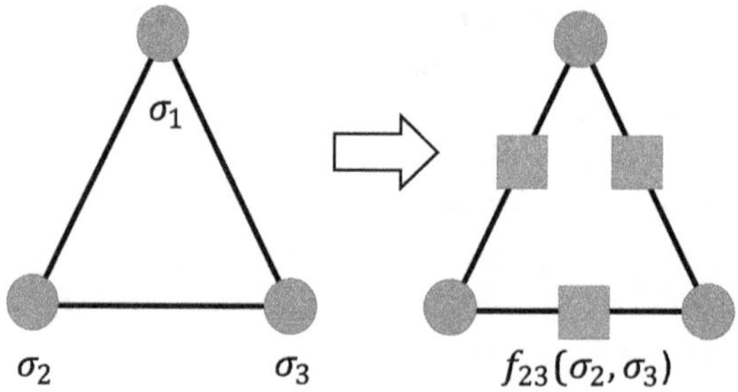

Figure 10.2. Factor graph representation for the simple case of pairwise factors. In the case of the Ising model, $f_{12}(x_1, x_2) = \exp(-J_{12}x_1x_2 + h_1x_1 + h_2x_2)$.

For example, if we consider a system with three spins all interacting with each other, the corresponding graph would be a triangle. The spins are the graph's nodes, and the pairwise interactions between them are the edges.

To handle such factorized problems more generally, we introduce a more flexible representation: factor graphs, where both factors and variables are associated with nodes. The transformation into this factor-graph representation for the three-spin example is shown in Fig. 10.2.

The Ising model, as well as other models discussed later in these lectures, can be formulated in terms of the general factor-graph framework:

$$P(x) = Z^{-1} \prod_{a \in \mathcal{V}_f} f_a(x_a), \quad x_a := (x_i \mid i \in \mathcal{V}_n,\ (i, a) \in \mathcal{E}), \tag{10.7}$$

where $(\mathcal{V}_f, \mathcal{V}_n, \mathcal{E})$ is a bipartite graph consisting of factors and variables.

This factor-graph language is more general and versatile. In the following section, we explore this in the context of a problem from information theory, specifically the decoding of graphical codes.

10.2.2 *Decoding of graphical codes as a factor-graph problem*

We now discuss the decoding of a graphical code. Our discussion here will be brief, but for more details, we recommend the book by Richardson and Urbanke [58].

In this context, a message consisting of L information bits is encoded into an N-bit codeword, where $N > L$. In the case of binary, linear coding (which we focus on here), the code is represented by $M \geq N - L$ parity-check constraints, or simply, checks. Formally, a bit configuration $\boldsymbol{\varsigma} = (\varsigma_i = 0, 1 \mid i = 1, \ldots, N)$ is a valid codeword if and only if it satisfies all the parity-check equations:

$$\sum_{i \sim \alpha} \varsigma_i = 0 \pmod 2 \quad \forall \alpha = 1, \ldots, M,$$

where $i \sim \alpha$ indicates that bit i contributes to check α. The relation between bits and checks is typically represented by the $M \times N$ parity-check matrix $\boldsymbol{H}$, where $H_{i\alpha} = 1$ if $i \sim \alpha$, and $H_{i\alpha} = 0$ otherwise. The set of valid codewords is therefore

$$\Xi^{(cw)} = \{\boldsymbol{\varsigma} \mid \boldsymbol{H}\boldsymbol{\varsigma} = \mathbf{0} \pmod 2\}.$$

The bipartite graph representation of $\boldsymbol{H}$, where bits are marked as circles, checks as squares and edges represent nonzero elements of $\boldsymbol{H}$, is called the Tanner graph of the code.

For example, the parity-check matrix of a code with $N = 10$ bits and $M = 5$ checks, corresponding to the Tanner graph in Fig. 10.3, is

$$\boldsymbol{H} = \begin{pmatrix} 1 & 1 & 1 & 1 & 0 & 1 & 1 & 0 & 0 & 0 \\ 0 & 0 & 1 & 1 & 1 & 1 & 1 & 1 & 0 & 0 \\ 0 & 1 & 0 & 1 & 0 & 1 & 0 & 1 & 1 & 1 \\ 1 & 0 & 1 & 0 & 1 & 0 & 0 & 1 & 1 & 1 \\ 1 & 1 & 0 & 0 & 1 & 0 & 1 & 0 & 1 & 1 \end{pmatrix}. \tag{10.8}$$

Assume that during transmission, each bit of the codeword is corrupted independently with some known probability, $p(x|\sigma)$, where $\sigma = 0, 1$ is the bit's original value and $x \in \mathbb{R}$ is the corrupted value. After receiving the corrupted vector $\boldsymbol{x} = (x_i \mid i = 1, \ldots, N)$, the task of maximum-a-posteriori (MAP) decoding is to reconstruct the most probable codeword that could have resulted in the observed vector:

$$\boldsymbol{\sigma}^{(\text{MAP})} = \arg \min_{\boldsymbol{\sigma} \in \Xi^{(cw)}} \prod_{i=1}^{N} p(x_i|\sigma_i). \tag{10.9}$$

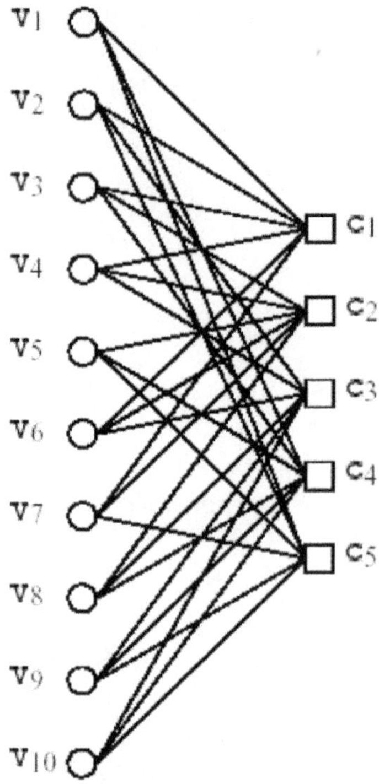

Figure 10.3. Tanner graph of a linear code with $N = 10$ bits, $M = 5$ checks and $L = N - M = 5$ information bits. This code selects 2^5 codewords from 2^{10} possible patterns. The adjacency (parity-check) matrix of the code is given by Eq. (10.8).

More generally, the probability of a codeword $\varsigma \in \Xi^{(cw)}$ being the pre-image of $\boldsymbol{x}$ is given by

$$\mathcal{P}(\varsigma|\boldsymbol{x}) = \frac{1}{Z(\boldsymbol{x})} \prod_{i \in \mathcal{G}_{0;v}} g^{(ch)}(x_i|\varsigma_i), \quad Z(\boldsymbol{x}) = \sum_{\varsigma \in \Xi^{(cw)}} \prod_{i \in \mathcal{G}_{0;v}} g^{(ch)}(x_i|\varsigma_i),$$

where $Z(\boldsymbol{x})$ is the partition function, dependent on the observed vector $\boldsymbol{x}$.

One may also consider the bit-wise MAP decoder, which estimates each bit individually:

$$\forall i: \quad \varsigma_i^{(\text{s-MAP})} = \arg\max_{\varsigma_i} \sum_{\varsigma \setminus \varsigma_i}^{\varsigma \in \Xi^{(cw)}} \mathcal{P}(\varsigma|\boldsymbol{x}).$$

10.2.3 *Partition function, marginal probabilities and maximum likelihood*

The partition function in Eq. (10.7) serves as the normalization factor:

$$Z = \sum_{x} \prod_{a \in \mathcal{V}_f} f_a(x_a), \quad x_a := (x_i \mid i \in \mathcal{V}_n, \quad (i,a) \in \mathcal{E}),$$

where $x = (x_i \in \{0,1\} \mid i \in \mathcal{V}_n)$. Here, we assume that the alphabet of each elementary random variable is binary, though the generalization to higher-alphabet cases is straightforward.

One common task is to "marginalize" Eq. (10.7) over a subset of variables, such as all variables except one:

$$P(x_i) := \sum_{x \setminus x_i} P(x). \tag{10.10}$$

The expected value of x_i, computed using the probability from Eq. (10.10), is referred to as the "magnetization" of the variable in the context of physics.

Example 10.2.1. Is a partition function oracle sufficient for computing $P(x_i)$? What is the relationship between $P(x_i)$ and $Z(h)$ in the Ising model?

Solution. There are two ways to relate $P(x_i)$ to the partition function. First, introduce an auxiliary GM derived from the original by fixing the value at node i to x_i. Then, $P(x_i)$ is simply the ratio of the partition function of this new GM to that of the original. Alternatively, modify the original model by introducing a multiplicative factor $\exp(x_i h_i)$ and denote the resulting partition function as $Z(h)$. The logarithm of $Z(h)$ then becomes the moment-generating function for $P(x_i)$. □

Another key object of interest is the ML estimate. Formally, this is the most probable state among all those represented in Eq. (10.7):

$$x_* = \arg\max_{x} P(x).$$

Computing the partition function, marginal probabilities or the ML estimate can be challenging because the number of operations required typically grows exponentially with system size, such as the number of variables or spins in the Ising model. However, for some special cases or classes of problems, these computations may be significantly easier. For example, for the ferromagnetic (or submodular) Ising model, computing the ML estimate can be done in polynomial time, though computing the partition function remains exponential.

An interesting case occurs when the Ising model is ferromagnetic, anti-ferromagnetic or glassy with zero magnetic field ($h = 0$) and is defined on a planar graph. In such cases, even computing the partition function becomes efficient, as it can be expressed via the determinant of a matrix. Calculating the determinant of a matrix of size N typically has $O(N^3)$ complexity, though for planar graphs, this reduces to $O(N^{3/2})$.

In general, however, we often rely on approximations to make such computations scalable. Before diving into these approximations, we can reformulate the problem of computing the partition function as an optimization problem, which leads us to the concept of the Kullback–Leibler (KL) divergence.

10.2.4 *Kullback–Leibler divergence and the probability polytope*

To transition from counting (i.e., computing the partition function) to optimization, we change our description from states to probabilities over the states, which we also refer to as beliefs. Let $b(x)$ represent our belief or probabilistic guess about the probability of state x. For instance, in the triangle system in Fig. (10.2), there are 2^3 states, corresponding to configurations $(x_1 = \pm 1, x_2 = \pm 1, x_3 = \pm 1)$, with each state occurring with a probability of $b(x_1, x_2, x_3)$. These beliefs must be non-negative and sum to one.

To compare a particular belief assignment with the true distribution $P(x)$ (as defined by Eq. (10.7)), we use the KL divergence:

$$D(b\|P) = \sum_x b(x) \log\left(\frac{b(x)}{P(x)}\right).$$

The KL divergence is a convex function of the beliefs, constrained within the probability polytope, defined by the following conditions:

$$\forall x: \quad b(x) \geq 0,$$

$$\sum_x b(x) = 1.$$

The unique minimum of $D(b\|P)$ is achieved when $b = P$, where the KL divergence is zero:

$$P = \arg\min_b D(b\|P), \quad \min_b D(b\|P) = 0. \tag{10.11}$$

Substituting Eq. (10.7) into Eq. (10.11) gives

$$\log Z = -\min_b \mathcal{F}(b), \quad \mathcal{F}(b) := \sum_x b(x) \log \left(\frac{\prod_a f_a(x_a)}{b(x)} \right), \tag{10.12}$$

where $\mathcal{F}(b)$, known as the (configurational) free energy, is a function of the beliefs. This concept originates from statistical physics.

Thus, we have successfully reformulated the problem of computing the partition function as an optimization problem. While this is promising, the number of variational degrees of freedom (the beliefs) is still as large as the original sum. The usefulness of this reformulation will become clear when we explore approximations.

10.2.5 *Variational approximation: Mean field approach*

The main idea behind variational approximations is to reduce the search space, approximating the high-dimensional belief space with a lower-dimensional proxy. A natural assumption is that all variables are independent:

$$b(x) \to b_{MF}(x) = \prod_i b_i(x_i), \tag{10.13}$$

$$\forall i \in \mathcal{V}_i, \ \ \forall x_i : \quad b_i(x_i) \geq 0, \tag{10.14}$$

$$\forall i \in \mathcal{V}_i : \quad \sum_{x_i} b_i(x_i) = 1. \tag{10.15}$$

Here, $b_i(x_i)$ is interpreted as the single-node marginal belief, or an estimate of the single-node marginal probability.

Substituting b by b_{MF} in Eq. (10.12) gives the mean field (MF) approximation for the partition function:

$$\log Z_{MF} = -\min_{b_{MF}} \mathcal{F}(b_{MF}), \tag{10.16}$$

$$\mathcal{F}(b_{MF}) := \sum_a \sum_{x_a} \left(\prod_{i \sim a} b_i(x_i) \right) \log f_a(x_a) - \sum_i \sum_{x_i} b_i(x_i) \log b_i(x_i).$$

To solve the variational problem in Eq. (10.16) under the constraints of Eqs. (10.13), (10.14) and (10.15), we search for the stationary point of the following MF Lagrangian:

$$\mathcal{L}(b_{MF}) := \mathcal{F}(b_{MF}) + \sum_i \lambda_i \sum_{x_i} b_i(x_i).$$

Example 10.2.2. Show that $Z \geq Z_{MF}$ and that $\mathcal{F}(b_{MF})$ is a strictly convex function of its argument. Write down the equations defining the stationary point of $\mathcal{L}(b_{MF})$.

Solution. The inequality $Z \geq Z_{MF}$ holds because we optimize over a restricted class of belief functions, $b_{MF}(x)$, which is strictly within the class of all valid beliefs $b(x)$. Convexity follows from the fact that the objective function in the optimization is a sum of convex functions. □

The fact that Z_{MF} provides a lower bound on Z is promising. However, the approximation can be quite crude, as it ignores significant correlations between variables. The following lecture will focus on a more accurate method – belief propagation – that often provides much better approximations for ML inference and partition function estimation.

In addition to discussing inference using belief propagation, we will also touch on the related inverse problem: learning in GMs.

10.2.6 *Dynamic programming for (exact) inference over trees*

Consider the Ising model over a linear chain of n spins, as shown in Fig. 10.4(a). The partition function for this model is given by

$$Z = \sum_{x_n} Z(x_n),$$

where $Z(x_n)$ represents the sum over all spin configurations except the last spin in the chain, labeled n.

The recursive nature of the partition function $Z(x_n)$ can be expressed as

$$Z(x_n) = \sum_{x_{n-1}} \exp(J_{n,n-1} x_n x_{n-1} + h_n x_n) Z_{(n-1)\to(n)}(x_{n-1}),$$

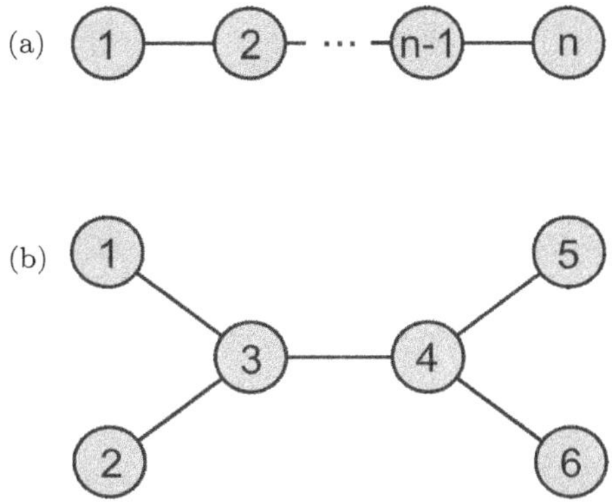

Figure 10.4. Examples of interaction/factor graphs that form a tree structure.

where $Z_{(n-1)\to(n)}(x_{n-1})$ is the partial partition function for the sub-chain rooted at $n-1$, excluding the branch toward n. This partial sum reduces the problem size by summing over one less spin compared to the full chain. In fact, the partial sum can be defined recursively as

$$Z_{(i-1)\to(i)}(x_{i-1}) = \sum_{x_{i-2}} \exp(J_{i-1,i-2}x_{i-1}x_{i-2} + h_{i-1}x_{i-1}) \times Z_{(i-2)\to(i-1)}(x_{i-2}),$$

where each partially summed object is computed based on the result of the previous step. The advantage of this recursive approach is clear: It allows the replacement of summation over exponentially many spin configurations with a much simpler summation over just two terms at each step of the recursion.

This method is an adaptation of the dynamic programming (DP) techniques, previously discussed in optimization, applied here to the problem of statistical inference.

The approach generalizes naturally from linear chains to more complex tree structures. For a general tree, $Z(x_i)$ represents the partition function of the entire tree with the spin at node i fixed. The partition function for such a tree is

$$Z(x_i) = e^{h_i x_i} \prod_{j\in\partial i} \left(\sum_{x_j} e^{J_{ij}x_i x_j} Z_{j\to i}(x_j) \right),$$

where ∂i denotes the set of neighbors of spin i, and

$$Z_{j\to i}(x_j) = e^{h_j x_j} \prod_{k\in\partial j\setminus i} \left(\sum_{x_k} e^{J_{kj}x_k x_j} Z_{k\to j}(x_k) \right)$$

is the partition function of the subtree rooted at node j.

To illustrate this general scheme, consider the tree shown in Fig. 10.4(b). The full partition function can be written as

$$Z = \sum_{x_4} Z(x_4),$$

where $Z(x_4)$, the partial partition function conditioned on the value of spin x_4, is

$$Z(x_4) = e^{h_4 x_4} \sum_{x_5} e^{J_{45}x_4 x_5} Z_{5\to 4}(x_5) \sum_{x_6} e^{J_{46}x_4 x_6} Z_{6\to 4}(x_6) \times \sum_{x_3} e^{J_{34}x_3 x_4} Z_{3\to 4}(x_3),$$

where the partial partition function for subtree rooted at node 3 is

$$Z_{3\to 4}(x_3) = e^{h_3 x_3} \sum_{x_1} e^{J_{13}x_1 x_3} Z_{1\to 3}(x_1) \sum_{x_2} e^{J_{23}x_2 x_3} Z_{2\to 3}(x_2).$$

Exercise 10.3. Consider the Ising model on a graph $\mathcal{G} = (\mathcal{V}, \mathcal{E})$, with spins $\boldsymbol{x}$. Let $\mathcal{V}_0 \subset \mathcal{V}$, and define the boundary set $\bar{\mathcal{V}}_0 = \{i \in \mathcal{V} \setminus \mathcal{V}_0 : (i,j) \in \mathcal{E}$ for some $j \in \mathcal{V}_0\}$. Prove that the spins on $\mathcal{V}_0$ are conditionally independent of all other spins, given the values of spins on $\bar{\mathcal{V}}_0$.

10.2.7 *Properties of tree-structured graphical models*

For pairwise GMs over trees, the joint distribution can be expressed solely via single-node marginals and pairwise marginals between neighboring nodes. Let's explore this factorization property through a few examples, as shown in Fig. 10.5.

In the simple two-node example in Fig. 10.5(a), the factorization follows directly from Bayes' rule:

$$P(x_1, x_2) = P(x_1)P(x_2|x_1),$$

or, equivalently, $P(x_1, x_2) = P(x_2)P(x_1|x_2)$.

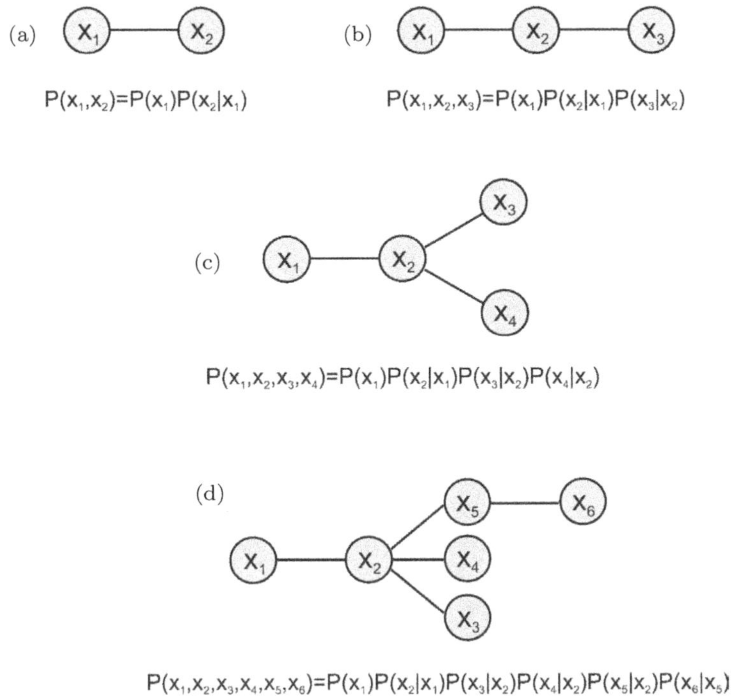

Figure 10.5. Examples of undirected tree – structured graphical models.

For the three-node model in Fig. 10.5(b), we have

$$P(x_1, x_2, x_3) = P(x_1, x_2)P(x_3|x_1, x_2) = P(x_1, x_2)P(x_3|x_2),$$

where the conditional independence of x_3 from x_1 given x_2, i.e., $P(x_3|x_1, x_2) = P(x_3|x_2)$, is used. Rearranging this, we get

$$P(x_1, x_2, x_3) = \frac{P(x_1, x_2)P(x_2, x_3)}{P(x_2)}.$$

Extending this logic to the four-node tree in Fig. 10.5(c), we find

$$P(x_1, x_2, x_3, x_4) = P(x_1, x_2)P(x_3|x_2)P(x_4|x_2),$$

which can also be written as

$$P(x_1, x_2, x_3, x_4) = \frac{P(x_1, x_2)P(x_2, x_3)P(x_2, x_4)}{P(x_2)^2}.$$

Finally, for the six-node example in Fig. 10.5(d), the joint probability distribution is

$$P(x_1, x_2, x_3, x_4, x_5, x_6) = \frac{P(x_1, x_2)P(x_2, x_3)P(x_2, x_4)P(x_2, x_5)P(x_5, x_6)}{P(x_2)^3 P(x_5)}.$$

Exercise 10.4. In the case of a general tree – structured GM, the joint probability distribution can be expressed in terms of pairwise and singleton marginals as follows:

$$P(x_1, x_2, \ldots, x_n) = \frac{\prod_{(i,j)\in\mathcal{E}} P(x_i, x_j)}{\prod_{i\in\mathcal{V}} P^{q_i-1}(x_i)}, \tag{10.17}$$

where q_i is the degree of node i. Prove this expression using mathematical induction.

10.2.8 *Bethe free energy and belief propagation*

As discussed earlier, DP is an exact method for inference on tree-structured graphs. For more complex graphs that contain loops, it still provides a good approximation. This approximation is commonly known as *Bethe–Peierls* or *belief propagation* (BP). When applied to loopy graphs, this method is often referred to as *loopy BP*. For foundational reading, see the original paper by Yedidia *et al.* [59], a comprehensive review by Wainwright and Jordan [55] and their detailed lecture notes.

In contrast to the MF approach (Eq. (10.13)), BP utilizes a different factorization. Specifically, BP assumes the following form:

$$b(x) \to b_{\mathrm{BP}}(x) = \frac{\prod_a b_a(x_a)}{\prod_i (b_i(x_i))^{q_i-1}}, \tag{10.18}$$

where $b_a(x_a)$ denotes beliefs associated with factor nodes and $b_i(x_i)$ represents beliefs for individual variables. The term q_i stands for the degree (or number of neighbors) of node i, and the exponent $q_i - 1$ corrects for over-counting contributions from variable nodes that appear in multiple factors.

The BP beliefs are subject to the following conditions:

$$\forall a \in \mathcal{V}_f, \quad \forall x_a : \; b_a(x_a) \geq 0, \tag{10.19}$$

$$\forall i \in \mathcal{V}_n, \quad \forall a \sim i : \; b_i(x_i) = \sum_{x_a \backslash x_i} b_a(x_a), \tag{10.20}$$

$$\forall i \in \mathcal{V}_n : \quad \sum_{x_i} b_i(x_i) = 1. \tag{10.21}$$

Substituting Eq. (10.18) into the KL divergence minimization framework leads to the *Bethe free energy* (BFE):

$$\mathcal{F}_{\text{BP}} := E_{\text{BP}} - \mathcal{H}_{\text{BP}}, \tag{10.22}$$

$$E_{\text{BP}} := -\sum_a \sum_{x_a} b_a(x_a) \log f_a(x_a), \tag{10.23}$$

$$\mathcal{H}_{\text{BP}} = \sum_a \sum_{x_a} b_a(x_a) \log b_a(x_a) - \sum_i \sum_{x_i} (q_i - 1) b_i(x_i) \log b_i(x_i), \tag{10.24}$$

where E_{BP} is the *self-energy*, a term borrowed from statistical physics, and $\mathcal{H}_{\text{BP}}$ represents the *BP entropy*, capturing the uncertainty in the beliefs. The task becomes that of minimizing $\mathcal{F}_{\text{BP}}$ over the beliefs, subject to the constraints in Eqs. (10.19)–(10.21).

Thus, the optimization problem for BP can be formulated as

$$\arg\min_{b_a, b_i} \mathcal{F}_{\text{BP}} \quad \text{subject to Eqs. (10.19), (10.20) and (10.21).} \tag{10.25}$$

Is $\mathcal{F}_{\text{BP}}$ a convex function of its arguments? The answer is not straightforward: While convexity holds in some cases, it is not guaranteed universally. The convexity of $\mathcal{F}_{\text{BP}}$ depends on the graph structure and the specific form of the factor functions.

The *maximum likelihood* or *zero-temperature* version of the problem simplifies to

$$\min_{b_a, b_i} E_{\text{BP}} \quad \text{subject to Eqs. (10.19), (10.20) and (10.21).} \tag{10.26}$$

This leads to a *linear programming* (LP) problem where we minimize a linear objective under linear constraints.

Belief propagation and message passing

To turn Eq. (10.25) into an unconditional optimization, we introduce *Lagrange multipliers*. The resulting *Lagrangian* is

$$\mathcal{L}_{\mathrm{BP}}(b,\eta,\lambda) := \sum_a \sum_{x_a} b_a(x_a)\log f_a(x_a) - \sum_a \sum_{x_a} b_a(x_a)\log b_a(x_a)$$
$$+ \sum_i \sum_{x_i} (q_i - 1) b_i(x_i)\log b_i(x_i)$$
$$- \sum_i \sum_{a\sim i} \sum_{x_i} \eta_{ia}(x_i)\left(b_i(x_i) - \sum_{x_a\backslash x_i} b_a(x_a)\right) + \sum_i \lambda_i\left(\sum_{x_i} b_i(x_i) - 1\right),$$

where η and λ are the Lagrange multipliers associated with Eqs. (10.20) and (10.21), respectively.

This transforms Eq. (10.25) into a **min-max** problem:

$$\min_b \max_{\eta,\lambda} \mathcal{L}_{\mathrm{BP}}(b,\eta,\lambda).$$

By minimizing over η, we derive the beliefs in terms of *messages*:

$$\forall a,\ \forall x_a: \quad b_a(x_a) \sim f_a(x_a) \prod_{i\sim a} n_{i\to a}(x_i) \sim f_a(x_a) \prod_{i\sim a} \prod_{b\sim i}^{b\neq a} m_{b\to i}(x_i),$$

$$\forall i,\ \forall x_i: \quad b_i(x_i) \sim \prod_{a\sim i} m_{a\to i}(x_i),$$

where $\sim$ denotes equality up to a normalization constant and m and n are **message variables**, related to the Lagrange multipliers η as

$$n_{i\to a}(x_i) := \exp(\eta_{ia}(x_i)), \quad m_{a\to i}(x_i) := \exp\left(\frac{\eta_{ia}(x_i)}{q_i - 1}\right).$$

Substituting these into Eq. (10.20) gives the **BP message-passing equations**:

$$\forall i,\ \forall a\sim i,\ \forall x_i: \quad n_{i\to a}(x_i) = \prod_{b\sim i}^{b\neq a} m_{b\to i}(x_i),$$

$$\forall a,\ \forall i\sim a,\ \forall x_i: \quad m_{a\to i}(x_i) = \sum_{x_a\backslash x_i} f_a(x_a) \prod_{j\sim a}^{j\neq i} n_{j\to a}(x_j).$$

If the BFE $\mathcal{F}_{\text{BP}}$ is non-convex, multiple fixed points may arise. To find a solution, we use the *message passing* (MP) algorithm, outlined as follows.

Algorithm 10 Message Passing, Sum-Product Algorithm (Factor Graph Representation)

Input: The graph, the factors.

Initialize messages: $\forall i, \forall a \sim i, \forall x_i : m_{a\to i} = 1,\ n_{i\to a} = 1$

loop Until convergence within an error tolerance (or fixed number of iterations)

  Update $\forall i, \forall a \sim i, \forall x_i : \quad n_{i\to a}(x_i) \leftarrow \prod_{b\sim i}^{b\neq a} m_{b\to i}(x_i)$

  Update

  $\forall a, \forall i \sim a, \forall x_i : \quad m_{a\to i}(x_i) \leftarrow \sum_{x_a\setminus x_i} f_a(x_a) \prod_{j\sim a}^{j\neq i} n_{j\to a}(x_j)$

end loop

10.3 Theory of Learning: Sufficient Statistics and Maximum Likelihood Estimation

10.3.1 *Sufficient statistics: Infinitely many samples*

So far, we have discussed the direct problem of inference in GMs. In the remainder of this lecture, we briefly explore inverse problems, which will also be revisited in subsequent lectures with specific examples involving trees.

Informally, the inverse problem deals with *learning* a GM from data or samples. Imagine a two-room setup: In the first room, a GM is known and used to generate many samples. These samples – but not the GM itself – are then passed to the second room. The task in the second room is to reconstruct the GM using only the samples.

The first question we should ask is whether it is possible, in principle, to recover the GM, even with an infinite number of samples. The powerful notion of *sufficient statistics* helps answer this question.

Consider the Ising model (which we have encountered multiple times in this course):

$$
\begin{aligned}
P(x) &= \frac{1}{Z(\theta)} \exp\left\{ \sum_{i\in V} h_i x_i + \sum_{\{i,j\}\in E} J_{ij} x_i x_j \right\} \\
&= \exp\{\theta^T \phi(x) - \log Z(\theta)\}, \qquad (10.27)
\end{aligned}
$$

where $x_i \in \{-1, 1\}$, $\theta := h \cup J = (h_i|i \in \mathcal{V}) \cup (J_{ij}|\{i,j\} \in \mathcal{E})$ and the partition function $Z(\theta)$ normalizes the probability distribution.

In fact, Eq. (10.27) represents a member of the *exponential family* of distributions. It is worth noting that any pairwise GM over binary variables can be written in the form of an Ising model.

Now, consider the collection of all first and second moments of the spin variables: $\mu^{(1)} := (\mu_i = \mathbb{E}[x_i], i \in V)$ and $\mu^{(2)} := (\mu_{ij} = \mathbb{E}[x_i x_j], \{i,j\} \in E)$. The statement of *sufficient statistics* is that to fully reconstruct θ, and therefore the GM, it is *sufficient* to know $\mu^{(1)}$ and $\mu^{(2)}$.

10.3.2 *Maximum likelihood estimation/learning of graphical models*

We can now turn the notion of sufficiency into a constructive statement through *ML estimation* over an exponential family of GMs.

First, note the following derivatives of the log-partition function with respect to θ:

$$\forall i: \quad \partial_{h_i} \log Z(\theta) = -\mu_i, \quad \forall i,j: \quad \partial_{J_{ij}} \log Z(\theta) = -\mu_{ij}.$$

This leads to the conclusion that, if we can compute the log-partition function for any θ, the task of reconstructing the correct θ becomes a convex optimization problem:

$$\theta^* = \arg\max_{\theta}\{\mu^T\theta - \log Z(\theta)\}. \tag{10.28}$$

If P is the empirical distribution of a set of independent, identically distributed (i.i.d.) samples $\{x^{(s)}, s = 1, \ldots, S\}$, then μ are the corresponding empirical moments, e.g., $\mu_{ij} = \frac{1}{S}\sum_s x_i^{(s)} x_j^{(s)}$.

General Remarks about GM Learning: The ML parameter estimation in Eq. (10.28) represents the best possible approach for learning, fundamental in machine learning tasks, and it extends beyond the Ising model to more general GMs.

Unfortunately, there are very few cases where the partition function can be computed efficiently for arbitrary θ. To make parameter estimation practical, we can employ one of the following approaches:

- Limit the model to a class where the partition function can be computed efficiently for all θ. For instance, this is the case in tree-structured models (e.g., the Chow–Liu tree) or Ising models over planar graphs. For further details, refer to Ref. [60].

- Use approximations, such as variational approximations (mean field and BP), MCMC or approximate methods such as DP.
- A recent innovative approach allows for efficient GM learning by using more information than what is implied by sufficient statistics. One researcher aptly phrased it: "sufficient statistics is not sufficient." This novel approach is beyond the scope of this course but is discussed in Ref. [61].

10.3.3 *Learning spanning trees*

Equation (10.17) suggests that, when the structure of a tree-based GM is known, the joint probability distribution can be expressed in terms of single-node and pairwise marginals. We now use this to pose and solve an inverse problem: reconstructing a tree based on correlations between multiple samples of the discrete random variables $x_1, x_2, \ldots, x_n$.

A straightforward approach is to first estimate the single-node and pairwise marginal distributions, $P(x_i)$ and $P(x_i, x_j)$, from the snapshots. Then, for each tree candidate, one could test if the relations in Eq. (10.17) hold. However, testing all possible trees would require checking exponentially many spanning trees, n^{n-2}. Thankfully, Chow and Liu proposed an efficient algorithm in 1968 to solve this problem [62].

Given a candidate distribution $P_T(x)$ over a tree $T = (\mathcal{V}, \mathcal{E})$, the tree-factorized form is

$$P_T(x_1, x_2, \ldots, x_n) = \frac{\prod_{(i,j)\in\mathcal{E}} P(x_i, x_j)}{\prod_{i\in\mathcal{V}} P(x_i)^{q_i - 1}}, \tag{10.29}$$

where q_i is the degree of node i.

The "distance" between the true (correct) joint probability distribution, P, and the candidate tree-factorized probability distribution, P_T, can be quantified using the KL divergence:

$$D(P \parallel P_T) = -\sum_x P(x) \log \frac{P(x)}{P_T(x)}. \tag{10.30}$$

As discussed in Section 8.5, the KL divergence is always non-negative: It is zero if P and P_T are identical and positive otherwise. Our goal is to find a tree structure that minimizes this KL divergence.

Substituting Eq. (10.29) into Eq. (10.30), we derive the following sequence of transformations:

$$\sum_x P(x)\left(\log P(x) - \sum_{(i,j)\in\mathcal{E}} \log P(x_i, x_j) + \sum_{i\in\mathcal{V}} (q_i - 1)\log P(x_i)\right)$$

$$= \sum_x P(x)\log P(x) - \sum_{(i,j)\in\mathcal{E}^{\mathcal{F}}} \sum_{x_i,x_j} P(x_i,x_j)\log P(x_i,x_j)$$

$$+ \sum_{i\in\mathcal{V}} (q_i - 1)\sum_{x_i} P(x_i)\log P(x_i)$$

$$= - \sum_{(i,j)\in\mathcal{E}^{\mathcal{F}}} \sum_{x_i,x_j} P(x_i,x_j)\log\frac{P(x_i,x_j)}{P(x_i)P(x_j)} + \sum_x P(x)\log P(x)$$

$$- \sum_{i\in\mathcal{V}^{\mathcal{F}}} \sum_{x_i} P(x_i)\log P(x_i),$$

where we have used the following marginalization relations for nodes and edges:

$$\forall i \in \mathcal{V}^{\mathcal{F}} :\ P(x_i) = \sum_{x\backslash x_i} P(x),\ \forall (i,j) \in \mathcal{E}^{\mathcal{F}} :\ P(x_i,x_j) = \sum_{x\backslash x_i,x_j} P(x).$$

Thus, the KL divergence becomes

$$D(P \parallel P_F) = - \sum_{(i,j)\in\mathcal{E}^{\mathcal{F}}} I(X_i,X_j) + \sum_{i\in\mathcal{V}^{\mathcal{F}}} S(X_i) - S(X),$$

where

$$I(X_i,X_j) := \sum_{x_i,x_j} P(x_i,x_j)\log\frac{P(x_i,x_j)}{P(x_i)P(x_j)}$$

represents the mutual information between the random variables X_i and X_j.

Since the entropies $S(X_i)$ and $S(X)$ are independent of the choice of tree, minimizing the KL divergence is equivalent to maximizing the following sum over the branches of a tree:

$$\sum_{(i,j)\in\mathcal{E}^{\mathcal{F}}} I(X_i,X_j).$$

Based on this, Chow and Liu proposed using Kruskal's greedy algorithm [63] to find the maximum spanning tree:

- **Step 1:** Sort the edges of the graph in decreasing order of mutual information, $I(X_i, X_j)$.
- **Step 2:** Add the first edge to the tree.
- **Step 3:** Add the next edge to the tree if it does not form a cycle.
- **Step 4:** Repeat Step 3 until the tree has $n-1$ edges.

This method reconstructs the best tree approximation even when the actual graph contains loops.

Exercise 10.5. Find the Chow–Liu optimal spanning tree approximation for the joint distribution of four random binary variables based on the data in Table 10.1.

Hint: Estimate the empirical pairwise mutual information and then apply the Chow–Liu–Kruskal algorithm.

Table 10.1. Information about an exemplary distribution of four binary variables for Exercise 10.5.

$x_1x_2x_3x_4$	$P(x_1, x_2, x_3, x_4)$	$P(x_1)P(x_2\|x_1)$ $P(x_3\|x_2)P(x_4\|x_1)$	$P(x_1)P(x_2)$ $P(x_3)P(x_4)$
0000	0.100	0.130	0.046
0001	0.100	0.104	0.046
0010	0.050	0.037	0.056
0011	0.050	0.030	0.056
0100	0.000	0.015	0.056
0101	0.000	0.012	0.056
0110	0.100	0.068	0.068
0111	0.050	0.054	0.068
1000	0.050	0.053	0.056
1001	0.100	0.064	0.056
1010	0.000	0.015	0.068
1011	0.000	0.018	0.068
1100	0.050	0.033	0.068
1101	0.050	0.040	0.068
1110	0.150	0.149	0.083
1111	0.150	0.178	0.083

10.4 Function Approximation with Neural Networks

The material presented in this section is both cutting-edge and rapidly evolving, driven by the recent breakthroughs in AI. While some foundational theoretical results date back over 30 years, their practical power, along with the surge of new approaches, has only become apparent in the past five years. As such, the field is developing rapidly, and we anticipate significant advancements in the years to come. At present, there are relatively few books focusing on the applied mathematics aspects of function approximation with neural networks (NNs). One notable exception is Gilbert Strang's book [64], *Linear Algebra and Learning from Data*, which we highly recommend. Part VII, *Learning from Data*, is particularly relevant to this section.

NNs, especially deep neural networks (DNNs), have emerged as the most powerful and versatile tool in applied mathematics for universal function approximation.

The mathematical foundation of this methodology is formalized by the following theorem.

Theorem 10.4.1 (Universal Approximation Theorem (Cybenko 1989 [65]; Hornik 1991 [66]; Pinkus 1999 [67])). *Let $\rho : \mathbb{R} \to \mathbb{R}$ be any continuous function (called the activation function). Let $\mathcal{NN}_n^\rho$ represent the class of feed-forward NNs with activation function ρ, consisting of n neurons in the input layer, one neuron in the output layer and a single hidden layer with an arbitrary number of neurons. Let $K \subset \mathbb{R}$ be compact. Then, $\mathcal{NN}_n^\rho$ is dense in $C(K)$ (the class of continuous functions on K) if and only if ρ is not a polynomial.*

This theorem was a catalyst for the revolution in NNs. Initially, it applied to networks with bounded depth and arbitrary width but was extended recently to networks with bounded width and arbitrary depth (see, in particular, Ref. [68]). Importantly, the term *deep* in deep learning refers to the large depth of these networks.

Although the universal approximation theorem is independent of the choice of the activation function, certain functions are more commonly used in practice. One activation function that has proven extremely effective is the rectified linear unit (ReLU), defined as $\text{ReLU}(x) = \max(x, 0)$.

NNs are not restricted to approximating a scalar function $\rho : \mathbb{R} \to \mathbb{R}$; they can be used to approximate vector-valued functions. For example, using ReLU, we can construct a piecewise linear function that maps a p-dimensional input vector v to an m-dimensional output:

- First, choose a (q, p) matrix A_1 and a q-dimensional vector b_1. Compute the activation by applying ReLU component-wise to $A_1 v + b_1$, resulting in $\mathrm{ReLU}(A_1 v + b_1) = (A_1 v + b_1)_+$.
- Next, apply a (m, q) matrix A_2 to the result: $A_2(A_1 v + b_1)_+$.

By introducing *depth* in an NN, we can construct more expressive piecewise linear functions, with an increasing number of linear regions. A standard method for enhancing expressivity is through composition:

$$\mathcal{NN}(v) = F_L(F_{L-1}(\cdots F_2(F_1(v)))) = (F_L \circ F_{L-1} \circ \cdots \circ F_2 \circ F_1)(v), \tag{10.31}$$

where $F_l(x) = (A_l x + b_l)_+$ for $l = 1, \ldots, L$.

While composition is the key operation in constructing DNNs, other crucial concepts include the *loss function*, the *chain rule*, *automatic differentiation* and *back-propagation*, all of which we discuss in the following sections.

Exercise 10.6 (Counting the Number of Pieces). Consider a $\mathcal{NN}(v)$ with a ReLU activation function, a one-dimensional input layer, a five-dimensional hidden layer and a one-dimensional output layer. Count the number of linear pieces in the resulting continuous piecewise linear function.

10.4.1 *Fitting a function with neural networks: An optimization approach*

A key element in fitting a function with an NN is the so-called *loss function*, a term commonly used in data science to describe the objective of the underlying optimization problem. A standard choice for the loss function is a norm, such as l_1, l_2 or l_∞, applied to the error between the target function, $F(v)$, and its NN approximation evaluated at a set of available samples, $s = 1, \ldots, S$. For example, minimizing the l_p-norm of the error leads to the following optimization problem:

$$\min_{\theta} \sum_{s=1}^{S} \|F(v_s) - \mathcal{NN}(v_s|\theta)\|_p, \tag{10.32}$$

where θ is the vector of NN parameters, for instance, $\theta = (A_L, b_L, \ldots, A_1, b_1)$ in the case of a continuous piecewise linear NN, as described in Eq. (10.31).

A standard method to solve this optimization problem is through gradient descent, which involves calculating the partial derivatives (gradient components) of the loss function with respect to the parameters and finding the point where all gradients are zero. Various forms of the gradient descent algorithm can be used to achieve this.

In order to compute the necessary derivatives of $\mathcal{NN}(v_s|\theta)$ with respect to the components of the parameter vector θ, a technique called *automatic differentiation* is employed. A widely used form of automatic differentiation in this context is *back-propagation.*

10.4.2 *Automatic differentiation, back-propagation and the chain rule*

The computational efficiency of automatic differentiation is based on the *chain rule* of calculus, which can be illustrated with the following example:

$$\frac{dg}{dx} = \frac{d}{dx}\left(g_3(g_2(g_1(x)))\right) = \left(\frac{dg_3}{dg_2}(g_2(g_1(x)))\right)\left(\frac{dg_2}{dg_1}(g_1(x))\right)\left(\frac{dg_1}{dx}\right).$$

Automatic differentiation can be carried out in two modes: *forward mode* and *backward mode.* Forward mode is preferable when there are many functions depending on a few input variables, while backward mode is more efficient when there is one function (such as the loss function) depending on many variables.

In the context of deep learning, where a single loss function depends on many weights, back-propagation is the natural choice. This backward mode of automatic differentiation propagates gradients from the output layer back to the input layer, following the structure of the network.

For example, consider a simple NN with $L = 2$ layers, where the activation function in the second layer is the identity

$$w = v_2 = b_2 + A_2 v_1 = b_2 + A_2 R(b_1 + A_1 v_0).$$

To compute the gradient with respect to A_1, we apply the chain rule:

$$\frac{\partial w}{\partial A_1} = A_2 R'(b_1 + A_1 v_0) \frac{\partial (b_1 + A_1 v_0)}{\partial A_1}.$$

Note that this computation proceeds from the output $w = v_2$ back to the input v_0, which is the essence of the back-propagation algorithm.

10.4.3 *Avoiding over-fitting*

Over-fitting occurs when a trained network performs well on the training data but generalizes poorly to new, unseen data. In contrast, if an NN performs poorly even on the training set, it is said to be under-fitted. Over-fitting typically arises when the model is overly complex and too closely follows the noise or specific features of the training data.

To prevent over-fitting, one standard recommendation is to introduce *regularization* in the loss function, which penalizes overly complex models. Another effective strategy is *early stopping*, where training is halted before the model over-fits.

Example 10.4.2. Consider an NN with $L = 2$, one hidden layer with a single neuron and tanh activation functions. The model is defined as

$$w = v_2 = \tanh(a_2 \tanh(a_1 v + b_1) + b_2),$$

with the weights initialized as $(a_1, b_1) = (1.0, 0.5)$ and $(a_2, b_2) = (-0.5, 0.3)$.

What is the gradient of the mean square error (MSE) loss function for the observation $(x, y) = (2, -0.5)$? What is the optimal MSE and the corresponding optimal parameter values?

Solution. First, evaluate the function (forward pass) with the initial parameters:

$$w = \tanh(-0.5 \tanh(1.0 \cdot 2 + 0.5) + 0.3) = -0.1909.$$

Now, define and compute the intermediary variables z_1, a_1 and z_2:

$$z_1 := a_1 v + b_1 = 2.500, \quad v_1 := \tanh(a_1 v + b_1) = 0.9866,$$
$$z_2 := a_2 v_1 + b_2 = -0.1933, \quad \hat{w} = v_2 = \tanh(a_2 v_1 + b_2) = -0.1909,$$

where $\hat{w}$ is the NN's predicted output and w is the actual sample value.

The MSE loss function is $\mathcal{L} = (\hat{w} - w)^2$. The gradient of the loss function with respect to the parameters is

$$\nabla \mathcal{L} = \begin{pmatrix} \frac{\partial \mathcal{L}}{\partial a_2} \\ \frac{\partial \mathcal{L}}{\partial b_2} \\ \frac{\partial \mathcal{L}}{\partial a_1} \\ \frac{\partial \mathcal{L}}{\partial b_1} \end{pmatrix} = \begin{pmatrix} \frac{\partial L}{\partial \hat{w}} \frac{\partial \hat{w}}{\partial z_2} \frac{\partial z_2}{\partial a_2} \\ \frac{\partial L}{\partial \hat{w}} \frac{\partial \hat{w}}{\partial z_2} \frac{\partial z_2}{\partial b_2} \\ \frac{\partial L}{\partial \hat{w}} \frac{\partial \hat{w}}{\partial z_2} \frac{\partial z_2}{\partial a_1} \frac{\partial a_1}{\partial z_1} \frac{\partial z_1}{\partial a_1} \\ \frac{\partial L}{\partial \hat{w}} \frac{\partial \hat{w}}{\partial z_2} \frac{\partial z_2}{\partial a_1} \frac{\partial a_1}{\partial z_1} \frac{\partial z_1}{\partial b_1} \end{pmatrix}.$$

Evaluating each partial derivative,

$$\frac{\partial L}{\partial \hat{w}} = 2(\hat{w} - w) = 2(-0.1909 + 0.5) = 0.6182,$$
$$\frac{\partial \hat{w}}{\partial z_2} = 1 - (-0.1933)^2 = 0.9626,$$
$$\frac{\partial z_2}{\partial a_2} = v_1 = 0.9866,$$
$$\frac{\partial z_2}{\partial b_2} = 1,$$
$$\frac{\partial z_2}{\partial a_1} = a_2 = -0.5,$$
$$\frac{\partial a_1}{\partial z_1} = 1 - (0.9866)^2 = 0.0266,$$
$$\frac{\partial z_1}{\partial a_1} = v = 2.0,$$
$$\frac{\partial z_1}{\partial b_1} = 1.$$

Putting this together, we obtain

$$\nabla\mathcal{L} = \begin{pmatrix} 0.5951 \\ 0.5871 \\ -0.0079 \\ -0.0158 \end{pmatrix}.$$

□

Exercise 10.7. Consider the following two-layer ($L = 2$) NN mapping $v \in \mathbb{R}$ to $w \in \mathbb{R}$, built using three ReLU neurons:

$$v_{1i} = \text{ReLU}(a_i v + b_i), \quad \forall i = 1, 2, \quad \boldsymbol{v}_1 = (v_{11}, v_{12}) \in \mathbb{R}^2,$$
$$w = v_2 = \text{ReLU}(\boldsymbol{A}_3 \cdot \boldsymbol{v}_1^T + b_3),$$

where $\boldsymbol{A}_3 \in \mathbb{R}^{1\times 2}$, and $a_1, a_2, b_1, b_2, b_3 \in \mathbb{R}$ are the parameters of the network.

(a) Describe the complexity of the class of functions represented by this NN.
(b) What is the minimal number P of non-degenerate samples $(v^{(p)}, w^{(p)})$, $p = 1, \ldots, P$, needed for an exact (!) reconstruction of the network's parameters?
(c) Construct an example where this NN outputs a continuous piecewise linear function with exactly two linear segments.

10.5 Reinforcement Learning

In this section, we return to Markov decision processes (MDPs), as introduced in Section 9.5, but now with a learning perspective. This perspective, called **reinforcement learning** (RL), is grounded in the concept of learning optimal behavior from interaction with an environment. To explain RL, we revisit the *gedanken* (thought) two-room experiment mentioned earlier in Section 10.3.1 in the context of graphical model learning.

Assume that in room #1, an MDP of the type described in Section 9.5, is used to generate a sequence of state–action–reward tuples: $(s_0, a_0, r_0), (s_1, a_1, r_1), \ldots$. The data stream is passed to room #2, but the underlying MDP model is not revealed (!!). In room #2, the RL task is to learn or estimate the MDP purely from the data stream.

RL is a rich field, covered extensively in various books and lecture notes. Our discussion will be concise, focusing on core applied mathematics concepts in RL. For a comprehensive study, we recommend the seminal work by Sutton and Barto, *Reinforcement Learning: An Introduction* [50]. Additionally, quality online courses, such as Berkeley's 2021 [69], Stanford's 2022 [70] and Columbia's 2019 [71], offer excellent supplementary resources.

The task of learning an MDP from data could involve one or more of the following objectives[a]:

- Estimate $\hat{\mathcal{P}}(s,a,s')$ (transition probability) and $\hat{r}(s,a,s')$ (reward function).
- Estimate the action-value function $\hat{Q}^{\pi}(s,a)$ under a potentially non-optimal (or random) policy.
- Estimate the optimal action-value function $\hat{Q}^{*}(s,a)$.
- Estimate the optimal state-value function $\hat{V}^{*}(s)$.
- Estimate the optimal policy $\hat{\pi}^{*}(s,a)$.

In the following sections, we explore several techniques to accomplish these RL tasks.

10.5.1 *Model-based Monte Carlo*

Given a data stream $SAR_N = (s_0, a_0, r_0, \ldots, s_N, a_N, r_N)$, we can estimate the transition probabilities and rewards of an MDP via Monte Carlo sampling. The estimators are defined as

$$\forall (s,a,s') \in SAR_N : \quad \hat{\mathcal{P}}(s,a,s') = \frac{\sum_{t=0}^{N-1} \mathbb{1}\left((s_t,a_t,s_{t+1}) = (s,a,s')\right)}{N}, \tag{10.33}$$

$$\hat{r}(s,a,s') = \frac{\sum_{t=0}^{N-1} r_{t+1}\mathbb{1}\left((s_t,a_t,s_{t+1}) = (s,a,s')\right)}{N}, \tag{10.34}$$

where $\mathbb{1}$("statement") is an indicator function equal to 1 if the statement is true, and 0 otherwise.

---

[a]We use the notations introduced in Section 9.5 and denote data-based estimates with a hat, e.g., $\hat{C}$.

These estimations can be carried out with either a fixed or random policy. However, exploring both the state space and policy space is expensive and requires a long data stream, even for relatively small MDPs. Furthermore, once $\hat{\mathcal{P}}(s,a,s')$ and $\hat{r}(s,a,s')$ are estimated, we still need to solve the corresponding MDP to find optimal policies and values.

This motivates a shift to Monte Carlo methods that directly estimate the action-value function $\hat{Q}^\pi(s,a)$, bypassing model learning altogether.

10.5.2 *Model-free Monte Carlo*

By analogy with Eqs. (10.33) and (10.34), the empirical estimate of the action-value function $\hat{Q}^\pi(s,a)$ can be written as

$$\forall(s,a) \in SAR_N: \quad \hat{Q}^\pi(s,a) = \frac{\sum_{t=0}^{N-1} u_t \mathbb{1}\left((s_t,a_t)=(s,a)\right)}{N},$$
$$u_t = r_t + \gamma r_{t+1} + \gamma^2 r_{t+2} + \cdots,$$

where u_t is the utility (or return) from time step t and actions are selected according to a prescribed policy π.

For real-time data streams (i.e., online learning), we initialize $\hat{Q}^\pi(s,a) = 0$ and update the estimates in an online fashion by interpolating between prior values and new corrections:

$$\forall t = 1, \cdots: \quad \hat{Q}^\pi(s_t,a_t) \leftarrow (1-\eta)\hat{Q}^\pi(s_t,a_t) + \eta u_t,$$

where η is a predefined learning rate.

Although this model-free approach improves on the model-based method by directly estimating values, it still suffers from requiring long sequences to accurately compute utilities. This leads to high variance and slow exploration.

10.5.3 *Temporal difference learning*

An alternative improvement to address the memory issue is to shift from (s,a,r) events to (s,a,r,s',a') events (often referred to as

SARSA), using the dynamic programming update rule for the action-value function:

$$\forall t = 1, \cdots : \quad \hat{Q}^{\pi}(s_t, a_t) \leftarrow (1-\eta)\hat{Q}^{\pi}(s_t, a_t) + \eta\left(r_t + \gamma \hat{Q}^{\pi}(s_{t+1}, a_{t+1})\right).$$

This approach is a form of *temporal difference* (TD) learning, which combines aspects of dynamic programming and Monte Carlo methods.

However, SARSA remains dependent on a predefined policy. To resolve this issue, the *Q-learning* algorithm [72] introduces an "off-policy" update rule:

$$\forall t = 1, \cdots : \quad \hat{Q}^{*}(s_t, a_t) \leftarrow (1-\eta)\hat{Q}^{*}(s_t, a_t) + \eta\left(r_t + \gamma \max_{a'} \hat{Q}^{*}(s_{t+1}, a')\right). \tag{10.35}$$

See Algorithm 11 for the implementation of Eq. (10.35).

Algorithm 11 Tabular Q-learning

Require: $P_0(\cdot)$; Initialize $Q^*(s,a) = 0$ for all (s,a); $e = 0$
  while Change in Q^* over consecutive episodes is small **do**
    $e \leftarrow e + 1$
    $t \leftarrow 1$, $s_1 \sim P_0(\cdot)$
    while episode e is running **do**
      $\delta_t = (r_t + \gamma \max_{a'} Q^*(s_{t+1}, a')) - Q^*(s_t, a_t)$
      $Q^*(s_t, a_t) \leftarrow Q^*(s_t, a_t) + \eta\delta_t$
      $t \leftarrow t + 1$
    end while
  end while

Q-learning with function approximation

The tabular approaches described so far do not scale well with increasing state space size. To address this, we use a *function approximation* technique, where we parameterize the Q-function as $Q^*_\theta(s,a)$, with θ representing a vector of parameters. For example, we might use linear regression or NNs to approximate $Q^*_\theta(s,a)$.

The goal is to learn θ such that the parametric Bellman equation

$$Q^*_\theta(s,a) = \mathbb{E}_{s'\sim P(\cdot|s,a)}\left[r(s,a,s') + \gamma \max_{a'} Q^*_\theta(s',a')\right]$$

holds for all state–action pairs (s, a).

We can formalize this as an optimization problem, minimizing the squared error between the target values and the learned Q-function:

$$\min_\theta \sum_{(s,a)} l_\theta(s,a), \quad l_\theta(s,a) = \mathbb{E}_{s'\sim P(\cdot|s,a)}\left[\left(Q^*_\theta(s,a) - r(s,a,s') - \gamma \max_{a'} Q^*_\theta(s',a')\right)^2\right].$$

To implement this, modify the Q-learning algorithm to update the parameter vector θ instead of directly updating $Q^*(s,a)$:

$$\theta \leftarrow \theta + \eta \delta_t \nabla_\theta Q^*_\theta(s_t, a_t).$$

Exploration vs. exploitation

When training RL models, there is a trade-off between *exploration* (trying new actions to discover their effects) and *exploitation* (choosing the best-known action). To balance this, we often use an ϵ-greedy policy:

$$\pi^\epsilon(s) = \begin{cases} \arg\max_{a'} Q^*(s,a') & \text{with probability } 1-\epsilon, \\ \text{random action from } A(s) & \text{with probability } \epsilon, \end{cases} \tag{10.36}$$

where $A(s)$ is the set of available actions from state s.

Example 10.5.1. (a) Simulate Monte Carlo and SARSA methods in World #1. Run it for a few steps and over longer episodes. (b) Compare SARSA and Q-learning in World #1b (which includes negative reward states). Does either method converge? Add a step limit if necessary. (c) Test SARSA and Q-learning in World #2, adjusting ϵ. Do they converge to the same policy? (d) Compare World #1b and World #2. Which method is more efficient and why?

Solution. The solution is available on the author's living book webpage at https://sites.google.com/site/mchertkov/living-books/applied-math-book. Refer to the notebook RL.ipynb for detailed explanations.

Appendix A

Midterm and Final Exams

This appendix contains midterm and final exams from the Math 581 course. As the final exams for Math 581A (December) and Math 581B (May) are part of the Applied Mathematics GIDP qualification process, they were collaboratively prepared by a group of program professors. We extend our gratitude to Professors Ibrahim Fatkullin, Ildar Gabitov, Kevin Lin, Laura Miller, Marek Rychlik and Charles Wolgemuth for their contributions to the exam problems.

A.1 Final Fall 2019 – Math 581a

Complex and Fourier Analysis

Submit three out of five problems

1. Find all values of $z \in \mathbb{C}$ satisfying the equation

$$\sin(z) = 3$$

.

2. Use contour integration to show that

$$\int_{-\infty}^{\infty} \frac{x}{\sinh(\pi x)}\, dx = \frac{1}{2}.$$

   Hint: Consider integration along a rectangular contour.

3. Prove that in the three-dimensional case,

$$\left(\frac{\partial^2}{\partial x^2}+\frac{\partial^2}{\partial y^2}+\frac{\partial^2}{\partial z^2}\right)\frac{1}{r}=-4\pi\delta(\boldsymbol{r}),$$

where $\boldsymbol{r} := (x, y, z) \in \mathbb{R}^3$ and $r := |\boldsymbol{r}| = \sqrt{x^2+y^2+z^2}$.

4. Consider the following sequence of functions in $L^2(0,\pi)$:

$$f_0(x)=\sqrt{\frac{1}{\pi}},\quad f_n(x)=\sqrt{\frac{2}{\pi}}\cos(nx),\quad n=1,\ldots,\infty.$$

   (a) Is the sequence orthonormal, i.e., is $\int_0^\pi f_n(x)f_m(x) = \delta_{nm}$ valid?
   (b) Does it form a basis in $L^2(0,\pi)$, i.e., is $\sum_{n=0} f_n(x)f_n(y) = \delta(x-y)$ valid?

5. An operator, U, is called unitary in $L^2(\mathbb{R})$ if, for any two functions $u, v \in L^2(\mathbb{R})$,

$$(Uf(x), Ug(x)) = (f(x), g(x)),$$

where

$$(f(x), g(x)) := \int_{-\infty}^{\infty} dx\ f(x)g(x).$$

Use the representation of functions in the Fourier integral to show explicitly that the operator $U := \exp\left(ia\frac{d^2}{dx^2}\right)$ is unitary in $L^2(R)$ for $a \in \mathbb{R}$. (Notation: $U := \exp\left(ia\frac{d^2}{dx^2}\right)$ means $Uf(x) = \sum_{n=0}^{\infty}\frac{(ia)^n}{n!}\frac{d^{2n}}{dx^{2n}}f(x)$.)

Differential Equations

Submit three out of five problems

1. Consider the three-dimensional dynamical system

$$\dot{x} = -\alpha x - \beta y - x(x^2 + y^2 + z^2),$$
$$\dot{y} = \beta x - \alpha y - y(x^2 + y^2 + z^2),$$
$$\dot{z} = \gamma z - z(x^2 + y^2 + z^2),$$

where $\alpha, \beta, \gamma \in \mathbb{R}^+$.

   (a) What is the local phase space dynamics of the system near the origin? Carefully sketch the behavior of the solution (in $\mathbb{R}^3$) near the origin.
   (b) Can orbits escape to ∞?
   (c) Might it be possible for the system to have an orbit that repeatedly passes near the origin? If so, sketch how you think it might look.

2. Let $u(x)$, $0 \leq x \leq 1$, represent the temperature of a rod with variable diffusion coefficient given by $k(x) = e^x$. If $f(x)$ is an external heat source, the temperature satisfies the equation

$$\frac{d}{dx}\left(k(x)\frac{du}{dx}\right) = f(x), \quad 0 \leq x \leq 1.$$

For this problem, the boundary conditions are chosen to be

$$u(0) = 0, \quad u'(1) = 0,$$

where $'$ denotes differentiation with respect to x. Find the Green function that satisfies this equation.

3. Find the general solution to the following quasi-linear equation,

$$\partial_t u + u\partial_x u = -x,$$

where $u(t;x) : \mathbb{R}^2 \to \mathbb{R}$, via the method of characteristics.
Hint: If $\xi(t;x;u)$ and $\eta(t;x;u)$ are two first integrals in the extended t, x, u space of a differential equation which can be integrated by the method of characteristics, then $g(\xi, \eta) = 0$, is describing solution implicitly, where g is an arbitrary function of ξ and η.

4. Find the value of the Green function at $r = |\boldsymbol{r}| = 0$ for the equation
$$(\partial_t^2 + \boldsymbol{\nabla}^4)u = \chi,$$
where $\boldsymbol{\nabla} = (\partial_{r_i}; i = 1, \ldots, 3)$.
5. Solve the following boundary value problem using the Fourier method (separation of variables):
$$\partial_{xx}u + \partial_{yy}u = 0, \quad 0 < x < L, \quad 0 < y < \infty$$
$$u(0,y) = u(L,y) = 0, \quad u(x,0) = A\frac{(L-x)x}{L^2}, \quad u(x,\infty) = 0.$$

A.2 Midterm Fall 2020 – Math 581a

Problem #1: Evaluate the integral
$$\int_C (z - \frac{\pi}{2}) \tan z dz,$$
where contour C is a clockwise circle of radius 3 around $z = 1$.

Problem #2: $f(z)$ is defined by
$$f(z) = \int_C \frac{e^w \sin(w) dw}{(w-z)^2},$$
where C is the clockwise circle of radius 2 around $w = 1$. Evaluate $f'(0)$.

Problem #3: Compute the integrals
$$I_1 = \int_0^\infty \frac{\log(x)}{x^2+1} dx,$$
$$I_2 = \int_0^\infty \frac{(\log(x))^2}{x^2+1} dx.$$

Problem #4: Consider the functions $f(x) = \sin(x/2)$ and $g(x) = \cos(x/2)$ on the interval $-\pi < x < \pi$. Let f_n and g_n be their respective Fourier coefficients:
$$f_n = \frac{1}{2\pi}\int_{-\pi}^{\pi} f(x)e^{-inx}\, dx \quad \text{and} \quad g_n = \frac{1}{2\pi}\int_{-\pi}^{\pi} g(x)e^{-inx}\, dx.$$
Without computing the coefficients, answer the following questions:

(a) Do you expect f_n to be purely real? purely imaginary? neither? Explain.
(b) Do you expect g_n to be purely real? purely imaginary? neither? Explain.
(c) Compare qualitatively (or quantitatively) the decay of $|f_n|$ and $|g_n|$ in the limit $|n| \to \infty$.

Problem #5: Find the Fourier transform of the Bessel function $J_0(x)$. Use the integral representation (which is one of many)

$$J_0(x) = \frac{1}{\pi} \int_0^{\pi} \cos(x \cos(\theta)) d\theta.$$

You do not need to justify interchanges of integration.

Hint: As a check on your answer, the Fourier transform has compact support.

Problem #6: What function, $f(x)$, has the Laplace transform $\tilde{f}(k) = k^{-1/2}$?

A.3 Final Fall 2020 – Math 581a

Problem #1: Let $a > 0$. Compute the integral

$$I_k = \int_{-\infty}^{\infty} \frac{e^{ikz}}{(z + ia)^{1/2}} dz$$

for $k > 0$ and for $k < 0$. The branch of the square root function is chosen such that $\mathrm{Arg}(z + ia) \in [-\pi/2, 3\pi/2)$ for any $z \in \mathbb{C}$.

Problem #2: Check if the following sequence of functions are orthonormal and if it forms a (complete) basis in the Hilbert spaces:

(a) The sequence of functions in the space $L^2([0, \pi])$[a]

$$f_0(x) = \sqrt{\frac{1}{\pi}}, \quad f_n(x) = \sqrt{\frac{2}{\pi}} \cos(nx), \quad n = 1, \ldots.$$

[a] $L^2([0, \pi])$ means square integrable, $L^2([0, \pi]) := \{f \;:\; \int_0^{\pi} |f(x)|^2 dx < \infty\}$.

(b) The sequence of functions in $L^2([0, 2\pi])$:

$$f_n(x) = \sqrt{\frac{1}{\pi}} \sin(nx), \quad n = 1, \ldots.$$

If the sequence does not form a basis, extend the sequence to make it form an orthonormal basis.

To check completeness, the following hint is useful:
The sum $\sum_{n=1}^{\infty} \cos(nx)\cos(ny)$ should be computed as the limit

$$\lim_{\varepsilon \to 0+} \sum_{n=1}^{\infty} \exp(-\varepsilon n) \cos(nx) \cos(ny).$$

Problem #3: Three positively charged particles are placed on the line. Two of them are stationary and located at $x = 0$ and $x = 1$, while the third one is free to move between the two stationary particles and its position is $x \in (0, 1)$. Its dynamic is given by

$$\ddot{x} = F(x), \quad \text{where } F = \frac{A}{x^2} - \frac{B}{(1-x)^2}$$

and A and B are positive constants.

(a) Find the potential V from which these forces derive.
(b) Identify the equilibrium position and determine its stability.
(c) Draw the phase portrait (particle position versus velocity).
(d) How is stability affected if friction is introduced in the problem? That is, we now consider

$$\ddot{x} = F(x) - \epsilon \dot{x}$$

where ϵ is small and positive. Describe the behavior of all trajectories as $t \to \infty$.

Problem #4: Find the Green function of the operator

$$x \in [0, +\infty) : \hat{L} = \frac{d^2}{dx^2} - \frac{2}{x^2}, \quad u(0) = 0, \quad |u(+\infty)| < \infty,$$

i.e., find the solution of

$$x, y \in [0, \infty) : \hat{L}G(x; y) = \delta(x - y), \; G(0, y) = 0, \; |G(\infty, y)| < +\infty.$$

Note: The boundary condition $u(0) = 0$ should be interpreted as the limit $\lim_{x \to 0+} u(x) = 0$.

Problem #5: Consider the boundary value problem for the following linear differential equation:

$$x^2y'' + xy' + y = x, \quad y(1) = 0, \quad y(e) = 0. \tag{A.1}$$

(a) Find $p(x)$, $q(x)$ and $r(x)$ such that Eq. (A.1) can be written in the form

$$(p(x)y')' + q(x)y = r(x)x, \quad y(1) = 0, \quad y(e) = 0. \tag{A.2}$$

(b) Using the $p(x)$, $q(x)$ and $r(x)$ you just found, solve the eigenvalue problem

$$(p(x)\phi')' + q(x)\phi = -\lambda r(x)\phi, \quad \phi(1) = 0, \quad \phi(e) = 0,$$

for eigenfunctions $\phi_n(x)$ and corresponding eigenvalues λ_n. Normalize the eigenfunctions.

(c) Express y in the eigenfunction basis (that is, $y(x) = \sum_{n=1}^{\infty} a_k\phi_k(x)$, where the coefficients a_k are to be determined), and substitute it into Eq. (A.2). Use your answer from part (b) to find an expression for the coefficients a_k.

Hint #1: It is known that the eigenfunctions for a Sturm–Liouville problem form a complete, countable basis and are orthogonal with respect to the weight function $r(x)$.

Problem #6: Use the Fourier (variable separation) method to solve the following PDE on the interval $0 \le x \le L$:

$$\frac{\partial y(t,x)}{\partial t} = -\frac{\partial^4 y(t,x)}{\partial x^4} - \alpha\frac{\partial^2 y(t,x)}{\partial x^2}, \tag{A.3}$$

with the boundary conditions

$$y(t,0) = \partial^2 y/\partial x^2|_{x=0} = y(t,L) = \partial^2 y/\partial x^2|_{x=L} = 0$$

and the initial condition $y(0,x) = \sin\left(\frac{\pi}{L}x\right)$.

On your way to deriving the result, state the general solution which satisfies the boundary conditions (but not necessarily satisfies the initial conditions).

Find the minimum value of α such that the solution grows in time.

Note: This is the overdamped beam equation with hinged boundary conditions. The parameter α represents a force applied at the end of the beam. You are effectively solving for the force that causes the beam to buckle.

A.4 Midterm Fall 2021 – Math 581a

Problem #1: The Γ-function is defined for all z satisfying $\mathrm{Re}(z) > 0$ by

$$\Gamma(z) = \int_0^\infty \exp(-t)t^{z-1}dt.$$

Integration by parts shows that the Γ-function satisfies the recurrence relation

$$\Gamma(z) = \frac{\Gamma(z+n+1)}{z(z+1)\cdots(z+n)},$$

for any integer n. The domain of Γ can be extended to all $z \in \mathbb{C}$ as follows: For any z with $\mathrm{Re}(z) < 0$, pick n such that $\mathrm{Re}(z+n+1) > 0$ and define $\Gamma(z)$ according to the recurrence relation.

Determine the locations of the poles of Γ, their orders and their residues.

Problem #2: Find the Laurent expansions of

$$g(z) = \frac{1}{z^3(z-1)^2}$$

near each pole. Describe the respective radius of convergence (the annulus for which the Laurent series converges).

Problem #3: Evaluate

$$\int_0^\infty \frac{\sqrt{x}dx}{x^2+4x+3}$$

via transformation to a contour integral.

Problem #4: A square wave with a jump of $\pm\pi/2$ at $x = n\pi$ can be represented by the following Fourier Series expansion:

$$f(x) = \sum_{n=0}^{\infty} \frac{\sin((2n+1)x)}{2n+1}.$$

(a) The truncation of $f(x)$,

$$S_N(x) = \sum_{n=0}^{N-1} \frac{\sin((2n+1)x)}{2n+1},$$

exhibits the Gibbs phenomenon. Explain the Gibbs phenomenon by comparing $S_N(x)$ and $f(x)$.

(b) Consider the following regularization of $S_N(x)$:

$$f_N(x) = \sum_{n=0}^{N-1} \operatorname{sinc}\left(\frac{2n+1}{2N+1}\pi\right) \frac{\sin((2n+1)x)}{2n+1},$$

where $\operatorname{sinc}(x) = \sin(x)/x$. Does $f_N(x) \to f(x)$? Does $f_N(x)$ show the Gibbs phenomenon?

Support your answers by numerical and/or theoretical arguments.

Problem #5: (a) Find the Fourier transform of the following function:

$$J_0(x) = \sum_{n=-\infty}^{\infty} \delta(x - an).$$

(b) What is the result of the convolution

$$(f * J_0)(x),$$

where $f(x)$ is a given function?
Let f be a given function. What is the result of

$$(f * J_0)(x).$$

You may assume that f is sufficiently smooth and that $|f(x)| \to 0$ sufficiently fast as $|x| \to \infty$. (c) Compute

$$\lim_{a\to 0} a(f * J_0)(x).$$

Problem #6: Given the function

$$f(x) = \frac{x\cos x - \sin x}{x^2},$$

evaluate its Fourier transform.

A.5 Final Fall 2021 – Math 581a

Problem #1: For the complex function

$$f(z) = \frac{\exp(z)}{z^{2021}},$$

- find all its singularities in $\mathbb{C}$;
- write the negative-power portion of the Laurent series of the function (also called principal part of the Laurent series) at each singularity;
- for each singularity, determine whether it is a pole, a removable singularity or an essential singularity;
- compute the residue of the function at each singularity.

Problem #2: Evaluate the contour integral of the function

$$\frac{1}{\exp(2z) - \exp(z)}$$

around the circle $|z| = 2021$, oriented counterclockwise.

Problem #3: For a function $f(x)$ with Fourier transform $\hat{f}(k)$, the Poisson summation formula states that

$$\sum_{n=-\infty}^{\infty} f(n) = \sum_{k=-\infty}^{\infty} \hat{f}(2\pi k)$$

and can be used to accelerate the computation of a sum. For the function

$$f(x) = \frac{\exp(i\pi x)}{1 + x^2},$$

compute its Fourier transform, $\hat{f}(k)$, and use the Poisson summation formula to show that

$$\sum_{n=0}^{\infty} \frac{(-1)^n}{1 + n^2} = \frac{\pi}{2\sinh(\pi)} + \frac{1}{2}.$$

Problem #4:

(a) Sketch solutions of the equation $dy(x)/dx = x^2 - y^2$ as flows in the (x, y) plane. Use the sketch to

- identify all the qualitatively different cases;
- find the stationary points (i.e., points, x, where $dy(x)/dx = 0$) of the solutions and whether all solutions have the same number of critical points.

(b) Show that every possible flow crosses the line $y = x$.

Problem #5: On the planet Barsoom, there is a very long, straight road, which can be assumed infinite in both directions. A stretch of the road collapsed, with the resulting sinkhole extending from $x = a$ miles to $x = b$ miles, measured from $x = 0$. The road sunk uniformly to depth h. Let $y = f(x)$ be the profile of the road, including the collapsed part (see the following figure).

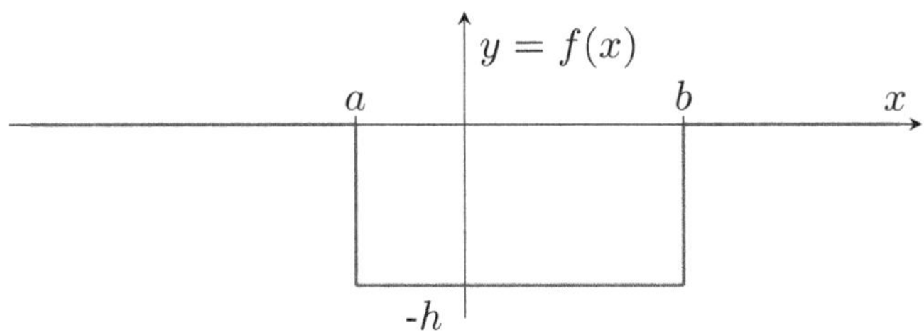

A certain kind of sensor allows one to measure the amplitude of the Fourier transform of the profile, $A_k = |\hat{f}(k)|$, for two different values of k, specifically for $k = 0$ and for $k = q \neq 0$. Call these measurements of the amplitude A_0 and A_q, respectively.

(a) Show that one cannot determine the position of the sinkhole, e.g., $(a+b)/2$, from the measurements of the amplitudes A_0 and A_q alone.

(b) For the given q, it is found that A_0 and A_q satisfy

$$A_0 > A_q \geq 0.22 \cdot A_0.$$

Find the width of the hole $b - a$, and explain why this condition is sufficient for finding $b - a$.

Problem #6:

(a) For $-\infty < x < \infty$, $0 \le y < \infty$, consider the first-order, homogeneous PDE

$$-yu_x + \beta x u_y = 0, \quad u(x,0) = x^n,$$

where β and n are parameters with $-\infty < \beta < \infty$ and $n \in \mathbb{N}$. Determine the values of β and n for which the PDE is well-posed.

(b) For $-\infty < x < \infty$, $0 \le y < \infty$, solve the first-order, nonlinear PDE

$$-yu_x + xu_y = -u^2, \quad u(0,y) = y.$$

A.6 Final Spring 2020 – Math 581b

Problem #1: *Economics of Stocks.* An investor wants to build a portfolio (distribution) of stocks, $\rho(c) \ge 0$, which is a function of stocks' cost, c, investing I_T dollars, where thus

$$I_T = \int_0^\infty dc\, c\, \rho(c). \tag{A.4}$$

The investor's model for the rate of return, aiming to avoid cheap and expensive stocks, is

$$r(c) = \alpha c e^{-kc}.$$

The investor's goal is to maximize the total rate of return, also penalizing non-smooth (non-diverse) portfolios, overall expressed in the maximization of the following objective:

$$R = \int_0^\infty dc \left(r(c)\rho(c) - \frac{\beta}{2}\left(\frac{\partial \rho}{\partial c}\right)^2 \right),$$

where β is a predefined regularization constant. Assuming that $\rho(0) = 0$ and $\partial\rho/\partial c|_{c=0} = 0$, formulate and solve the problem of maximizing R over $\rho(c)$, subject to the constraint (A.4).

Hint: The optimal $\rho(c)$ is non-negative but is not necessarily smooth at all c, and it may be zero at sufficiently large c.

Problem #2: *Quadratic Programming.* For the optimization problem

$$\min_{x_1,x_2} \left(-8x_1 - 16x_2 + x_1^2 + 4x_2^2\right)$$
$$\text{s.t. } x_1 + x_2 \le 5,$$

(a) introduce the Lagrangian, (b) state the stationarity conditions, (c) state the complementary slackness conditions, (c) state the dual problem and (d) find the optimal solution and optimal objective.

Problem #3: *Factory Owner.* A factory produces $x(t) \ge 0$ products per unit of time. Assume that you are the owner and can reinvest some fraction, $0 \le \alpha(t) \le 1$, of the factory output, in which case you boost the production rate according to

$$\begin{cases} \dot{x}(t) = k\alpha(t)x(t), \\ x(0) = x_0 > 0, \end{cases} \tag{A.5}$$

where x_0 is the initial value of $x(t)$ and the constant $k > 0$ defines the efficiency of the reinvestment. The products are sold making a net profit of

$$P = \int_0^T (1 - \alpha(t))x(t)dt. \tag{A.6}$$

Your goal is to maximize the profit (A.6) accumulated over time T.

(a) Find the optimal control, $\alpha^*(t)$, $0 \le t \le T$.
(b) Find the optimal production rate $x^*(t)$, $0 \le t \le T$. Explain different regimes (short T vs. long T).
(c) Assuming optimal control, find the profit (A.6) accumulated in time T.

Hint: Think bang-bang control.

Problem #4: *Z channel.* A Z channel takes a binary signal, $x = 0, 1$, as an input and outputs a binary signal, $y = 0, 1$. If the input is $x = 0$, then the output is $y = 0$ with a probability of one. When the input is $x = 1$, the output is $y = 0$ or $y = 1$ with probabilities of μ and $1 - \mu$, respectively. Consider the Z channel with $\mu = 0.1$ and the following probability distribution of the input symbols: $P(x = 0) = 0.8$ and $P(x = 1) = 0.2$.

(a) Compute the probability distribution of output, $P(y)$.
(b) Compute the probability of $x = 1$ given $y = 0$.
(c) Compute the mutual information between input and output, $I(X; Y)$.

Problem #5: *Markov chain.* Let $S = \{x_0, x_1, \ldots, x_{K-1}\}$ be K equidistant points on the circle, i.e., $x_k = e^{2\pi ik/K}$. Let $p, q, r \in (0, 1)$ satisfy $p+q+r = 1$, and consider the random walk (X_n) defined by

$$P(X_{n+1} = x_{k+1}|X_n = x_k) = p, \tag{A.7a}$$

$$P(X_{n+1} = x_{k-1}|X_n = x_k) = q, \tag{A.7b}$$

$$P(X_{n+1} = x_k|X_n = x_k) = r. \tag{A.7c}$$

We view p and q as free parameters. Observe that the chain is irreducible for all $p, q \in (0, 1)$. Let π be the unique stationary distribution.

(a) For what values of p and q does the chain satisfy detailed balance?
(b) Let T denote the transition matrix. Find exact expressions for the eigenvalues of T.
Hint: the linear transformation represented by T is a convolution operator, i.e., there is a g such that $(Tv)_k = (g \star v)_k = \sum_\ell g(x_{k-\ell})v(x_\ell)$ for all n-vectors v,[b] *and thus can be diagonalized by the discrete Fourier transform.*
(c) The *spectral gap* of T is $1 - |\lambda'|$, where λ' is the second-largest (in absolute value) eigenvalue of T. The size of the spectral gap determines how fast an irreducible aperiodic chain converges to its stationary distribution: The larger the gap, the faster the convergence. Suppose $r = 0.99$ and $p = q$. Use the result of the previous part to find the spectral gap of T to leading order in $1/K$ as $K \to \infty$.

[b]Identifying n vectors with functions on S.

(d) Is the chain aperiodic for all $p, q, r \in (0,1)$ (so long as $p+q+r=1$)? Explain.
(e) Are there initial distributions that converge to the stationary distribution at a rate faster than the second-largest eigenvalue? If so, give an example. If not, explain why not.

Problem #6 *Implementing Social Distancing.* In the context of social distancing, a restaurant has to solve the problem of computing the probability of COVID-19 infection. The restaurant is to host a business meeting with $2n$ participants, placing seats equally spaced from each other on both sides of a long table, as shown in Fig. A.1. A patron $i \in Ind$, where $Ind := \{1a, \ldots, na; 1b, \ldots, nb\}$, can be in either of the two states $\sigma_i = \{S, I\}$, corresponding to "Susceptible" and "Infected" (two states). We call a pair of patrons neighbors if they sit next to each other on the same side of the table or if they sit across from each other. Assume that the probability for the entire group to be in the state $\sigma = (\sigma_i | i \in Ind)$ by the end of the dinner is estimated according to the following neighborhood-factorized formula:

$$P(\sigma) \propto \left(\prod_{k=1,\ldots,n-1} \rho_{ka,(k+1)a}(\sigma_{ka}, \sigma_{(k+1)a}) * \rho_{kb,(k+1)b}(\sigma_{kb}, \sigma_{(k+1)b}) \right) * \prod_{m=1,\ldots,n} \rho_{ma,mb}(\sigma_{ma}, \sigma_{mb}), \tag{A.8}$$

where $\rho_{i,j}(\sigma_i, \sigma_j)$ are known functions which are different for different pairs of neighbors. (A neighborhood function depends on various factors, such as patrons' temperature measurements, quality of individual protection, strength of immune system, number of patrons on premises since last disinfection, etc).

(a) What is the number of possible states σ?
(b) Suggest a dynamic programming (DP) algorithm of $O(n)$ complexity computing the normalization factor (partition function) in Eq. (A.8).

Figure A.1. Sitting arrangement over the restaurant table.

(c) Adapt the algorithm in part (b) to compute efficiently, i.e., in $O(n)$ steps, the probability for a customer occupying position ka to be infected at the end of the meeting.
(d) Suggest an algorithm for generating i.i.d. samples from the probability distribution (A.8) efficiently.

A.7 Midterm Spring 2021 – Math 581b

Problem #1: Find the extremal of the functional

$$\int_0^\pi y^2 + y_x^2 - 2xy\, dx$$

of $y(x)$ subject to the following boundary conditions (two different cases):

(a) $y(0) = 0,\ y_x(\pi) = 2$,
(b) $y(0) = 0,\ y(\pi) = \text{free}$.

Notation: Here, y_x means $\frac{dy}{dx}$.

Problem #2: $f(x)$ is given by

$$f(x) = \begin{cases} a(x\log(x) + (1-x)\log(1-x)), & x \in [0,1]; \\ +\infty, & x \notin [0,1]. \end{cases}$$

Consider two cases, $a = +1$ and $a = -1$, and compute (a) the Legendre–Fenchel transform of f and (b) the double Legendre–Fenchel transform.

Problem #3: Solve the following constrained optimization problem:

$$\begin{cases} \text{Minimize} & x_1^2 + x_2^2 \\ \text{subject to} & x_2 \geq \frac{1}{2}x_1^2 - 3x_1 + 6, \\ & x_2 \leq -x_1 + 6. \end{cases}$$

Explain all the steps.

Problem #4: Solve the optimal control problem

$$\{u^*(\tau)\} = \arg \max_{\{q(\tau),u(\tau)\}} \int_0^2 (2q(\tau) - 3u(\tau)) \bigg|_{\substack{\dot{q} = q + u, \ \forall \tau \in [0,2] \\ q(0) = 4 \\ u(\tau) \in [0,2], \ \forall \tau \in [0,2].}}$$

Give details. Plot the optimal $q_*(\tau)$ and $u_*(\tau)$.

Problem #5: Solve the optimal control problem

$$\min_{\{x(\tau);u(\tau)\}} \left(\frac{\gamma}{2}(x(t))^2 + \int_{t_i}^{t} d\tau \, (u(\tau))^2 \right) \Bigg|_{\substack{\forall \tau \in [t_i, t] : \ \dot{x}(\tau) = u(\tau) \\ x(t_i) = x_0}},$$

using Pontryagin's minimum principle (variational calculus). State the result as a feedback policy, i.e., express, $\forall \tau \in [t_i, t]$, the optimal control $u^*(\tau)$ via $x_*(\tau)$ (i.e., express, $\forall \tau \in [t_i, t]$, the optimal control $u^*(\tau)$ in terms of $x_*(\tau)$).

Problem #6: Consider the compression of gas done in N stages, from $p_0 = 1$ to $p_N = 2$ (in some properly chosen dimensionless units). Assume that the energy required is a function of the intermediate pressures and is proportional to

$$E(p_0, p_1, \ldots, p_N) = \left(\frac{p_1}{p_0}\right)^\alpha + \left(\frac{p_2}{p_1}\right)^\alpha + \cdots + \left(\frac{p_N}{p_{N-1}}\right)^\alpha,$$

where α is a positive constant. Use dynamic programming to define intermediate energies, $E_k(p_k)$, dependent on p_k and k only and such that $E_0 = 0$ and $\min_{p_N} E_N(p_N) = \min_{p_0,\ldots,p_N} E(p_0, \ldots, p_N)$, and find a formula for $E_k(p_k)$ for the cases where $\alpha = 1/2, 2$ and $N = 2, 10$ (four different cases). Use your formula to find the minimum value of E, and state the intermediate pressures p_k that achieve this minimum. Complement analytic computations with numerical computations if needed.

A.8 Final Spring 2021 – Math 581b

Problem #1: Find the extremal of the functional

$$S\{y_1(x), y_2(x)\}$$
$$= \int_0^{\pi/2} \left(\left(\frac{d}{dx} y_1(x) \right)^2 + \left(\frac{d}{dx} y_2(x) \right)^2 - \left(y_1(x) \right)^2 - \left(y_2(x) \right)^2 \right) dx$$

satisfying the boundary conditions $y_1(0) = 1$, $y_2(0) = -1$, $y_1(\pi/2) = 1$, $y_2(\pi/2) = 1$ and the constraint equation $y_1(x) - y_2(x) - 2\cos x = 0$.

Problem #2:

(a) A measure of the uncertainty of a continuous random variable X with the known probability density function $p(x)$ is the entropy $H(X)$, defined as follows:

$$H(X) = -\int_{-\infty}^{\infty} p(x) \ln p(x) dx,$$

where $p(x) \ln p(x) = 0$ whenever $p(x) = 0$. Among all probability density functions on $\mathbb{R}$ of a continuous random variable X that satisfy

$$\int_{-\infty}^{\infty} \ln(1 + x^2) p(x) dx = \ln 4,$$

find the probability density function with the maximal entropy. You may use the following fact if necessary:

$$\int_{-\infty}^{\infty} \frac{\ln(1 + x^2)}{1 + x^2} dx = \pi \ln(4).$$

(b) Assume that you have access to a function that can generate i.i.d. samples of a random variable that is uniformly distributed on $(0, 1)$ and that you want to generate random samples from the distribution you found in part (a). Explain how you would implement a sampling algorithm of your choice.

(c) Determine which moments of $p(x)$ are finite. Does the central limit theorem apply to $p(x)$? If you used your algorithm to generate N samples from $p(x)$, would the average of the N samples converge as N gets large?

Problem #3: A firm has the right to extract oil from a well over the interval $[0, T]$. The oil can be sold at price \$$p$ per unit. To extract oil at rate u when the remaining quantity of oil in the well is x incurs cost at rate \$$u^2/x$. Thus, the problem is one of maximizing

$$\int_0^T \left(pu(t) - \frac{u(t)^2}{x(t)} \right) dt$$

subject to $dx(t)/dt = -u(t)$, $u(t) > 0$, $x(t) > 0$.

(a) Explain why $\lambda(t)$, the adjoint variable, has the terminal condition $\lambda(T) = 0$.
(b) Use Pontryagin's maximum principle to find an expression for the optimal controller in terms of both the adjoint variable λ and the state variable x.
(c) Assuming that the control is optimal, find the oil remaining in the well at time T as a function of $x(0)$, p and T.

Problem #4: Mark n i.i.d. uniformly distributed points in $(0, 1)$. What is the probability density function (PDF) of the position of the kth point from the left (X_k)?

Problem #5: Let $\mathcal{F}$ be the family of four-state Markov chains (MCs). Each MC in $\mathcal{F}$ has a 4×4 transition matrix with entries between zero and one.

(a) Among all Markov chains in $\mathcal{F}$ that are ergodic, find an example of one that has the least possible number of nonzero entries in the transition matrix. Verify that the MC you found is ergodic. Find the resulting stationary distribution of your example.
(b) Among all MCs in $\mathcal{F}$ that are both ergodic and satisfy detailed balance, find an example of one that has the least possible number of nonzero entries in the transition matrix. Verify that the MC you found is ergodic and that is satisfies detailed balance. Find the resulting stationary distribution of your example.

Problem #6: Consider a Brownian particle in $\mathbb{R}^2$. The trajectory of the particle, $(x(t), y(t))$, is described by

$$\dot{x}(t) = \sqrt{2D}\,\xi_x(t), \quad \dot{y}(t) = \sqrt{2D}\,\xi_y(t), \tag{A.9}$$

where D is the diffusion constant of the medium and $\xi_x(t), \xi_y(t)$ satisfy

$$\begin{aligned}\langle \xi_x(t)\rangle = \langle \xi_y(t)\rangle = 0, \quad \langle \xi_x(t)\xi_x(t')\rangle &= \langle \xi_y(t)\xi_y(t')\rangle \\ = \delta(t-t'), \quad \langle \xi_x(t)\xi_y(t')\rangle &= 0.\end{aligned}$$

At $t = 0$ the particle is distributed uniformly within the square $-1 \le x, y \le 1$.

(a) Compute the probability that at time $T > 0$, the particle is found within a circle of radius R. (You can also express your answer as an integral.)
(b) Describe the asymptotic dependence of your answer on T in the regime where $R \gg 1$.

A.9 Midterm Spring 2022 – Math 581b

Problem #1: Consider a metal rod of length L that is clamped at one end and left free at the other end. Let $y(x)$ be the height of the rod at location x. The total potential energy of the metal rod is the sum of the elastic potential energy (caused by bending) and the gravitational potential energy. In the limit where $|y'(x)| \ll 1$, the total potential energy of the metal rod is approximately

$$\tilde{E}[y(x)] = \int_0^L \varepsilon^{-2}\left(y''(x)\right)^2 + \rho g y(x)\,dx,$$

where ε and ρ are parameters that describe the stiffness and density of the metal rod, respectively, and g represents acceleration due to gravity (i.e., ε, ρ and g are constants.)

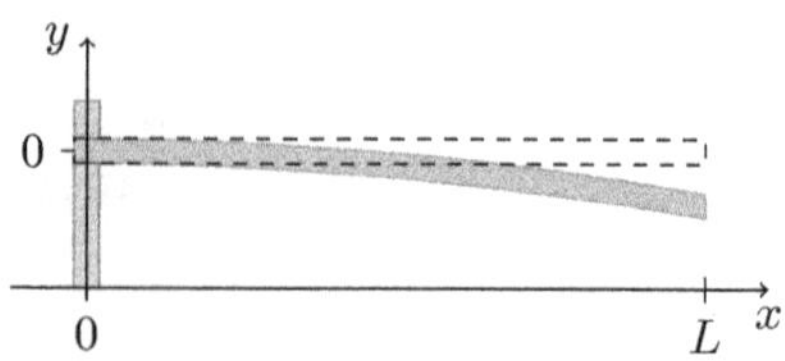

A mathematical model proposes that the metal rod will take the profile $y_*(x)$ that minimizes its total potential energy.

(a) Derive the Euler–Lagrange equations for Lagrangians with second derivatives, paying particular attention to the free end-point condition at $x = L$.

(b) Find the profile of the metal rod by solving the minimization problem

$$y_*(x) = \arg \min_{\{y \in \mathcal{A}\}} \int_0^L \left(\varepsilon^{-1} y''(x)\right)^2 + \rho g y(x)\, dx$$

where

$$\mathcal{A} = \{y : [0, L] \to \mathbb{R} \mid y(0) = 0, y'(0) = 0,$$
$$y(L) = \text{"free"}, y'(L) = \text{"free"}\}.$$

Problem #2: Let $x(t) = (x_1(t), x_2(t)) : [0,1] \to \mathbb{R}^2$ be a C^2 (twice differentiable) curve which is closed, $x(0) = x(1)$. The area and perimeter enclosed by the curve are given by the following functionals:

$$S\{x(t)\} = \frac{1}{2}\int_0^1 dt\,(x_1(t)\dot{x}_2(t) - \dot{x}_1(t)x_2(t)), \tag{A.10}$$

$$P\{x(t)\} = \int_0^1 dt\,|x(t)| = \int_0^1 dt\sqrt{(\dot{x}_1(t))^2 + (\dot{x}_2(t))^2}. \tag{A.11}$$

Maximize the area, $S\{x(t)\}$, under the condition that the perimeter is fixed, $P\{x(t)\} = l$, i.e., find and describe the resulting (minimal) curve.

Problem #3: Find an expression for the optimal controller $u_*(t)$ and for the controlled dynamics $x_*(t)$ for the following optimal control problem:

$$\begin{cases} \text{Minimize} & \int_0^1 \frac{1}{2}\left(x_2(t)\right)^2 + \frac{1}{2}\left(u(t)\right)^2 dt \\ \text{subject to} & \dot{x}_1(t) = x_2(t), \\ & \dot{x}_2(t) = u(t), \\ & x_1(0) = 0,\ x_1(1) = 1, \\ & x_2(0) = 0,\ x_2(1) = \text{"free."} \end{cases}$$

Problem #4: A circular disk with center x_0 and radius $r_n = 1$ is partitioned into n regions by n circles with centers at x_0 and radii $0 \leq r_1 \leq \cdots \leq r_n = 1$. (For notational convenience, also define $r_0 = 0$). The following figure shows an example for $n = 4$.

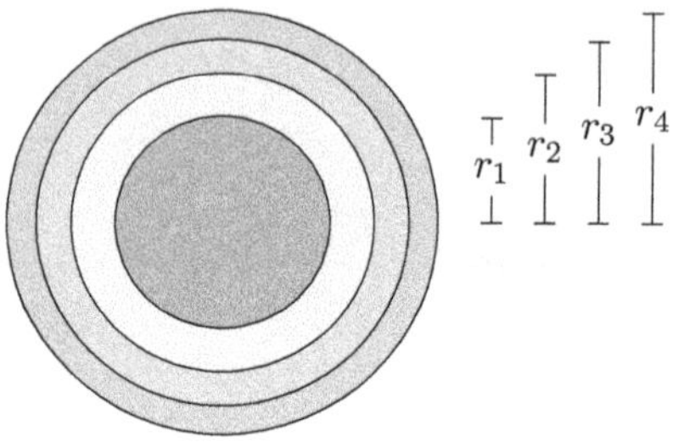

Use dynamic programming to find values for the radii such that the sum of the squares of the areas of the regions is minimized. That is, solve the following optimization problem:

$$\min_{r_1,\ldots,r_n} \left(\sum_{k=1}^{n} \left(\pi r_k^2 - \pi r_{k-1}^2\right)^2 \right)_{0=r_0\leq r_1\leq\cdots\leq r_n=1}.$$

Problem #5: Consider n i.i.d. random variables drawn from the standard normal distribution (i.e., with zero mean and unit variance) so that $\forall i = 1,\ldots,n: \; X_i \sim \mathcal{N}(0,1)$, and let $Y = \sum_{i=1}^{n} X_i^2$. Use the Chernoff bound to prove that the probability that Y is larger than ny, where $y \geq 1$ is less than or equal to $y^{n/2}\exp(-n(y-1)/2)$.

Hint: Follow the logic of the sketch used in the proof of Theorem 8.4.8 from the lecture notes.

Problem #6: Consider the random vector (X_1, X_2, X_3, X_4) where each of the components $X_i, i = 1,\ldots,4$ takes values ± 1. The probability mass function for the random vector is determined by the graph with vertices $\mathcal{V}$ and edges $\mathcal{E}$, as shown in the following figure.

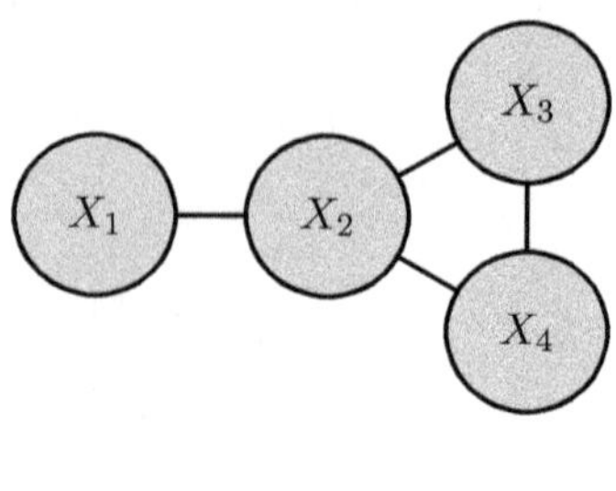

Specifically, the probability mass function is given by

$$p(x_1, x_2, x_3, x_4) \propto \exp\left(\sum_{(i,j)\in\mathcal{E}} Jx_ix_j\right), \tag{A.12}$$

where J is a constant. The constant of proportionality is called the partition function and is often denoted by the symbol $\mathcal{Z}^{-1}$.

(a) Set $J = \ln(2)$. Complete the table (where some entries have been filled for you.)

X_1	X_2	X_3	X_4	$\exp\left(\sum_{(i,j)\in\mathcal{E}} Jx_ix_j\right)$	$P(x_1, x_2, x_3, x_4)$
+1	+1	+1	+1	16	
+1	+1	+1	−1	1	
+1	+1	−1	+1	1	
+1	+1	−1	−1	1	
+1	−1	+1	+1	1/4	
+1	−1	+1	−1	1/4	
+1	−1	−1	+1	1/4	
+1	−1	−1	−1	4	
−1	+1	+1	+1		
−1	+1	+1	−1		
−1	+1	−1	+1		
−1	+1	−1	−1		
−1	−1	+1	+1		
−1	−1	+1	−1		
−1	−1	−1	+1		
−1	−1	−1	−1		

(a) Define the random variable $W := X_1 + X_2 + X_3 + X_4$ and set $J = \ln(2)$.

(i) Compute the probability mass function of W conditioned on $X_1 = +1$. That is, enter the values for $P(W|X_1 = 1)$ on the following table.

		W		
-4	-2	0	$+2$	$+4$
Please	Enter	Your	Values	Here

(ii) Compute the conditional entropy of W given that $X_1 = 1$, $H(W|X_1 = 1)$.

(iii) It is observed that for this graphical model, $H(W|X_1 = 1) \to 0$ as $J \to \infty$ and that $H(W|X_1 = 1) \not\to 0$ as $J \to -\infty$. Explain why this must be so.

(c) Define the random variables $Y := X_2 + X_3$ and $Z := X_4$. Set $J = \ln(2)$.

(i) Find the joint probability distribution of Y and Z, conditioned on $X_1 = +1$ and enter the values on the following table. That is, find $P(Y, Z|X_1 = 1)$.

			Y		
		-2	0	$+2$	$P_Z(z)$
Z	-1	Enter	Your	Values	Here
	$+1$	Enter	Your	Values	Here
	$P_Y(y)$	Enter	Your	Values	

(ii) Find the mutual information of Y and Z conditioned on $X_1 = 1$, $I(Y; Z|X_1 = 1)$.

(iii) It is observed that for this graphical model, $I(Y; Z|X_1 = 1) \to 0$ as $J \to \infty$ and that $I(Y; Z|X_1 = 1) \not\to 0$ as $J \to -\infty$. Explain why this must be so.

A.10 Final Spring 2022 – Math 581b

Problem #1: The shape of the nose cone of a rocket is designed to minimize the total friction on the rocket as it flies through the atmosphere. If the air density is low and the velocity of the rocket is much less than the speed of sound, then the angle of incidence of the air molecules is assumed to be equal to the angle of reflection. Hence, the pressure of the air against the rocket can be modeled as $p = 2\rho V^2 \sin^2 \theta$, where p is the pressure and θ is the angle between the tangential to the body surface direction of the rocket and is a function of x.

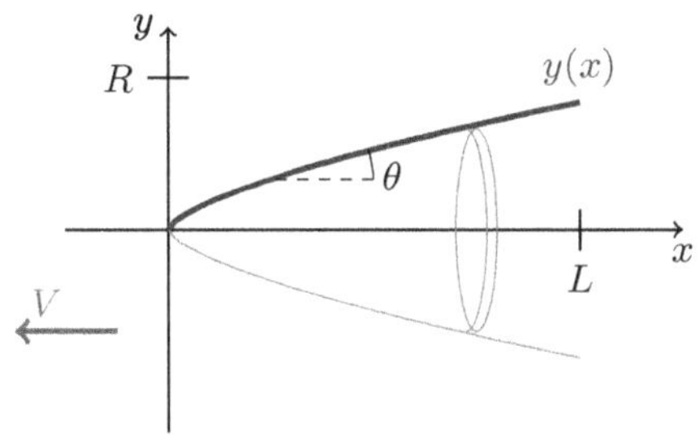

For a nose cone of length L and maximum radius R, the total frictional force is

$$F[y(x)] = \int_0^L 2\rho V^2 \sin^2 \theta \left(2\pi y \sqrt{1 + y'^2} \sin \theta\right) dx.$$

For the case where $R \ll L$, the "small slopes" approximation $\sin(\theta) \approx y'$ gives

$$F[y(x)] \approx 4\pi\rho V^2 \int_0^L yy'^3 dx. \tag{A.13}$$

Find the optimal shape of the nose cone by minimizing Eq. (A.13) subject to the boundary conditions $y(0) = 0$ and $y(L) = R$.

Problem #2: (a) Consider N independent events with probabilities given by p_i for $i = 1, \ldots, N$. Show that the information gained from observing the combination of these events is equal to the sum of the information gained from observing each of these events separately and in any order.

(b) At an animal rescue, the number of cats that are black outnumber the number of dogs that are black by five to one. There are only dogs and cats at the rescue. How much information, in bits, is gained by learning that an animal at the rescue is a black dog?

(c) Recall that the entropy in bits for a discrete random variable distributed over states with probabilities p_i is given as $H = -\sum_i p_i \log_2(p_i)$. Let X and Y be two random, integer-valued, independent variables. The random variable X takes on the values 1, 2, 3, 4 with equal probability. The random variable Y can take on any positive integer value k with probability $p_k = 1/2^k$. What are the entropies $H(X)$ and $H(Y)$? What is the joint entropy $H(X, Y)$ and the mutual information $I(X; Y)$?
(d) Consider N discrete random variables $X_1, X_2, X_3, \ldots, X_N$. Assume that the entropy for each is given by $H(X_i)$. What is the upper bound on the joint entropy $H(X_1, X_2, X_3, \ldots, X_N)$? Explain.

Problem #3: Consider N reversible Markov chains, $p^{(n)}$, $n = 1, \ldots, N$, defined over the same space, converging to the same stationary distribution, π^*, i.e., for any of the Markov chains, the following detailed balance conditions are satisfied:

$$\forall n = 1, \ldots, N: \ \forall i, j: \quad p^{(n)}(i \to j)\pi_i^* = \pi_j^* p^{(n)}(j \to i),$$

Consider the Markov chain with the transition probability $q^{(N)}$ defined as the matrix product, $q^{(N)} = p^{(N)} \cdot p^{(N-1)} \cdots p^{(2)} \cdot p^{(1)} \cdot p^{(2)} \cdots p^{(N)}$. Prove or disprove the following statement:

- The Markov chain $q^{(N)}$ is reversible and converges to the same stationary distribution π^*.

Problem #4: The low-resolution LCD display of an old device is using three letters, "D", "J" or "C", to show its status. The status values are equiprobable.

Each letter is displayed as a 5×5 image consisting of 25 pixels operating independently of each other, as shown in the following:

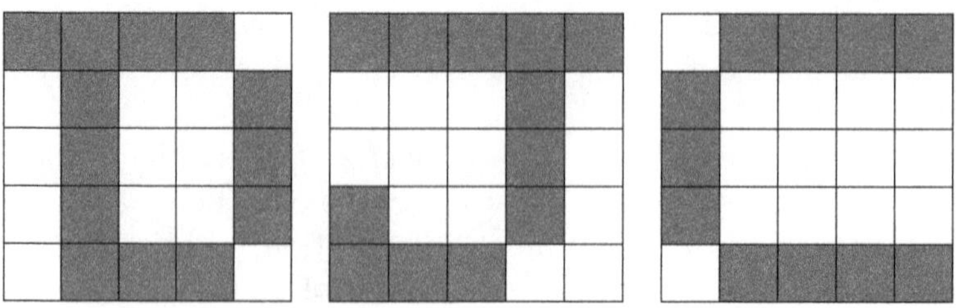

However, after many years of use, there is only a 70% chance that any particular pixel is displayed correctly, else the pixel is flipped. Upon turning on the device, the actual image shown is

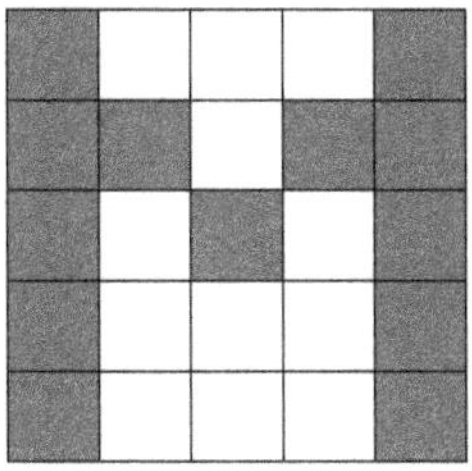

which looks like the letter "M". What is the probability that the actual letter is "D"?

Problem #5: A truncated geometric distribution is a discrete probability distribution with the probability mass function

$$\pi(x) = \begin{cases} c\,\rho^x & \text{for } x = 0, 1, \dots, N, \\ 0 & \text{for } x < 0 \text{ or } x > N. \end{cases}$$

where N is a given natural number and $0 < \rho < 1$.

(a) Describe a Metropolis–Hastings MCMC method to find an ergodic Markov chain with π as a stationary distribution, so that π satisfies the detailed balance condition. (No coding is expected.)
(b) For $N = 2$ and $\rho = 2/3$, find the spectral gap for the probability transition matrix you have designed.

Problem #6: Consider the following piecewise affine function with four linear pieces

$$f(x) = \begin{cases} 3x, & 0 \le x \le 1/2 \\ 3(1-x), & 1/2 \le x \le 1 \\ 0, & \text{otherwise.} \end{cases}$$

(a) Suggest a representation of $f(x)$ via an NN of $\mathrm{ReLu}(x) = \max\{0, x\}$ functions. Is the representation unique?

(b) Consider the nth order self-composition of $f(x)$, $f^{(n)}(x)$, where $f^{(2)}(x) = f(f(x))$, $f^{(3)}(x) = f(f(f(x)))$, etc. What is the number of linear pieces in $f^{(n)}(x)$? You may experiment first, but to receive a full credit, we expect you to derive a closed-form (analytic) answer for any $n \in \mathbb{N}_+$.

A.11 Midterm Spring 2023 – Math 581b

Problem #1: Find the solution to the following optimization problem:

$$\max_{\{y(t);t\in[0,T]\}} \int_0^T e^{-\beta t} \log(1 + y(t))\, dt$$

$$\text{s.t.} \quad \int_0^T e^{-\alpha t} y(t)\, dt = Q.$$

Problem #2: Your task is to navigate from point $A := (r_1 \cos\theta_1, r_1 \sin\theta_1)$ to point $B = (r_2 \cos\theta_2, r_2 \sin\theta_2)$ such that the amount of a pollutant accumulated along the path is minimal. Let $(r(t)\cos\theta(t), r(t)\sin\theta(t))$ denote your location at time t. The total amount of pollutant collected along the path is

$$E\{r(t), \theta(t)\} = \int_0^T I\big(r(t), \theta(t)\big)\, v(t)\, dt$$

where $v(t)$ is your speed (which is constant) and the density of the pollutant I is given by

$$I(r, \theta) = \frac{1}{r}.$$

Find the optimal path.

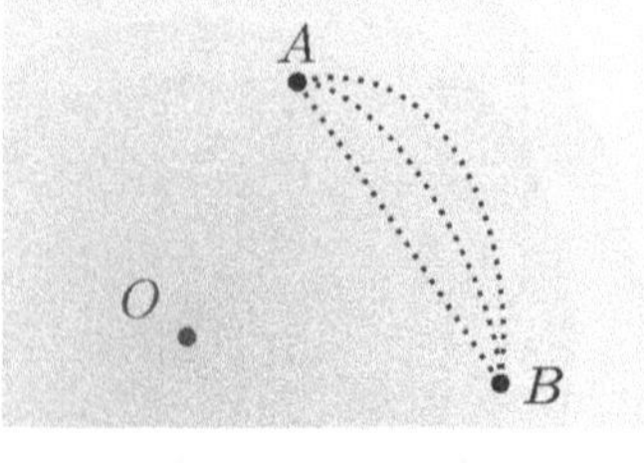

Hint: You may find it useful to write $r = r(\theta)$ and then minimize a functional in the form

$$\int_{\theta_1}^{\theta_2} L(\theta, r(\theta), r'(\theta))\, d\theta.$$

Problem #3: The wealth of a company evolves according to

$$\ddot{x}(t) = -\alpha^2 x(t) + u(t) \qquad \text{for } 0 \le t \le T,$$

where $x(0) = x_0$, $\dot{x}(0) = 0$ and $u(t)$ is the effort that the company's CEO invests in the company at time t. The CEO is greedy and lazy, so he wants to maximize the wealth of the company at the time of the next board meeting (and hence the size of his bonus) with the minimal amount of effort possible. His strategy is to maximize

$$x(T) - \frac{1}{2}\int_0^T \Big(u(t)\Big)^2 dt.$$

Find the optimal control, $u^*(t, x(t))$, and the corresponding state vector, $x^*(t)$.

Problem #4: A nation with total population P is divided into s states. The nation's legislature has R representatives, where $R > s$. Under strictly proportional representation, if state j has population p_j, it would receive r_j representatives, where $r_j/R = p_j/P$. This allocation is not feasible, however, because r_j may not be integer-valued.

The objective is to allocate an integer number of representatives to each state in a way that minimizes the maximum difference between the actual allocation, y_j, and the proportional allocation, r_j, over all states:

$$\min_{y_1,\ldots,y_s} \left(\max\left(|y_1 - r_1|, \ldots, |y_s - r_s|\right)\right)\Bigg|_{\substack{\sum_{j=1}^{s} y_j = R \\ \forall j:\ y_j \ge 0}}.$$

Note: The solution may not be unique.

(a) Formulate a dynamic programming algorithm that finds a solution to this problem by optimizing for only one of the s variables at each step.

(b) Apply your method to the data $R = 4$, $s = 3$, and $r_1 = 0.4$, $r_2 = 2.4$, $r_3 = 1.2$.

Problem #5: Consider $Y_n = n^{-1}\sum_{k=1}^{n} X_k$, where $X_1, \ldots, X_n$ are i.i.d. random numbers generated from the uniform probability distribution $p(x) = 1$ if $-1/2 \le x \le 1/2$ and $p(x) = 0$ otherwise.

(a) Find the mean, μ, variance, σ^2, and moment-generating function, $M_{Y_n}(t)$, of Y_n.

(b) According to the strong version of the central limit theorem (Theorem 8.4.6 of the lecture notes),

$$\forall y > \mu : \quad \lim_{n\to\infty} P(Y_n \ge y) = \exp(-n\,\Phi^*(y)),$$

where $\Phi^*(y)$ is the Crámer function of Y_n, defined by

$$\Phi^*(y) = \sup_t (y\,t - \log M_X(t)),$$

where $M_X(t)$ is the moment-generating function of $X_1, \ldots, X_n$. Find the asymptotic of the Crámer function at $Y_n \approx 1/2$ (Y_n close to $1/2$).

Note: By construction, Y_n cannot be larger than $1/2$.

Problem #6: Given the probability density function

$$p(x \in [-1, 1]) = 1 - |x|,$$

consider the problem of approximating $p(x)$ by the "belief" or "proxy" probability density function $b(x)$:

$$b(x \in [-1, 1]) = \begin{cases} a_0, & |x| \le x_0 < 1 \\ a_1, & x_0 < |x| \le 1 \end{cases}.$$

Find the parameters (a_0^*, a_1^*, x_0^*) which minimize the KL divergence, $D(b\|p)$, and give the corresponding value of $D(b^*\|p)$ (the best fit).

A.12 Final Spring 2023 – Math 581b

Problem #1: Consider a river of width b with straight parallel banks. Assume that one bank of the river coincides with the y-axis and the velocity of the river is $v = v(x)$. A boat moving with speed

c, which is constant relative to the moving water ($c > \max v(x)$, $x \in [0, b]$), should cross the river in the shortest time, sailing from the point $(0, 0)$ and arriving at the point (b, y_1). We also assume that the river flows from south to north and, due to the Coriolis force, has an asymmetric depth profile. As a result, the river velocity can be approximately considered a linear function of the distance from the bank $v(x) = ax$.

(a) State the problem as a variational calculus problem.
(b) Reduce the variational formulation to a transcendental equation.
(c) Solve the transcendental equation numerically and illustrate your solution with plots.

Problem #2: Consider the control dynamics of a particle in one dimension:

$$\dot{x}(s) = a(s), \; 0 \leq t \leq s \leq 1, \quad x(t) = y, \tag{A.14}$$

where the control $a(s)$ is limited to only three values: $a(s) \in \{-1, 0, 1\}$.

(a) Solve the optimal control problem

$$V(t, x) = \min_{\{a(s)|s\in[t,1]\}} \int_t^1 ds|x(s)|\Big|_{\text{Eq. (A.14)}}, \tag{A.15}$$

i.e., find the optimal control using your intuition about the dynamics.
(b) Derive the Hamilton–Jacobi–Bellman (HJB) equation for $V(t, x)$.
(c) Show that $V(t, x)$, evaluated according to the optimal control solution found in (a), solves the HJB equation found in (b). *Hint*: Consider three regions distinguished by how $|x|$ compares with $1 - t$.

Problem #3: Consider a population of algae with exponential growth that is being grown for biofuel. The control cost for the population is to be kept down during the production. The optimal control problem is formulated as

$$\max_u \int_0^2 (2x(t) - 3u(t))\, dt$$

$$\frac{dx(t)}{dt} = x(t) + u(t), \ x(0) = 5, \ 0 \leq u(t) \leq 2.$$

Find the optimal solution.

Problem #4: The probability density function of T, the length of time that a customer waits in line at the grocery store in minutes, is given by

$$f(t) = \begin{cases} k(100 - t^2) & \text{if } 0 \leq t \leq 10, \\ 0 & \text{otherwise,} \end{cases} \tag{A.16}$$

where k is a constant.

(a) Show that $k = 3/2000$.
(b) Find the cumulative distribution function.
(c) Find the probability that the customer will wait for longer than five minutes.
(d) Assuming that the customer has already waited for five minutes, find the probability that the customer will wait for at least eight minutes.

Problem #5: A mouse lives in the labyrinth shown in Fig. A.2. At each time step, the mouse chooses at random one of the doors and leaves the room through this door. The process repeats. Formally,

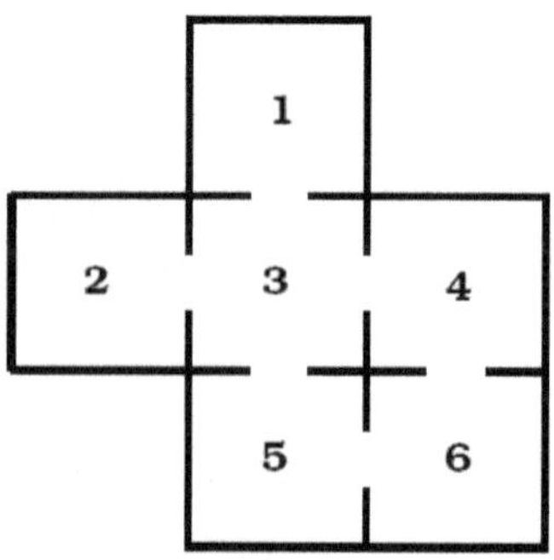

Figure A.2.

mouse dynamics is described fully by a Markov chain of transitions between six states.

(a) Write down the transition matrix P for this Markov chain. Is it irreducible, aperiodic, ergodic? Explain you answer.
(b) By definition, the stationary distribution π^* is an eigenvector of P, which corresponds to the eigenvalue $\lambda = 1$, i.e., it satisfies the equation $P\pi^* = \pi^*$. Find the stationary distribution. Does the detailed balance hold?
(c) Assume that, initially at $t = 0$, the mouse was in the room 1. What is the probability to find the mouse in the room 5 in four steps? What about in five steps?
(d) Do the probabilities of finding the mouse in different rooms converge to the stationary distribution π^*?

Problem #6: We are studying weather patterns in three cities: A, B and C. Every day is classified either as "sunny" (0: the absence of rain) or "rainy" (1). It is known that the probability of a rainy day in cities A, B and C is 0.2, 0.5 and 0.7, respectively.

Furthermore, the probability of any combination of a rainy and sunny day in two cities was found for every pair of cities, as follows:

a	b	$P(A=a, B=b)$	a	c	$P(A=a, C=c)$	b	c	$P(B=b, C=c)$
0	0	0.4	0	0	0.2	0	0	0.15
1	0	0.1	1	0	0.1	1	0	0.15
0	1	0.4	0	1	0.6	0	1	0.35
1	1	0.1	1	1	0.1	1	1	0.35

We are required to build the best spanning tree pair-wise model (as measured by the Kulback–Leibler divergence) of the weather in the three cities in the form of the joint distribution $P(A = a, B = b, C = c)$.

Construct **all** optimal joint distributions. The formulas for the distributions do not have to be explicit, but it needs to be clear about how to calculate

$$P(A = a, B = b, C = c).$$

Appendix B

Convex and Non-Convex Optimization (*)

This appendix[a] is divided into four sections. Sections B.1 and B.2 focus on the fundamentals of convex and non-convex optimization in finite dimensions. Subsequently, Sections B.3 and B.4 discuss iterative optimization methods, based on the formulations introduced in Sections B.1 and B.2, for both constrained and unconstrained problems. We refer the books of Boyd and Vandenberghe [73], Ben-Tal and Nemirovski [74] and Polyak [75] for in-depth study of convex and non-convex optimization methods. Several examples from these books are used in these lectures.

The general problem introduced in Section B.1 involves minimizing a function $f : S \subseteq \mathbb{R}^n \to \mathbb{R}$:

$$\begin{aligned} & f(x) \to \min \\ & \texttt{s.t.}\ \ x \in S \subseteq \mathbb{R}^n. \end{aligned} \tag{B.1}$$

It is worth noting that alternative notations are possible, such as

$$\min_{x \in S \subseteq \mathbb{R}^n} f(x).$$

[a]This appendix was originally prepared by Dr. Yury Maximov from Los Alamos National Laboratory (with editing by the author). It should be viewed as a $*$-section of the main text, intended for supplementary study, and serves as supporting material for Chapter 6.

Section B.1 serves as an introduction, establishing notations and laying the foundation for the discussion of optimization duality in Section B.2.

In Sections B.3 and B.4, we explore iterative algorithms designed to solve Eq. (B.1). Each iteration updates the current estimate x_k based on previous values x_j and $f(x_j)$ (with $j \leq k$), possibly its gradient $\nabla f(x)$ and potentially the Hessian matrix $\nabla^2 f(x)$. The goal is to reach the optimum in the limit[b]:

$$\lim_{k \to +\infty} f(x_k) = \inf_{x \in S \subseteq \mathbb{R}^n} f(x).$$

Iterative algorithms can be categorized based on the type of information available at each step:

- *Zero-order algorithms*: At each iteration, only the value of $f(x)$ at a given point x is accessible (without any information about $\nabla f(x)$ or $\nabla^2 f(x)$).
- *First-order algorithms*: In addition to the function value $f(x)$, the gradient $\nabla f(x)$ is also available at each iteration.
- *Second-order algorithms*: At each step, the function value $f(x)$, its gradient $\nabla f(x)$ and its Hessian matrix $\nabla^2 f(x)$ are accessible.
- *Higher-order algorithms*: These involve access to the function value and all its derivatives up to a given order.

In these notes, we will primarily focus on first-order algorithms, with brief mentions of second-order methods in Sections B.3 and B.4, while higher-order methods will not be discussed.

B.1 Convex Functions, Sets and Optimizations

Calculus of convex functions and sets

An important class of functions that can be efficiently minimized are convex functions, previously introduced in Definition 6.7.3. For convenience, we restate it here.

[b]In this appendix, we assume that all infima are attained.

Definition B.1.1 (Definition 6.7.3). A function $f : \mathbb{R}^n \to \mathbb{R}$ is convex if

$$\forall x, y \in \mathbb{R}^n,\ \lambda \in (0, 1): \quad f(\lambda x + (1 - \lambda)y) \leq \lambda f(x) + (1 - \lambda)f(y).$$

For smooth functions, an equivalent definition of convexity can be stated as follows:

Definition B.1.2. A smooth function $f(x) : \mathbb{R}^n \to \mathbb{R}$ is convex iff

$$\forall x, y \in \mathbb{R}^n: \ f(y) \geq f(x) + \nabla f(x)^\top (y - x).$$

Additionally, for functions with a smooth gradient, convexity can be characterized by the Hessian matrix:

Definition B.1.3. Let $f : \mathbb{R}^n \to \mathbb{R}$ be a function with a smooth gradient. Then, f is convex if and only if

$$\forall x: \ \nabla^2 f(x) \succeq 0,$$

where the Hessian matrix $\nabla^2 f(x)$ is positive semi-definite at every point. Recall that a real symmetric $n \times n$ matrix H is positive semi-definite if $x^\top H x \geq 0$ for all $x \in \mathbb{R}^n$.

Lemma B.1.4. *The definitions above are equivalent for sufficiently smooth functions.*

Proof. Assume the function is convex according to Definition B.1.1. For any $h \in \mathbb{R}^n$ and $\lambda \in [0, 1]$, we have

$$f(\lambda(x+h)+(1-\lambda)x) - f(x) = f(x+\lambda h) - f(x) \leq \lambda(f(x+h) - f(x)).$$

Thus,

$$f(x+h) - f(x) \geq f(x+\lambda h) - f(x) = \nabla f(x)^\top h + O(\lambda), \quad \forall \lambda \in [0, 1].$$

Taking the limit as $\lambda \to 0$, we obtain

$$\nabla f(x)^\top h \leq f(x+h) - f(x), \quad \forall h \in \mathbb{R}^n,$$

which matches Definition B.1.2.

Conversely, assume $\forall x, y: \ f(y) \leq f(x) + \nabla f(x)^\top (y-x)$. Setting $z = \lambda x + (1-\lambda)y$ for any $\lambda \in [0,1]$, we get

$$f(y) \geq f(z) + \nabla f(z)^\top (y-z) = f(z) + \lambda \nabla f(z)^\top (y-x),$$
$$f(x) \geq f(z) + \nabla f(z)^\top (x-z) = f(z) + (1-\lambda) \nabla f(z)^\top (x-y).$$

Summing these inequalities with weights $1-\lambda$ and λ, we get $f(\lambda x + (1-\lambda)y) \leq \lambda f(x) + (1-\lambda) f(y)$, which corresponds to Definition B.1.1.

Finally, if f is sufficiently smooth, using the Taylor expansion,

$$f(y) = f(x) + \nabla f(x)^\top (y-x) + \frac{1}{2}(y-x)^\top \nabla^2 f(x)(y-x) + o(\|y-x\|_2^2),$$

taking the limit as $y \to x$, we derive Definition B.1.3 from Definition B.1.2, and vice versa. □

Definition B.1.5. A function $f(x)$ is called *concave* if $-f(x)$ is convex.

Definition B.1.2 is particularly useful in practice. For non-smooth functions, we introduce the concept of the *sub-gradient.*

Definition B.1.6. A vector $g \in \mathbb{R}^n$ is a sub-gradient of a convex function $f : \mathbb{R}^n \to \mathbb{R}$ at a point x if

$$\forall y \in \mathbb{R}^n : \ f(y) \geq f(x) + g^\top (y-x).$$

The set of all sub-gradients at point x is called the *sub-differential* of f, denoted $\partial f(x)$.

To further analyze sub-gradients, we introduce the concept of convex sets.

Definition B.1.7. A set S is convex if, for any $x_1, x_2 \in S$ and $\theta \in [0,1]$, the point $x_1\theta + x_2(1-\theta) \in S$. In other words, a set is convex if it contains the line segment connecting any two points within it.

Theorem B.1.8. *For any convex function $f : \mathbb{R}^n \to \mathbb{R}$, the sub-differential $\partial f(x)$ is a convex set. That is, for any $g_1, g_2 \in \partial f(x)$, $\theta g_1 + (1-\theta) g_2 \in \partial f(x)$. Moreover, if f is smooth, then $\partial f(x) = \{\nabla f(x)\}$.*

Proof. Let $g_1, g_2 \in \partial f(x)$. Then,

$$f(y) \geq f(x) + g_1^\top (y - x), \quad \text{and} \quad f(y) \geq f(x) + g_2^\top (y - x).$$

Thus, for any $\lambda \in [0, 1]$, we have

$$f(y) \geq f(x) + (\lambda g_1 + (1 - \lambda) g_2)^\top (y - x),$$

which means $\lambda g_1 + (1 - \lambda) g_2$ is also a sub-gradient. Therefore, the set of all sub-gradients is convex. Moreover, if f is smooth, using the Taylor expansion,

$$f(x + h) = f(x) + \nabla f(x)^\top h + O(\|h\|_2^2),$$

we conclude that $\partial f(x)$ consists only of $\nabla f(x)$. □

- The sub-differential of $|x|$ is

$$\partial f(x) = \begin{cases} 1, & \text{if } x > 0, \\ -1, & \text{if } x < 0, \\ [-1, 1], & \text{if } x = 0. \end{cases}$$

- The sub-differential of $f(x) = \max\{f_1(x), f_2(x)\}$ is

$$\partial f(x) = \begin{cases} \nabla f_1(x) & \text{if } f_1(x) > f_2(x), \\ \nabla f_2(x) & \text{if } f_1(x) < f_2(x), \\ \{\theta \nabla f_1(x) + (1 - \theta) \nabla f_2(x) \mid \theta \in [0, 1]\} & \text{if } f_1(x) = f_2(x), \end{cases}$$

  assuming f_1 and f_2 are smooth.

Exercise B.1. Consider $f(x, y) = \sqrt{x^2 + 4y^2}$. Prove that f is convex. Sketch the level curves of f and find the sub-differential $\partial f(0, 0)$.

Example B.1.9. Examples of convex functions include the following:

(a) x^p, for $p \geq 1$ or $p \leq 0$, is convex; however, x^p for $0 \leq p \leq 1$ is concave.
(b) The exponential function, $\exp(x)$ for $x \in \mathbb{R}$, and the negative logarithm, $-\log(x)$ for $x \in \mathbb{R}_{++}$, are convex.

(c) A composition of functions $f(h(x))$, where $f : \mathbb{R} \to \mathbb{R}$ and $h : \mathbb{R} \to \mathbb{R}$, is convex if:

(i) $f(x)$ is convex and non-decreasing, and $h(x)$ is convex;
(ii) or, $f(x)$ is convex and non-increasing, and $h(x)$ is concave.

For smooth functions, this can be shown by considering the second derivative:

$$g''(x) = f''(h(x)) \cdot (h'(x))^2 + f'(h(x)) \cdot h''(x).$$

This result can also be extended to non-smooth and multidimensional functions.

(d) The LogSumExp (also known as soft-max) function, $\log\left(\sum_{i=1}^n \exp(x_i)\right)$, is convex in $x \in \mathbb{R}^n$. It plays an important role in optimization because it provides a smooth approximation to the maximum function, bridging smooth and non-smooth optimization:

$$\max(x_1, x_2, \ldots, x_n) \approx \frac{1}{\lambda} \log\left(\sum_{i=1}^n \exp(\lambda x_i)\right), \quad \lambda \to 0, \quad \lambda > 0. \tag{B.2}$$

(e) The ratio of a quadratic function in one variable to a linear function in another variable, such as $f(x, y) = x^2/y$, is jointly convex in x and y when $y > 0$.

(f) The vector norm

$$\|x\|_p = \left(\sum_{i=1}^n |x_i|^p\right)^{1/p}, \quad x \in \mathbb{R}^n,$$

also called the p-norm or ℓ_p-norm, is convex for $p \geq 1$.

(g) The dual norm $\|\cdot\|_*$ of a norm $\|\cdot\|$ is defined as

$$\|y\|_* = \sup_{\|x\| \leq 1} x^\top y.$$

The dual norm is always convex.

(h) The indicator function of a convex set, $I_S(x)$, is convex:

$$I_S(x) = \begin{cases} 0 & \text{if } x \in S, \\ +\infty & \text{if } x \notin S. \end{cases}$$

Example B.1.10. Examples of convex sets include the following:

1. The intersection of any number of convex sets $\{S_i\}_i$ is also convex: $\bigcap_i S_i$ is convex.
2. The affine image of a convex set is convex. For example, the set

$$\bar{S} = \{x : Ax + b,\ x \in S\},$$

is convex.

3. The image (and inverse image) of a convex set S under perspective mapping $P : \mathbb{R}^{n+1} \to \mathbb{R}^n$, defined as $P(x,t) = x/t$ with dom $P = \{(x,t) : t > 0\}$, is convex.

To illustrate, let $y_1 = x_1/t_1$ and $y_2 = x_2/t_2$ be points in $P(S)$. We need to show that for any $\lambda \in [0,1]$,

$$y = \lambda y_1 + (1-\lambda) y_2 = \lambda \frac{x_1}{t_1} + (1-\lambda)\frac{x_2}{t_2} = \frac{\lambda t_2 x_1 + (1-\lambda) t_1 x_2}{\lambda t_2 + (1-\lambda) t_1}.$$

This expression confirms that the set $P(S)$ is convex. The inverse statement can be similarly proven.

4. The image of a convex set under a linear-fractional function is convex. For example, the function $f : \mathbb{R}^{n+1} \to \mathbb{R}^n$, defined by

$$f(x) = \frac{Ax+b}{c^\top x + d}, \quad \mathrm{dom}\, f = \{x : c^\top x + d > 0\},$$

produces a convex image since $f(x)$ is a perspective transformation of an affine function.

Exercise B.2. Verify that all the functions and sets described above are convex using Definition B.1.1 (or the equivalent Definitions B.1.2 and B.1.3) for convex functions, and Definition B.1.7 for convex sets.

B.1.1 *Strongly convex functions*

A subclass of convex functions, known as strongly convex functions, guarantees faster convergence in optimization than general convex functions.

Definition B.1.11. A function $f : \mathbb{R}^n \to \mathbb{R}$ is μ-strongly convex with respect to the norm $\|\cdot\|$ for some $\mu > 0$ if

1. $\forall x, y : \ f(y) \geq f(x) + \nabla f(x)^\top (y - x) + \frac{\mu}{2}\|y - x\|^2$,
2. f is sufficiently smooth, strong convexity in the ℓ_2 norm is equivalent to $\forall x : \nabla^2 f(x) \succeq \mu$.

As we will see later, generalizing the concept of strong convexity to ℓ_p norms allows for the design of more efficient algorithms in various cases. (Concavity, strong concavity and convexity in ℓ_p norms are defined similarly.)

Exercise B.3. Find a subset of $\mathbb{R}^3$ containing $(0, 0, 0)$ such that $f(u) = \sin(x + y + z)$ is (a) convex and (b) strongly convex.

Exercise B.4. Determine if the following functions are convex and/or strongly convex:

1. $f(x) = \frac{x^2}{2} - \sin x$,
2. $g(x) = \sqrt{1 + x^\top x}$, $x \in \mathbb{R}^n$.

Exercise B.5. Examine whether the function $\sum_{i=1}^n x_i \log x_i$, defined on $\mathbb{R}^n_{++}$, is

- convex, concave, strongly convex or strongly concave?
- strongly convex/concave in the ℓ_1, ℓ_2 and ℓ_∞ norms?

Hint: To show strong convexity in the ℓ_p norm, it is sufficient to prove that

$$h^\top \nabla^2 f(x) h \geq \|h\|_p^2.$$

Convex optimization problems

The optimization problem

$$f(x) \to \min_{x \in S \subseteq \mathbb{R}^n}$$

is convex if $f(x)$ and S are convex. *Complexity* of an iterative algorithm initiated with x_0 to solve the optimization problem is measured in the number of iterations required to get a point x_k such that $|f(x_k) - \inf_{x \in S \subseteq \mathbb{R}^n} f(x)| < \varepsilon$. Each iteration means an update of x_k. Complexity classification is as follows:

- Linear: The number of iterations $k = O(\log(1/\varepsilon))$, and in other words $f(x_{k+1}) - \inf_{x \in S} f(x) \leq c(f(x_k) - \inf_{x \in S} f(x))$ for some constant c, $0 < c < 1$. Roughly, after iteration we increase the number of correct digits in our answer by one.

- Quadratic: $k = O(\log\log(1/\varepsilon))$, and $f(x_{k+1}) - \inf_{x\in S} f(x) \leq c(f(x_k) - \inf_{x\in S} f(x))^2$ for some constant c, $0 < c < 1$. That is, after iteration we double the number of correct digits in our answer.
- Sub-linear: It is characterized by the rate slower than $O(\log(1/\varepsilon))$. In convex optimization, it is often the case that the convergence rate for different methods is $k = O(1/\varepsilon)$, $O(1/\varepsilon^2)$ or $O(1/\sqrt{\varepsilon})$ depending on the properties of function f.

Consider an optimization problem:

$$
\begin{aligned}
f(x) &\to \min_{x\in\mathbb{R}^n} \\
\text{s.t. } g(x) &\leq 0 \\
h(x) &= 0.
\end{aligned}
$$

If the inequality constraint $g(x)$ is convex and the equality constraint is affine, $h(x) = Ax + b$, a feasible set of this problem, $S = \{x : g(x) \leq 0 \text{ and } h(x) = 0\}$, is convex that follows immediately from definitions of a convex set and a convex function. As we will see later in the lectures, in contrast to non-convex problems the convex ones admit very efficient and scalable solutions.

Exercise B.6. Let $\Pi_C^{\ell_p}(x)$ be a projection of a point x to a convex compact set C in ℓ_p norm if

$$
\Pi_C^{\ell_p}(x) = \arg\min_{y\in C} \|x - y\|_p.
$$

Find ℓ_1, ℓ_2, ℓ_∞ projections of $x = \{1, 1/2, 1/3, \ldots, 1/n\} \in \mathbb{R}^n$ on the unit simplex $S = \{x : \sum_{i=1}^n |x_i| = 1\}$. Which of the ℓ_1, ℓ_2, ℓ_∞ projections of an arbitrary point $x \in \mathbb{R}^n$ to a unit simplex is easier to compute?

B.2 Duality

Duality is very powerful tool which allows us to perform the following: design efficient (tractable) algorithms to approximate non-convex problems; build efficient algorithms to convex and non-convex problems with constraints (which are often of a much smaller dimensionality than the original formulations); and formulate necessary and sufficient conditions of optimality for convex and non-convex optimization problems.

Lagrangian

Consider the following constrained (not necessary convex) optimization problem:

$$\begin{aligned} & f(x) \to \min \\ \text{s.t. } & g_i(x) \le 0, \quad 1 \le i \le m \\ & h_j(x) = 0, \quad 1 \le j \le p \\ & x \in \mathbb{R}^n, \end{aligned} \tag{B.3}$$

with the optimal value p^* (which is possibly $-\infty$). Let S be the feasible set of this problem, that is, the set of all x for which all the constraints are satisfied.

Compose the so-called Lagrangian function $\mathcal{L} : \mathbb{R}^n \times \mathbb{R}^m \times \mathbb{R}^p \to \mathbb{R}$:

$$\begin{aligned} \mathcal{L}(x, \lambda, \mu) &= f(x) + \sum_{i=1}^{m} \lambda_i g_i(x) + \sum_{j=1}^{p} \mu_j h_j(x) \\ &= f(x) + \lambda^\top g(x) + \mu^\top h(x), \quad \lambda \ge 0 \end{aligned} \tag{B.4}$$

which is a weighted combination of the objective and the constraints. The Lagrange multipliers λ and μ can be viewed as penalties for violation of inequality and equality constraints.

The Lagrangian function (B.4) allows us to formulate the constrained optimization, Eq. (B.3), as a min-max (also called saddle-point) optimization problem:

$$p^* = \min_{x \in S \subseteq \mathbb{R}^n} \max_{\lambda \ge 0, \mu} \mathcal{L}(x, \lambda, \mu) \tag{B.5}$$

where the optimum of Eq. (B.3) is achieved at p_*.

Weak and strong duality

Let us consider the saddle-point problem (B.5) in greater detail. For any feasible point $x \in S \subseteq \mathbb{R}^n$ one has $f(x) \ge \mathcal{L}(x, \lambda, \mu)$, $\lambda \ge 0$.

Thus,

$$L(\lambda, \mu) = \min_{x \in S} \mathcal{L}(x, \lambda, \mu) \leq \min_{x \in S} f(x) = p^* \Rightarrow \underbrace{\max_{\lambda \geq 0, \mu} \min_{x \in S} \mathcal{L}(x, \lambda, \mu)}_{L(\lambda,\mu)} \leq p^*$$

$$= \min_{x \in S} \max_{\lambda \geq 0, \mu} \mathcal{L}(x, \lambda, \mu),$$

where $L(\lambda, \mu) = \inf_{x \in \mathbb{R}^n} \mathcal{L}(x, \lambda, \mu) = \inf_{x \in \mathbb{R}^n} \{f(x) + \lambda^\top g(x) + \mu^\top h(x)\}$ is called the Lagrange dual function. One can restate it as

$$d^* = \max_{\lambda \geq 0, \mu} \min_{x \in S} \mathcal{L}(x, \lambda, \mu) \leq \min_{x \in S} \max_{\lambda \geq 0, \mu} \mathcal{L}(x, \lambda, \mu) = p^*$$

The original optimization, $\min_{x \in S} f(x) = \min_{x \in S} \max_{\lambda \geq 0, \mu} \mathcal{L}(x, \lambda, \mu)$, is called the Lagrange primal optimization, while $\max_{\lambda \geq 0, \mu} L(\lambda, \mu) = \max_{\lambda \geq 0, \mu} \min_{x \in S} \mathcal{L}(x, \lambda, \mu)$ is called the Lagrange dual optimization.

Note that $\max_{\lambda \geq 0, \mu} \min_{x \in S} \mathcal{L}(x, \lambda, \mu) = \max_{\lambda \geq 0, \mu} \min_{x \in \mathbb{R}^n} \mathcal{L}(x, \lambda, \mu)$, regardless of what S is. This is because $\hat{x} \notin S$ one has $\max_{\lambda \geq 0, \mu} \mathcal{L}(\hat{x}, \lambda, \mu) = +\infty$, thus allowing us to perform unconstrained minimization of $\mathcal{L}(x, \lambda, \mu)$ over x much more efficiently.

Let us describe a number of important features of the dual optimization:

1. *Concavity of the dual function*: The dual function $L(\lambda, \mu)$ is always concave. Indeed, for $(\bar{\lambda}, \bar{\mu}) = \theta(\lambda_1, \mu_1) + (1 - \theta)(\lambda_2, \mu_2)$, one has

$$\begin{aligned} L(\bar{\lambda}, \bar{\mu}) &= \min_x \mathcal{L}(x, \bar{\lambda}, \bar{\mu}) = \min_x \{\theta \mathcal{L}(x, \lambda_1, \mu_1) \\ &\quad + (1 - \theta)\mathcal{L}(x, \lambda_2, \mu_2)\} \\ &\geq \theta \min_x \mathcal{L}(x, \lambda_1, \mu_1) + (1 - \theta) \min_x \mathcal{L}(x, \lambda_2, \mu_2) \\ &= \theta L(\lambda_1, \mu_1) + (1 - \theta) L(\lambda_2, \mu_2). \end{aligned}$$

   The dual (maximization) problem $\max_{\lambda \geq 0, \mu} L(\lambda, \mu)$ is equivalent to the minimization of the convex function $-L(\lambda, \mu)$ over the convex set $\lambda \geq 0$.
2. *Lower bound property*: $L(\lambda, \mu) \leq p^*$ for any $\lambda \geq 0$.

3. *Weak duality*: For any optimization problem $d^* \leq p^*$. Indeed, for any feasible (x, λ, μ), we have $f(x) \geq \mathcal{L}(x, \lambda, \mu) \geq L(\lambda, \mu)$, thus $p^* = \min_{x \in \mathbb{R}^n} f(x) \geq \max_{\lambda \geq 0, \mu} L(\lambda, \mu) = d^*$.
4. *Strong duality*: We say that strong duality holds if $p^* = d^*$. Convexity of the objective function and convexity of the feasible set S is neither sufficient nor necessary condition for strong duality (see the following example).

Example B.2.1. Convexity alone is not sufficient for the strong duality. Find the dual problem and the duality gap $p^* - d^*$ for the following optimization:

$$\exp(-x) \to \min_{y>0,x}$$

$$\text{s.t. } x^2/y \leq 0.$$

The optimal problem is $p^* = 1$, which is achieved at $x = 0$ and any positive y. The dual problem is

$$L(\lambda) = \inf_{y>0,x} (\exp(-x) + \lambda x^2/y) = 0.$$

That is, the dual problem is $\max_{\lambda \geq 0} 0 = 0$, and the duality gap is $p^* - d^* = 1$.

Theorem B.2.2 (Slater's (sufficient) conditions). *Consider the optimization* (B.3) *where all the equality constraints are affine and all the inequality constraints and the objective function are convex. The strong duality holds if there exists an x_* such that x_* is strictly feasible, i.e., all constraints are satisfied and the nonlinear constraints are satisfied with strict inequalities.*

The Slater conditions imply that the set of optimal solutions of the dual problem, therefore making the conditions sufficient for the strong duality of the optimization.

Optimality conditions

A key feature of the Lagrangian function is its role in establishing the necessary and sufficient conditions for a triplet, (x, λ, μ), to be the solution of the saddle-point optimization problem (B.5). Let us first formulate the necessary conditions of optimality for the following problem:

$$\begin{aligned} & f(x) \to \min \\ \text{s.t. } & g_i(x) \leq 0, \quad 1 \leq i \leq m \\ & h_j(x) = 0, \quad 1 \leq j \leq p \\ & x \in S \subseteq \mathbb{R}^n. \end{aligned}$$

According to Eq. (B.5), this optimization is equivalent to

$$\min_{x \in S} \max_{\lambda \geq 0, \mu} \mathcal{L}(x, \lambda, \mu),$$

where the Lagrangian function is defined by Eq. (B.4). The following conditions, known as the Karush–Kuhn–Tucker (KKT) conditions, are necessary for a triplet, (x^*, λ^*, μ^*), to be optimal:

1. *Primal feasibility*: $x^* \in S$.
2. *Dual feasibility*: $\lambda^* \geq 0$.
3. *Vanishing gradient*: $\nabla_x \mathcal{L}(x^*, \lambda^*, \mu^*) = 0$ for smooth functions, and $0 \in \partial \mathcal{L}(x^*, \lambda^*, \mu^*)$ for non-smooth functions. This holds because, for the optimal (λ^*, μ^*), the Lagrangian must attain its minimum at x^*.
4. *Complementary slackness*: $\lambda_i^* g_i(x^*) = 0$. If $g_i(x^*) < 0$ and $\lambda_i^* > 0$, one could reduce the Lagrange multiplier and improve the objective.

Note that the KKT conditions generalize (in finite dimensions) the Euler–Lagrange conditions introduced in variational calculus. Let us now discuss when these conditions are sufficient for optimality.

The KKT conditions are sufficient if strong duality holds, for which the Slater conditions (discussed earlier) are sufficient. Specifically, if strong duality holds and the point (x^*, λ^*, μ^*) satisfies the KKT conditions, then

$$L(\lambda^*, \mu^*) = f(x^*) + g(x^*)^\top \lambda^* + h(x^*)^\top \mu^* = f(x^*), \tag{B.6}$$

where the first equality holds due to stationarity and the second holds because of complementary slackness.

Example B.2.3. Find the duality gap and solve the dual problem for the following minimization:

$$\min\left((x_1-3)^2+(x_2-2)^2\right)$$
$$\text{s.t.}\quad x_1+2x_2=4,\quad x_1^2+x_2^2\le 5.$$

This is a convex problem, and since Slater's conditions are satisfied, the minimum is unique, and there is no duality gap. The Lagrangian is

$$\begin{aligned}\mathcal{L}(x,\lambda,\mu) &= (x_1-3)^2+(x_2-2)^2+\mu(x_1+2x_2-4)\\ &\quad+\lambda(x_1^2+x_2^2-5),\quad \lambda\ge 0.\end{aligned}$$

The dual problem becomes

$$L(\lambda,\mu)=\inf_{x\in\mathbb{R}^n}\mathcal{L}(x,\lambda,\mu).$$

The KKT conditions are

$$\nabla\mathcal{L}=\begin{pmatrix}2(x_1-3)+\mu+2\lambda x_1\\ 2(x_2-2)+2\mu+2\lambda x_2\end{pmatrix}=0.$$

Solving this system with the primal feasibility constraint, we get $x_1=\frac{12+4\lambda}{5(1+\lambda)}$ and $x_2=\frac{4+8\lambda}{5(1+\lambda)}$. The dual problem then becomes

$$L(\lambda)=5-\frac{9\lambda}{5}-\frac{16}{5(1+\lambda)}\to\max_{\lambda\ge 0}.$$

Solving for the stationary point, we find $\lambda^*=\frac{1}{3}$. Thus, the optimal solution is $(x_1^*,x_2^*,\lambda^*,\mu^*)=(2,1,\frac{2}{3},\frac{1}{3})$.

Example B.2.4. Consider the following linear programming problem:

$$\begin{aligned}&3x+7y+z\to\min\\ \text{s.t.}\quad &x+5y=2\\ &x+y\ge 3\\ &z\ge 0.\end{aligned}$$

Find the dual problem, the optimal values of the primal and dual objectives, and the optimal solutions for both the primal and dual variables.

Solution. 1. First, observe that the problem is equivalent to

$$3x + 7y \to \min \quad \text{s.t.} \quad x + 5y = 2, \quad x + y \geq 3,$$

since x and y are independent of z, and the minimum value of z is 0.

2. The Lagrangian is

$$\mathcal{L}(x, y, \mu, \lambda) = 3x + 7y + \mu(2 - x - 5y) + \lambda(3 - x - y).$$

3. The KKT conditions from $\nabla\mathcal{L}(x, y, \mu, \lambda)$ are

$$\frac{\partial}{\partial x}\mathcal{L}(x, y, \mu, \lambda) = 3 - \mu - \lambda = 0,$$
$$\frac{\partial}{\partial y}\mathcal{L}(x, y, \mu, \lambda) = 7 - 5\mu - \lambda = 0.$$

This gives $\mu = 1$ and $\lambda = 2$, showing that the Lagrange multipliers are feasible.

4. The complementary slackness condition $\lambda(3 - x - y) = 0$ implies that $x + y = 3$.
5. Solving the primal feasibility constraints $x+5y = 2$ and $x+y = 3$, we find $y = -0.25$ and $x = 3.25$.
6. The optimal values of the primal variables are $(x^*, y^*, z^*) = (3.25, -0.25, 0)$.

Exercise B.7. For each of the following primal optimization problems, find the dual problem, the optimal values of the primal and dual objectives and the optimal solutions for the primal and dual variables. Provide detailed steps in your solution.

1. $\min 4x + 5y + 7z$, s.t. $2x + 7y + 5z + d = 9$ and $x, y, z, d \geq 0$. *Hint*: Consider dropping an inequality constraint and verify after finding the optimal solution if the dropped inequality is satisfied.
2. $\min\{(x_1 - \frac{5}{2})^2 + 7x_2^2 - x_3^2\}$, s.t. $x_1^2 - x_2 \leq 0$ and $x_3^2 + x_2 \leq 4$.

Examples of duality

Example B.2.5 (Duality and Legendre–Fenchel Transform). Let us explore the relationship between transforming the Lagrangian function to the dual (Lagrange) function and the Legendre–Fenchel (LF) transform (also known as the conjugate function), defined as

$$f^*(y) = \sup_{x \in \mathbb{R}^n} (y^\top x - f(x)),$$

which was introduced in the variational calculus section (see Section 6). One of the main conclusions of LF analysis is that $f(x) \geq f^{**}(x)$. This inequality directly connects to duality theory, where the dual problem provides a lower bound to the primal optimization. To illustrate this, consider the following optimization problem:

$$\begin{aligned} &f(x) \to \min \\ &\text{s.t. } x = b, \end{aligned}$$

where b is a fixed parameter. The duality relationship can be expressed as

$$\begin{aligned} \min_x \max_\mu \{f(x) + \mu^\top (b - x)\} &\leq \max_\mu \min_x \{f(x) + \mu^\top (b - x)\} \\ &= \max_\mu \{-\mu^\top b - \max_x (\mu^\top x - f(x))\} \\ &= \max_\mu \{-\mu^\top b - f^*(\mu)\} = f^{**}(-b). \end{aligned}$$

Minimizing the expression over all $b \in \mathbb{R}^n$ gives

$$\min_{x \in \mathbb{R}^n} f(x) \geq \min_{x \in \mathbb{R}^n} f^{**}(x).$$

This shows that the dual problem provides a lower bound to the primal one, similar to how the LF transform does.

Example B.2.6 (Duality in Linear Programming (LP)). Consider the following linear programming problem:

$$\begin{aligned} &c^\top x \to \min \\ &\text{s.t. } Ax \leq b. \end{aligned}$$

The Lagrangian for this problem is given by

$$\mathcal{L}(x, \lambda) = c^\top x + \lambda^\top (Ax - b), \quad \lambda \geq 0.$$

The dual objective function becomes

$$\begin{aligned} L(\lambda) &= \inf_{x\in\mathbb{R}^n} \mathcal{L}(x,\lambda) \\ &= \inf_{x\in\mathbb{R}^n} \left\{ x^\top(c + A^\top\lambda) - b^\top\lambda \right\} \\ &= \begin{cases} -b^\top\lambda, & \text{if } c + A^\top\lambda = 0, \\ -\infty, & \text{otherwise.} \end{cases} \end{aligned}$$

Thus, the dual problem becomes

$$L(\lambda) = -b^\top\lambda \to \max_{\lambda\geq 0, c+A^\top\lambda=0}.$$

This is the dual of the original linear programming problem.

Example B.2.7 (Non-convex Problems with Strong Duality). Consider the following non-convex quadratic minimization problem:

$$\begin{gathered} x^\top Ax + 2b^\top x \to \min \\ \text{s.t. } x^\top x \leq 1, \end{gathered}$$

where $A \not\succeq 0$ (i.e., A is not positive semi-definite). The dual objective is

$$\begin{aligned} L(\lambda) &= \inf_{x\in\mathbb{R}^n} \mathcal{L}(x,\lambda) \\ &= \inf_{x\in\mathbb{R}^n} \left\{ x^\top(A+\lambda I)x - 2b^\top x - \lambda \right\} \\ &= \begin{cases} -\infty & \text{if } A + \lambda I \not\succeq 0, \\ -\infty & \text{if } A + \lambda I \succeq 0 \text{ and } b \notin \operatorname{Im}(A+\lambda I), \\ -b^\top(A+\lambda I)^+ b - \lambda & \text{otherwise.} \end{cases} \end{aligned}$$

The corresponding dual optimization problem is

$$-b^\top(A+\lambda I)^+ b - \lambda \to \max,$$

subject to

$$A + \lambda I \succeq 0, \quad b \in \operatorname{Im}(A+\lambda I).$$

To restate this as a convex optimization problem, introduce an auxiliary variable t:

$$-t - \lambda \to \max,$$

subject to

$$t \geq b^\top (A + \lambda I)^+ b, \quad A + \lambda I \succeq 0, \quad b \in \operatorname{Im}(A + \lambda I).$$

Finally, this can be written as

$$-t - \lambda \to \max,$$

subject to

$$\begin{pmatrix} A + \lambda I & b \\ b^\top & t \end{pmatrix} \succeq 0.$$

Example B.2.8 (Dual to Binary Quadratic Programming (QP)). Consider the following binary quadratic optimization problem:

$$x^\top A x \to \max$$

$$\text{s.t. } x_i^2 = 1, \quad 1 \leq i \leq n,$$

where $A \succeq 0$. The dual optimization is

$$\begin{aligned} &\min_{x \in \mathbb{R}^n} \left\{ -x^\top A x + \sum_{i=1}^n \mu_i (x_i^2 - 1) \right\} \\ &= \min_{x \in \mathbb{R}^n} \left\{ x^\top (\operatorname{Diag}(\mu) - A) x - \sum_{i=1}^n \mu_i \right\} \to \max_{\mu} . \end{aligned}$$

Thus, the dual problem is

$$\sum_{i=1}^n \mu_i \to \min,$$

subject to

$$\operatorname{Diag}(\mu) \succeq A.$$

This optimization problem is convex and provides a non-trivial lower bound to the original binary quadratic optimization. This bound is often referred to as the semi-definite programming (SDP) relaxation.

Example B.2.9. Show that $\min_x \lambda^\top x\Big|_{\|x\|_p\le 1} = -\|\lambda\|_{p/(p-1)}$, where $x \in \mathbb{R}^d$, $p \ge 1$ and $\|x\|_p := (|x_1|^p + \cdots + |x_d|^p)^{1/p}$ is the p-norm of x.

Solution. We start by formulating the dual problem for this minimization. The original problem is

$$\min_x \lambda^\top x \quad \text{subject to} \quad \|x\|_p \le 1.$$

The dual formulation introduces the Lagrange multiplier $\mu \ge 0$ for the constraint $\|x\|_p - 1 \le 0$, leading to the Lagrangian

$$\min_x \max_{\mu\ge 0} \left(\lambda^\top x + \mu(\|x\|_p - 1)\right).$$

Since the problem is convex, Slater's condition is satisfied (as any small x in the p-norm is feasible), so strong duality holds. Therefore, we can reverse the order of optimization:

$$\max_{\mu\ge 0} \min_x \left(\lambda^\top x + \mu(\|x\|_p - 1)\right).$$

Next, we apply the KKT conditions. The stationary point condition with respect to x is

$$\forall i = 1, \ldots, d: \quad \lambda_i + \mu \frac{|x_i^*|^{p-2} x_i^*}{\left(\sum_{j=1}^d |x_j^*|^p\right)^{1-1/p}} = 0.$$

The complementary slackness condition is

$$\mu(\|x^*\|_p - 1) = 0.$$

Assuming $\lambda \ne 0$ (if $\lambda = 0$, the result is trivially zero), we must have $\mu \ne 0$ by the stationarity condition, and $\|x^*\|_p = 1$. Thus, the stationary point condition simplifies to

$$\lambda_i = -\mu \frac{|x_i^*|^{p-2} x_i^*}{\left(\sum_{j=1}^d |x_j^*|^p\right)^{1/p}} = -\mu x_i^* |x_i^*|^{p-2}.$$

By combining these equations, we find

$$\mu = -\left(\sum_{i=1}^{d} |\lambda_i|^{p/(p-1)}\right)^{(p-1)/p} = -\|\lambda\|_{p/(p-1)}, \tag{B.7}$$

$$\lambda^\top x^* = -\mu \sum_{i=1}^{d} |x_i^*|^p = -\mu. \tag{B.8}$$

Therefore, substituting the optimal values back into the objective proves the desired result.

Exercise B.8. Consider the following quadratic optimization problem:

$$\min_{x\in\mathbb{R}^d} \; -\frac{1}{2}x^\top Lx + b^\top x \tag{B.9}$$

$$\text{s.t. } \|x\|_\infty \leq 1. \tag{B.10}$$

(a) Describe the conditions on L and b that guarantee the convexity of Eq. (B.9).
(b) Find the dual of Eq. (B.9), reformulating the ℓ_∞-constraint as a convex quadratic constraint. Is the duality gap zero when $L \preceq 0$?
(c) Show that if $bb^\top \succeq \varepsilon L \succeq 0$ for some $\varepsilon > 0$, then $L = cbb^\top$, where c is a constant.
(d) Assuming the conditions in (c) are satisfied, solve Eq. (B.9) analytically.
Hint: Transition to a scalar variable and show that the problem reduces to a one-dimensional quadratic concave optimization over a bounded domain.

Conic duality

A standard conic optimization problem can be formulated as

$$c^\top x \to \min_x \tag{B.11}$$

$$\text{s.t. } Ax = b,$$

$$x \in \mathcal{K},$$

where $\mathcal{K}$ is a proper cone, meaning it satisfies the following properties:

1. $\mathcal{K}$ is a convex cone: for any $x, y \in \mathcal{K}$ and $\alpha, \beta \geq 0$, $\alpha x + \beta y \in \mathcal{K}$;
2. $\mathcal{K}$ is closed;

3. $\mathcal{K}$ is solid, meaning it has a non-empty interior;
4. $\mathcal{K}$ is pointed, meaning that if $x \in \mathcal{K}$ and $-x \in \mathcal{K}$, then $x = 0$.

Conic optimization is important in various applications. For example, in Example B.2.8, we encountered a dual problem that is a conic optimization problem over the cone of positive semi-definite matrices.

The dual cone $\mathcal{K}^*$ of $\mathcal{K}$ is defined as

$$\mathcal{K}^* = \{c : c^\top x \geq 0 \text{ for all } x \in \mathcal{K}\}.$$

Exercise B.9. Show that the following sets are self-dual cones (i.e., $\mathcal{K}^* = \mathcal{K}$):

1. The set of positive semi-definite matrices, $\mathbb{S}^n_+$.
2. The positive orthant, $\mathbb{R}^n_+$.
3. The second-order cone, $Q^n = \{(x, t) \in \mathbb{R}^n \times \mathbb{R} : t \geq \|x\|_2\}$.

For the case of semi-definite matrices, note that the condition $c^\top x = \sum_{i,j=1}^n c_{ij} x_{ij}$ is equivalent to the Hadamard product of matrices.

The Lagrangian for the problem in Eq. (B.11) is

$$\mathcal{L}(x, \mu, \lambda) = c^\top x + \mu^\top (b - Ax) - \lambda^\top x,$$

where the term $-\lambda^\top x$ enforces the constraint $x \in \mathcal{K}$. Using the definition of the dual conc, wc dcrivc

$$\max_{\lambda \in \mathcal{K}^*} -\lambda^\top x = \begin{cases} 0, & x \in \mathcal{K}, \\ +\infty, & x \notin \mathcal{K}. \end{cases}$$

Thus, the dual problem becomes

$$p^* = \min_{x \in \mathcal{K}} \max_{\lambda \in \mathcal{K}^*, \mu} \mathcal{L}(x, \lambda, \mu) \geq d^* = \max_{\lambda \in \mathcal{K}^*, \mu} \min_{x \in \mathcal{K}} \mathcal{L}(x, \lambda, \mu).$$

The dual objective function is

$$L(\lambda, \mu) = \min_{x \in \mathcal{K}} \{c^\top x + \mu^\top (b - Ax) - \lambda^\top x\} = \begin{cases} \mu^\top b & \text{if } c - A^\top \mu - \lambda = 0, \\ -\infty & \text{otherwise.} \end{cases}$$

Thus, the dual problem simplifies to

$$d^* = \max \mu^\top b \quad \text{subject to} \quad c - A^\top \mu - \lambda = 0, \quad \lambda \in \mathcal{K}^*.$$

Finally, eliminating λ gives the form

$$\mu^\top b \to \max \quad \text{subject to} \quad c - A^\top \mu \in \mathcal{K}^*.$$

Exercise B.10. Find the dual problem (see Example B.2.8) to

$$1^\top \mu = \sum_{i=1}^{n} \mu_i \to \min$$

$$\text{s.t. } \mathrm{Diag}(\mu) \succeq A.$$

Show that the dual problem is equivalent to

$$\langle A, X \rangle \to \max \quad \text{subject to} \quad X \in \mathbb{S}_+^n, \quad X_{ii} = 1 \; \forall i.$$

In the remainder of this chapter, we will study iterative algorithms for solving the optimization problems discussed so far. It will be useful to think of iterations as occurring in "discrete (algorithmic) time," while also considering the "continuous time" limit, where the change in values per iteration becomes sufficiently small and the number of iterations becomes large. In the continuous-time analysis, we will use the language of differential equations, which offers both intuitive and rigorous tools for analyzing the behavior of the algorithms. However, we will also refer back to the discrete case when needed to derive specific conclusions.

B.3 Unconstrained First-Order Convex Minimization

In this section, we explore the unconstrained convex minimization problem,

$$f(x) \to \min_{x \in \mathbb{R}^n},$$

focusing on first-order optimization methods. These methods assume that both the objective function $f(x)$ and its gradient $\nabla f(x)$ can be computed efficiently. First-order methods are widely used to solve the majority of practical problems in machine learning, data science and applied mathematics.

We assume the objective function $f(x)$ is smooth, meaning that

$$\forall x, y \in \mathbb{R}^n : \|\nabla f(x) - \nabla f(y)\|_* \leq \beta \|x - y\|,$$

for some constant $\beta > 0$. When using the ℓ_2-norm, this smoothness condition can be expressed as

$$f(y) \leq f(x) + \nabla f(x)^\top (y - x) + \frac{\beta}{2}\|y - x\|_2^2, \quad \forall x, y \in \mathbb{R}^n.$$

To simplify notation, we omit further references to norms when assuming the ℓ_2-norm.

Smooth optimization

Gradient descent. Gradient descent (GD) is the simplest and one of the most popular algorithms for solving both convex and non-convex optimization problems. The GD iteration step is given by

$$\begin{aligned} x_{k+1} &= x_k - \eta_k \nabla f(x_k) \\ &= \arg\min_x \underbrace{\left\{ f(x_k) + \nabla f(x_k)^\top (x - x_k) + \frac{1}{2\eta_k} \|x - x_k\|_2^2 \right\}}_{h_{\eta_k}(x)}, \end{aligned}$$

where $\eta_k \leq 1/\beta$ is the step size. Here, we assume that $f(x)$ is β-smooth with respect to the ℓ_2-norm. Each step of GD can be interpreted as minimizing the quadratic upper bound $h_{\eta_k}(x)$ of $f(x)$.

Definition B.3.1. A function $f : \mathbb{R}^n \to \mathbb{R}$ is β-smooth with respect to a norm $\|\cdot\|$ if

$$\|\nabla f(x) - \nabla f(y)\|_* \leq \beta \|x - y\| \quad \forall x, y \in \mathbb{R}^n.$$

When $\|\cdot\| = \|\cdot\|_2$, we simply say that the function is β-smooth.

Theorem B.3.2. *Let $f : \mathbb{R}^n \to \mathbb{R}$ be a convex and β-smooth function. Then, after k iterations of the GD algorithm with a fixed step size $\eta \leq 1/\beta$, the function value $f(x_k)$ satisfies*

$$f(x_k) - f(x^*) \leq \frac{\|x_1 - x^*\|_2^2}{2\eta k}, \quad \eta \leq 1/\beta, \tag{B.12}$$

where x^* is the optimal solution.

We prove this theorem using both a continuous-time approach and a discrete-time approach, relying on the concept of a Lyapunov function.

Definition B.3.3. A Lyapunov function, $V(x(t))$, for a differential equation $\dot{x}(t) = f(x(t))$, is a function that satisfies the following properties:

1. It decreases monotonically along the trajectory, i.e., $\dot{V}(x(t)) < 0$.
2. It converges to zero as $t \to \infty$, i.e., $V(x(\infty)) = 0$, where $x^* = x(\infty)$.

For the continuous-time analysis, we use $X(t)$ to denote the continuous-time version of the iterates x_k.

Proof of Theorem B.3.2: Continuous Time. The GD algorithm can be viewed as a discretization of the first-order differential equation

$$\dot{X}(t) = -\nabla f(X(t)).$$

We define the Lyapunov function for this system as $V(X(t)) = \frac{1}{2}\|X(t) - x^*\|_2^2$. Taking its time derivative, we get

$$\begin{aligned}\frac{d}{dt}V(t) &= (X(t) - x^*)^\top \dot{X}(t)\\ &= -\nabla f(X(t))^\top (X(t) - x^*) \leq -(f(X(t)) - f^*),\end{aligned}$$

where the last inequality follows from the convexity of f. Integrating this inequality over time yields

$$V(X(t)) - V(X(0)) \leq tf^* - \int_0^t f(X(\tau))d\tau.$$

Using Jensen's inequality for convex functions,

$$f\left(\frac{1}{t}\int_0^t X(\tau)d\tau\right) \leq \frac{1}{t}\int_0^t f(X(\tau))d\tau,$$

and noting that $V(t) \geq 0$, we derive

$$f\left(\frac{1}{t}\int_0^t X(\tau)d\tau\right) - f^* \leq \frac{V(X(0))}{t}.$$

The proof is complete by setting $t \approx k/\beta$ since f is β-smooth. □

Proof of Theorem B.3.2: Discrete Time. For the discrete case, applying the smoothness condition to $y = x - \eta\nabla f(x)$ gives

$$\begin{aligned} f(y) &\le f(x) + \nabla f(x)^\top (y - x) + \frac{\beta}{2}\|y - x\|_2^2 \\ &= f(x) - \eta\|\nabla f(x)\|_2^2 + \frac{\beta}{2}\eta^2\|\nabla f(x)\|_2^2 \\ &= f(x) - \left(1 - \frac{\beta\eta}{2}\right)\eta\|\nabla f(x)\|_2^2. \end{aligned}$$

Since $\eta \le 1/\beta$, we have $1 - \beta\eta/2 \le 1/2$ and, therefore,

$$f(y) \le f(x) - \frac{\eta}{2}\|\nabla f(x)\|_2^2.$$

Now, using convexity,

$$f(x^*) \ge f(x) + \nabla f(x)^\top (x^* - x),$$

and plugging this into the smoothness inequality, we get

$$f(y) - f(x^*) \le \nabla f(x)^\top (x - x^*) - \frac{\eta}{2}\|\nabla f(x)\|_2^2.$$

This leads to

$$\sum_{j=1}^{k}(f(x_j) - f(x^*)) \le \frac{1}{2\eta}\left(\|x_1 - x^*\|_2^2 - \|x_{k+1} - x^*\|_2^2\right).$$

Thus,

$$f(\bar{x}) - f(x^*) \le \frac{\beta\|x_1 - x^*\|_2^2}{2k},$$

where $\bar{x} = \frac{1}{k}\sum_{j=1}^{k} x_j$, completing the proof. □

One naturally aims to choose a step size in GD that results in the fastest convergence. However, this problem – finding the best, or even only a good, step size – is challenging and remains an open question. This difficulty extends to determining an effective stopping criterion for iterations. In the following, we describe some practical and empirical strategies for selecting the step size in GD:

- **Exact line search:** In each iteration, choose the step size η_k such that

$$\eta_k = \arg\min_{\eta} \{f(x_k - \eta \nabla f(x_k))\}.$$

This strategy ensures the most progress is made along the gradient direction at each step but can be computationally expensive, especially in high-dimensional problems.

- **Backtracking line search:** Here, the step size η_k is chosen to satisfy

$$f(x_k - \eta_k \nabla f(x_k)) \leq f(x_k) - \frac{\eta_k}{2} \|\nabla f(x_k)\|_2^2.$$

Since the difference between the right- and left-hand sides of the inequality decreases monotonically with η_k, you can start with an initial η and iteratively reduce it using $\eta \to b\eta$, where $0 < b < 1$. This approach ensures sufficient decrease in function value while adjusting the step size dynamically.

- **Polyak's step-size rule:** If the optimal function value f^* is known, a more targeted step size can be used. By minimizing the right-hand side of

$$\|x_{k+1} - x^*\| \leq \|x_k - x^*\|_2^2 - 2\eta_k(f(x_k) - f(x^*)) + \eta_k^2 \|\nabla f(x_k)\|_2^2 \to \min_{\eta_k},$$

one arrives at Polyak's step-size rule:

$$\eta_k = \frac{f(x_k) - f(x^*)}{\|\nabla f(x_k)\|_2^2}.$$

This rule is particularly useful when solving systems of linear equations, such as $Ax = b$, as it often leads to rapid convergence in those cases.

Exercise B.11. Recall that GD minimizes a convex quadratic upper bound $h_{\eta_k}(x)$ of $f(x)$. Consider a modified GD algorithm where the step size is $\eta = (2 + \varepsilon)/\beta$, with $\varepsilon > 0$. (Note that in Theorem B.3.2, the step size condition was $\eta \leq 1/\beta$.) Derive the modified version of Eq. (B.12). Can you find a convex quadratic function for which this modified algorithm fails to converge?

Exercise B.12. Consider the minimization of the following (non-convex) function f:

$$\begin{aligned} &f(x) \to \min \\ &\text{s.t. } \|x - x^*\| \le \varepsilon, \\ &x \in \mathbb{R}^n, \end{aligned}$$

where x^* is the global and unique minimum of the β-smooth function f. Furthermore, assume that

$$\forall x \in \mathbb{R}^n : \frac{1}{2}\|\nabla f(x)\|_2^2 \ge \mu(f(x) - f(x^*)),$$

for some constant $\mu > 0$. Prove or disprove whether GD with a step size of $\eta_k = 1/\beta$ converges to the optimum for some small $\varepsilon > 0$. How does ε depend on β and μ?

Exercise B.13 (difficult). In many optimization problems, the gradient is often noisy, meaning that only a perturbed version is available. This leads to the "inexact oracle" optimization problem: $f(x) \to \min, x \in \mathbb{R}^n$, where for each x, one can compute $\hat{f}(x)$ and $\hat{\nabla} f(x)$ such that

$$\forall x : |f(x) - \hat{f}(x)| \le \delta \quad \text{and} \quad \|\nabla f(x) - \nabla \hat{f}(x)\|_2 \le \varepsilon,$$

where $\delta, \varepsilon > 0$ are the tolerances. Propose and analyze a modification of GD that addresses this "inexact oracle" optimization.

Gradient descent in ℓ_p-norm. GD can be generalized to other norms, such as the ℓ_p-norm:

$$x_{k+1} = \underset{x \in S \subset \mathbb{R}^n}{\arg\min} \left\{ f(x_k) + \nabla f(x_k)^\top (x - x_k) + \frac{1}{2\eta_k}\|x - x_k\|_p^2 \right\},$$

where $\eta_k \le 1/\beta_p$ and $\beta_p \ge \sup_x \|\nabla f(x)\|_p$. Depending on the choice of p, this method can converge significantly faster than in the ℓ_2-norm. GD in the ℓ_1-norm, in particular, is quite popular for promoting sparsity.

Exercise B.14. Restate and prove the discrete-time version of Theorem B.3.2 for GD in the ℓ_p-norm.

Hint: Consider the following Lyapunov function: $\|x - x^*\|_p^2$.

Gradient descent for strongly convex, smooth functions.

Theorem B.3.4. *GD applied to a strongly convex function f with a fixed step size,*

$$x_{k+1} = x_k - \eta \nabla f(x_k), \quad \eta = 1/\beta,$$

converges to the optimal solution as

$$f(x_{k+1}) - f(x^*) \leq c^k (f(x_1) - f(x^*)),$$

where $c \leq 1 - \mu/\beta$.

Exercise B.15. Extend the proof of Theorem B.3.2 to cover Theorem B.3.4.

Fast gradient descent. While GD is simple and efficient in practice, it can suffer from slow convergence when the gradient is small. Additionally, it may oscillate near the optimum if the gradient is nearly orthogonal to the optimal direction. To address these issues, two major modifications of GD have been introduced:

- **Polyak's heavy-ball method** [25]:

$$x_{k+1} = x_k + \eta_k \nabla f(x_k) + \mu_k (x_k - x_{k-1}), \tag{B.13}$$

  where the additional μ_k term introduces momentum from the previous step, analogous to inertia in classical mechanics. This momentum term helps accelerate the movement toward the optimum but may also cause overshooting due to insufficient damping.
- **Nesterov's fast gradient method (FGM)** [26]:

$$x_{k+1} = x_k + \eta_k \nabla f(x_k + \mu_k (x_k - x_{k-1})) + \mu_k (x_k - x_{k-1}), \tag{B.14}$$

  In FGM, the momentum term is adjusted by applying the gradient to a linear combination of the current and previous steps. This modification often leads to faster convergence and less overshooting compared to the heavy-ball method.

In both methods, the term involving μ_k is referred to as the "momentum" or "inertia" term, which aims to reduce oscillations and accelerate convergence by maintaining a portion of the previous step's direction. Despite the similarity between the two methods, the convergence behavior can differ significantly, with FGM generally exhibiting superior convergence rates, while Polyak's heavy-ball method can sometimes lead to divergence due to overshooting.

Exercise B.16. Construct a convex function f with a piecewise linear gradient such that the heavy-ball algorithm (B.13) fails to converge for some fixed values of μ and η.

Consider the following simplified version of FGM with a two-step recurrence:

$$x_k = y_{k-1} - \eta\nabla f(y_{k-1}), \qquad y_k = x_k + \frac{k-1}{k+2}(x_k - x_{k-1}), \quad \text{(B.15)}$$

which can be analyzed in continuous time by the differential equation

$$\ddot{X}(t) + \frac{3}{t}\dot{X}(t) + \nabla f(X(t)) = 0. \quad \text{(B.16)}$$

This differential equation is derived by interpreting $t \approx k\sqrt{\eta}$ and rescaling the discrete-time recurrence relation from Eq. (B.15). Expanding the differences $x_{k+1} - x_k$ and $x_k - x_{k-1}$ using a Taylor series, we obtain

$$\frac{x_{k+1} - x_k}{\sqrt{\eta}} = \dot{X}(t) + \frac{1}{2}\ddot{X}(t)\sqrt{\eta} + o(\sqrt{\eta}),$$

and similarly for $x_k - x_{k-1}$. This leads to Eq. (B.16), a second-order differential equation, which describes the continuous-time behavior of the FGM algorithm.

Convergence analysis of FGM (sketch)

Following Ref. [27], to analyze the convergence rate of FGM, we introduce the following Lyapunov function:

$$V(X(t)) = t^2(f(X(t)) - f^*) + 2\|X(t) + \frac{t}{2}\dot{X}(t) - x^*\|_2^2.$$

Taking the derivative of $V(X(t))$, we get

$$\begin{aligned}\dot{V}(X(t)) &= 2t(f(X(t)) - f^*) + t^2\nabla f(X(t))^\top\dot{X}(t)\\ &\quad + 4\left(X(t) + \frac{t}{2}\dot{X}(t) - x^*\right)^\top\left(\frac{3}{2}\dot{X}(t) + \frac{t}{2}\ddot{X}(t)\right).\end{aligned}$$

Using the fact that $\dot{X}(t) + \frac{t}{2}\ddot{X}(t) = -\frac{t}{2}\nabla f(X(t))$ and applying the convexity of f, we simplify to

$$\dot{V}(X(t)) \leq 0.$$

Since $V(X(t))$ is non-increasing and $V(X(t)) \geq 0$, we have

$$f(X(t)) - f^* \leq \frac{V(0)}{t^2} = \frac{2\|x_0 - x^*\|_2^2}{t^2}.$$

Finally, since $t \approx k\sqrt{\eta}$, we conclude that

$$f(x_k) - f^* \leq \frac{2\|x_0 - x^*\|_2^2}{\eta k^2}, \qquad \eta \leq \frac{1}{\beta}.$$

This yields the following theorem.

Theorem B.3.5. *FGD, applied to the minimization of a β-smooth convex function $f(x)$, with the update rule*

$$x_k = y_{k-1} - \eta\nabla f(y_{k-1}), \quad y_k = x_k + \frac{k-1}{k+2}(x_k - x_{k-1})$$

converges to the optimum at a rate of

$$f(x_k) - f^* \leq \frac{2\|x_0 - x^*\|_2^2}{\eta k^2}.$$

Exercise B.17. Consider the differential equation:

$$\ddot{X}(t) + \frac{r}{t}\dot{X}(t) + \nabla f(X(t)) = 0,$$

for some $r > 0$. Derive the corresponding discrete-time algorithm, analyze its convergence and show that if $r \leq 2$, the convergence rate is $O(1/k^2)$.

Exercise B.18. Show that the FGM method described by Eq. (B.15) transitions to the more general update rule given by Eq. (B.14) for some choice of η_k.

Non-smooth problems

Sub-gradient method. We begin our discussion of sub-gradient (SG) methods with the simplest and arguably most popular algorithm:

$$x_{k+1} = x_k - \eta_k g_k, \quad g_k \in \partial F(x_k), \tag{B.17}$$

which mirrors GD, except that the gradient is replaced by a SG to accommodate non-smooth functions. It is important to note that it is not entirely appropriate to call this algorithm SG descent because, unlike GD, the function value $f(x_{k+1})$ can be larger than $f(x_k)$ during iterations. To mitigate this issue, one can track the best point encountered so far or use an averaged result over past iterations (for instance, using a finite horizon of previous points). A common approach is to augment Eq. (B.17) with

$$f_{\text{best}}^{(k)} = \min\{f_{\text{best}}^{(k-1)}, f(x_k)\}.$$

We assume the SG of $f(x)$ is bounded, meaning that

$$\forall x: \ \|g(x)\| \leq L, \quad g(x) \in \partial f(x),$$

which follows from the Lipschitz condition $|f(x) - f(y)| \leq L\|x - y\|$ for some constant L. Let x^* be the optimal point of the problem $f(x) \to \min_{x \in \mathbb{R}^n}$. Then, we have the following bound on the distance to the optimum:

$$\begin{aligned} \|x_{k+1} - x^*\|_2^2 &= \|x_k - \eta_k g_k - x^*\|_2^2 \\ &= \|x_k - x^*\|_2^2 - 2\eta_k g_k^\top (x_k - x^*) + \eta_k^2 \|g_k\|_2^2 \\ &\leq \|x_k - x^*\|_2^2 - 2\eta_k (f(x_k) - f(x^*)) + \eta_k^2 \|g_k\|_2^2, \end{aligned} \tag{B.18}$$

where the last inequality uses the convexity of f, i.e., $f(x^*) \geq f(x_k) + g_k^\top (x^* - x_k)$. Recursively applying this inequality, we get

$$\|x_{k+1} - x^*\|_2^2 \leq \|x_1 - x^*\|_2^2 - 2\sum_{j \leq k} \eta_j (f(x_j) - f(x^*)) + \sum_{j \leq k} \eta_j^2 \|g_j\|_2^2,$$

which yields

$$\left(2\sum_{j\le k}\eta_j\right)(f_{\text{best}}^{(k)} - f(x^*)) \le \|x_1 - x^*\|_2^2 + \sum_{j\le k}\eta_j^2\|g_j\|_2^2.$$

Assuming that the SGs are bounded by L_2, we have

$$f_{\text{best}}^{(k)} - f(x^*) \le \frac{\|x_1 - x^*\|_2^2 + L_2^2\sum_{j\le k}\eta_j^2}{2\sum_{j\le k}\eta_j}.$$

For the step size $\eta_k = \frac{R}{L\sqrt{k}}$, where $R_2^2 \ge \|x_1 - x^*\|_2^2$, this simplifies to

$$f_{\text{best}}^{(k)} - f(x^*) \le \frac{RL}{\sqrt{k}}. \tag{B.19}$$

Note that the $\sim 1/\sqrt{k}$ convergence rate is significantly slower than the $\sim 1/k^2$ rate achieved for smooth functions. We will discuss this in more detail and explore methods to improve the convergence rate.

Proximal gradient method. In many applications, particularly in machine learning and statistics, the objective function is often composed as a sum over samples. A typical example of this is the *composite optimization* problem:

$$f(x) = g(x) + h(x) \to \min_{x\in\mathbb{R}^n}, \tag{B.20}$$

where $g : \mathbb{R}^n \to \mathbb{R}$ is a convex, smooth function and $h : \mathbb{R}^n \to \mathbb{R}$ is a convex, closed, but possibly non-smooth function. A well-known example of composite optimization is the *Lasso* problem:

$$f(x) = \|Ax - b\|_2^2 + \lambda\|x\|_1 \to \min_{x\in\mathbb{R}^n}, \tag{B.21}$$

where the ℓ_1-norm introduces non-smoothness at $x = 0$.

To handle non-smooth components, we introduce the *proximal operator*:

$$\text{prox}_h(x) = \arg\min_{u\in\mathbb{R}^n}\left(h(u) + \frac{1}{2}\|u - x\|_2^2\right).$$

Common examples of the proximal operator include the following:

1. For $h(x) = I_C(x)$, where $I_C(x)$ is the indicator function of a convex set C, the proximal operator corresponds to the projection onto C:

$$\operatorname{prox}_h(x) = \arg\min_{u \in C} \|x - u\|_2^2.$$

2. For $h(x) = \lambda\|x\|_1$, the proximal operator performs soft-thresholding:

$$\operatorname{prox}_h(x)_i = \begin{cases} x_i - \lambda & \text{if } x_i \geq \lambda, \\ x_i + \lambda & \text{if } x_i \leq -\lambda, \\ 0 & \text{otherwise.} \end{cases}$$

The proximal operator allows us to handle non-smooth terms effectively. This leads to the *proximal gradient descent* (PGD) algorithm:

$$\begin{aligned} x_{k+1} &= \operatorname{prox}_{\eta_k h}(x_k - \eta_k \nabla g(x_k)) \\ &= \arg\min_u \left(\frac{1}{2}\|x_k - \eta_k \nabla g(x_k) - u\|_2^2 + \eta_k h(u) \right), \end{aligned}$$

where $\eta_k \leq 1/\beta$ and g is β-smooth with respect to the ℓ_2 norm.

Similar to GD, PGD minimizes a convex upper bound of the objective at each iteration. The convergence rate of PGD matches that of GD for smooth functions, as captured by the following theorem.

Theorem B.3.6. *The PGD algorithm,*

$$x_{k+1} = \operatorname{prox}_h(x_k - \eta \nabla g(x_k)), \quad \eta \leq 1/\beta,$$

with a fixed step size, converges to the optimal solution x^* of the composite optimization problem (B.20) *at the rate of*

$$f(x_{k+1}) - f(x^*) \leq \frac{\|x_0 - x^*\|_2^2}{2\eta k}.$$

The proof of this theorem follows the same logic as the proof for Theorem B.3.2 for the GD algorithm. Furthermore, PGD can be accelerated in a manner similar to GD. The accelerated version of PGD is given by

$$x_k = \text{prox}_{\eta_k}(y_{k-1} - \eta_k \nabla f(y_{k-1})), \quad y_k = x_k + \frac{k-1}{k+2}(x_k - x_{k-1}).$$

This leads to the following accelerated convergence result.

Theorem B.3.7. *The accelerated PGD method for solving a convex optimization problem,*

$$f(x) \to \min_{x \in \mathbb{R}^n},$$

with the update rule

$$x_k = \text{prox}_{h\eta}(y_{k-1} - \eta \nabla f(y_{k-1})), \quad y_k = x_k + \frac{k-1}{k+2}(x_k - x_{k-1}),$$

converges to the optimum as

$$f(x_{k+1}) - f(x^*) \leq \frac{2\|x_0 - x^*\|_2^2}{\eta k^2},$$

for any β-smooth convex function f.

PGD is one of the key methods for dealing with non-smooth optimization problems. We explore other approaches in the following.

Smoothing out non-smooth objectives

Consider the following min-max optimization problem:

$$\max_{1 \leq i \leq n} f_i(x) \to \min_{x \in \mathbb{R}^n},$$

which represents one of the most common types of non-smooth optimization problems. A smooth and convex approximation to the maximum function can be provided by the *soft-max* function, as defined in Eq. (B.2). This approximation allows the problem to be tackled using accelerated GD, which achieves a convergence rate of $O(1/\sqrt{\varepsilon})$, in contrast to the slower $O(1/\varepsilon^2)$ rate typical for non-smooth functions. A careful choice of the parameter λ within the soft-max function can further accelerate the algorithm, often improving convergence to $O(1/\varepsilon)$.

B.4 Constrained First-Order Convex Minimization

Projected gradient descent

The PGD algorithm solves convex optimization problems with constraints. The update step for PGD is

$$\begin{aligned} x_{k+1} &= \Pi_C(x_k - \eta_k \nabla f(x_k)) \qquad \text{(B.22)} \\ &= \arg\min_{y \in C} \left(f(x_k) + \nabla f(x_k)^\top (y - x_k) + \frac{1}{2\eta_k} \|x_k - y\|_2^2 \right), \end{aligned}$$

where Π_C represents the Euclidean projection onto the convex set C:

$$\Pi_C(y) = \arg\min_{x \in C} \|x - y\|_2^2.$$

PGD enjoys the same convergence rate as standard GD. The proof is analogous, leveraging the fact that projections do not increase the distance to the optimal solution, i.e.,

$$\|x_{k+1} - x^*\|_2^2 \leq \|x_k - \eta_k \nabla f(x_k) - x^*\|_2^2, \quad \text{for } x^* \in C.$$

Exercise B.19 (Alternating Projections). Consider two convex sets $C, D \subseteq \mathbb{R}^n$ and the problem of finding $x \in C \cap D$. Starting from $x_0 \in C$, apply alternating projections:

$$y_k = \Pi_C(x_k), \quad x_{k+1} = \Pi_D(y_k).$$

How many iterations are required to guarantee that

$$\max \left\{ \inf_{x \in C} \|x_k - x\|, \inf_{x \in D} \|x_k - x\| \right\} \leq \varepsilon?$$

Frank–Wolfe algorithm (conditional gradient method)

The Frank–Wolfe (FW) algorithm solves constrained convex optimization problems by avoiding explicit projections, which can be computationally expensive. It solves the following problem:

$$f(x) \to \min \quad \text{s.t. } x \in C. \qquad \text{(B.23)}$$

Instead of projecting onto the set C, as in PGD, FW solves a linear approximation of the objective at each step:

$$y_k = \arg\min_{y \in C} \nabla f(x_k)^\top y, \quad x_{k+1} = (1 - \gamma_k)x_k + \gamma_k y_k, \quad \gamma_k = \frac{2}{k+2}. \tag{B.24}$$

For instance, if C is the unit simplex,

$$f(x) \to \min \quad \text{s.t. } x \in S = \{x \geq 0,\ x^\top 1 = 1\},$$

the update y_k corresponds to a unit vector in the direction of the maximum coordinate of the gradient. The time complexity per iteration is $O(n)$, making it significantly more efficient than PGD for large-scale problems. A comprehensive analysis of the FW method is presented in Ref. [76].

Stopping criterion. A key advantage of the FW algorithm is its reliable stopping criterion. Due to the convexity of f, we have

$$f(y) \geq f(x_k) + \nabla f(x_k)^\top (y - x_k),$$

for any $y \in C$. Minimizing both sides over $y \in C$, we get

$$f^* \geq f(x_k) + \min_{y \in C} \nabla f(x_k)^\top (y - x_k),$$

where f^* is the optimal value. This leads to the following duality gap:

$$\max_{y \in C} \nabla f(x_k)^\top (x_k - y) \geq f(x_k) - f^*. \tag{B.25}$$

The left-hand side of Eq. (B.25) provides a straightforward stopping criterion for the algorithm.

Theorem B.4.1. *If $f(x)$ is a convex β-smooth function and C is a bounded, convex set, then the FW algorithm converges to the optimal solution f^* of Eq.* (B.23) *at the rate of*

$$f(x_k) - f^* \leq \frac{2\beta D^2}{k+2},$$

where $D^2 \geq \max_{x,y \in C} \|x - y\|_2^2$.

Proof. Convexity of f implies

$$f(x) \geq f(x_k) + \nabla f(x_k)^\top (x - x_k), \quad \forall x \in C.$$

Minimizing both sides over $x \in C$, we obtain

$$f(x^*) \geq f(x_k) + \nabla f(x_k)^\top (y_k - x_k).$$

Thus,

$$f(x_k) - f(x^*) \leq \nabla f(x_k)^\top (x_k - x^*).$$

Using the update rule $x_{k+1} = (1 - \gamma_k)x_k + \gamma_k y_k$, we derive

$$\begin{aligned} f(x_{k+1}) - f(x_k) &\leq \gamma_k \nabla f(x_k)^\top (y_k - x_k) + \frac{\beta \gamma_k^2}{2} \|y_k - x_k\|_2^2 \\ &\leq (1 - \gamma_k)(f(x_k) - f(x^*)) + \frac{\beta \gamma_k^2 D^2}{2}. \end{aligned}$$

By induction on k, starting with $k = 1$, we obtain the desired convergence rate:

$$f(x_k) - f^* \leq \frac{2\beta D^2}{k + 2}.$$

Comparison with other methods. While the FW algorithm has a slower iteration-based convergence rate compared to FGMs, it often performs better in practice, particularly when the feasible set C is simple (e.g., a norm ball or polytope). This is because FW avoids the need for explicit projections, which can be computationally expensive in PGD or other constrained methods.

Primal-dual gradient algorithm

Consider the following smooth convex optimization problem:

$$\begin{aligned} &f(x) \to \min \\ &\text{s.t. } Ax = b, \quad x \in \mathbb{R}^n. \end{aligned}$$

To handle the constraints effectively, it is often useful to work with an *augmented problem*:

$$\begin{aligned} &f(x) + \frac{\rho}{2} \|Ax - b\|_2^2 \to \min \\ &\text{s.t. } Ax = b, \end{aligned}$$

where $\rho > 0$ is a penalty parameter. This formulation introduces a penalty term to enforce the constraint softly.

The *augmented Lagrangian* for this problem is defined as

$$\mathcal{L}(x,\mu) = f(x) + \mu^\top(Ax-b) + \frac{\rho}{2}\|Ax-b\|_2^2,$$

where μ is the vector of Lagrange multipliers associated with the equality constraint $Ax = b$.

A point (x,μ) is called *primal-dual optimal* if it satisfies the following optimality conditions:

$$\begin{aligned} 0 &= \nabla_x \mathcal{L}(x,\mu) = \nabla f(x) + A^\top \mu + \rho A^\top(Ax-b), \\ 0 &= -\nabla_\mu \mathcal{L}(x,\mu) = b - Ax. \end{aligned}$$

These conditions can be compactly expressed using a primal-dual operator $T(x,\mu)$ as

$$T(x,\mu) = \begin{pmatrix} \nabla_x \mathcal{L}(x,\mu) \\ -\nabla_\mu \mathcal{L}(x,\mu) \end{pmatrix}.$$

Thus, (x,μ) is primal-dual optimal if $T(x,\mu) = 0$.

Primal-dual gradient algorithm (PDG). The primal-dual gradient (PDG) algorithm is an iterative method that updates both the primal variable x and the dual variable μ using the primal-dual operator:

$$\begin{pmatrix} x \\ \mu \end{pmatrix}_{k+1} = \begin{pmatrix} x \\ \mu \end{pmatrix}_k - \eta_k T(x_k,\mu_k),$$

where η_k is the step size at iteration k.

Incorporating inequality constraints. Now, consider the case where the optimization problem includes inequality constraints:

$$\begin{aligned} & f(x) \to \min \\ & \text{s.t. } g_i(x) \le 0, \quad 1 \le i \le m. \end{aligned}$$

The corresponding *augmented problem* becomes

$$f(x) + \frac{\rho}{2}\sum_{i=1}^{m}(g_i(x))_+^2 \to \min,$$

where $(g_i(x))_+ = \max\{0, g_i(x)\}$ denotes the positive part of $g_i(x)$. The inequality constraints $g_i(x) \le 0$ are softly enforced through the penalty term.

The *augmented Lagrangian* in this case is

$$\mathcal{L}(x, \lambda) = f(x) + \lambda^\top F(x) + \frac{\rho}{2}\|F(x)\|_2^2,$$

where $F(x)_i = (g_i(x))_+$ and λ is the vector of Lagrange multipliers for the inequality constraints.

A point (x, λ) is primal-dual optimal if it satisfies

$$0 = \nabla_x \mathcal{L}(x, \lambda) = \nabla f(x) + \sum_{i=1}^{m} (\lambda_i + \rho g_i(x)_+) \nabla g_i(x),$$

$$0 = -\nabla_\lambda \mathcal{L}(x, \lambda) = -F(x).$$

Primal-dual gradient algorithm with inequality constraints. The PDG algorithm for problems with inequality constraints follows a similar structure:

$$\begin{pmatrix} x \\ \lambda \end{pmatrix}_{k+1} = \begin{pmatrix} x \\ \lambda \end{pmatrix}_k - \eta_k T(x_k, \lambda_k),$$

where the update is applied to both the primal variable x and the dual variable λ and $T(x, \lambda)$ is the primal-dual operator associated with the inequality-constrained problem.

Convergence analysis. The convergence analysis for the PDG algorithm mirrors that of standard GD but in an extended space of primal and dual variables. The Lyapunov function used to establish convergence is

$$V(x, \lambda) = \|x_k - x^*\|_2^2 + \|\lambda_k - \lambda^*\|_2^2.$$

This function measures the combined distance of the primal and dual variables from their optimal values. By showing that $V(x_k, \lambda_k)$ decreases monotonically, one can prove convergence to the optimal solution.

Exercise B.20. Analyze the convergence of the PDG algorithm for convex optimization with inequality constraints, assuming that all functions involved (both the objective f and the constraint functions g_i) are convex and β-smooth.

Mirror descent algorithm

Our previous analysis primarily focused on cases where the objective function f is smooth in the ℓ_2 norm, and the distance from the initial point to the optimal point is also measured using the ℓ_2 norm. From the perspective of GD, optimizing over a unit simplex and a unit Euclidean sphere appears equivalent in terms of computational complexity. However, the geometry of these domains is drastically different: The volume of a unit simplex is exponentially smaller than that of a unit sphere. The mirror descent (MD) algorithm exploits the geometry of the optimization domain, providing faster convergence for specific domains, such as the simplex, with acceleration up to a factor of $\sim \sqrt{d}$, where d is the dimension of the space.

Let's begin with an unconstrained convex optimization problem:

$$f(x) \to \min \quad \text{s.t. } x \in S \subseteq \mathbb{R}^n.$$

An elementary iteration of the GD algorithm is given by

$$x_{k+1} = x_k - \eta_k \nabla f(x_k).$$

In this form, we are summing objects from different spaces: x belongs to the primal space, while the gradient $\nabla f(x)$ resides in the dual (conjugate) space, which may be geometrically distinct. To address this inconsistency, Nemirovski and Yudin (1978) introduced the MD algorithm, which proceeds as follows:

$$y_k = \nabla \phi(x_k) \quad \text{(map the primal point to the dual space)}$$

$$y_{k+1} = y_k - \eta_k \nabla f(x_k) \quad \text{(update in the dual space)}$$

$$\bar{x}_{k+1} = (\nabla \phi)^{-1}(y_{k+1}) = \nabla \phi^*(y_{k+1}) \quad \text{(map back to the primal space)}$$

$$x_{k+1} = \Pi_C^{D_\phi}(\bar{x}_{k+1}) = \operatorname*{arg\,min}_{x \in C} D_\phi(x, \bar{x}_{k+1}) \quad \text{(project to the feasible set)},$$

where $\phi(x)$ is a strongly convex function called the *mirror map*, and $\phi^*(y)$ is its LF conjugate, defined as

$$\phi^*(y) = \sup_{x \in \mathbb{R}^n} (y^\top x - \phi(x)).$$

The function $D_\phi(u, v) = \phi(u) - \phi(v) - \nabla \phi(v)^\top (u - v)$, known as the *Bregman divergence*, measures the distance between u and v in terms of the mirror map ϕ. For strongly convex ϕ, the Bregman divergence satisfies several key properties:

- **Non-negativity:** $D_\phi(u, v) \geq 0$ for any convex function ϕ.
- **Convexity in the first argument:** The Bregman divergence is convex in its first argument, although not necessarily in its second.
- **Linearity:** For any non-negative coefficients λ and μ, the Bregman divergence satisfies the linearity property:

$$D_{\lambda\phi+\mu\psi}(u, v) = \lambda D_\phi(u, v) + \mu D_\psi(u, v).$$

- **Duality:** For a function ϕ with a convex conjugate ϕ^*, the Bregman divergences of ϕ and ϕ^* are related by

$$D_{\phi^*}(u^*, v^*) = D_\phi(u, v), \quad \text{where } u^* = \nabla\phi(u),\ v^* = \nabla\phi(v).$$

Examples of Bregman divergence

- **Euclidean norm:** If $\phi(x) = \frac{1}{2}\|x\|_2^2$, then the Bregman divergence reduces to the squared Euclidean distance

$$D_\phi(x, y) = \frac{1}{2}\|x - y\|_2^2.$$

- **Negative entropy:** If $\phi(x) = \sum_{i=1}^n x_i \ln x_i$, defined on the positive orthant $\mathbb{R}_{++}^n$, then the Bregman divergence corresponds to the Kullback–Leibler (KL) divergence

$$D_\phi(x, y) = \sum_{i=1}^n x_i \ln\left(\frac{x_i}{y_i}\right) - x_i + y_i = D_{KL}(x\|y).$$

- **Bounds:** If ϕ is μ-strongly convex with respect to a norm $\|\cdot\|$, then the Bregman divergence satisfies the bounds

$$\frac{\mu}{2}\|x - y\|^2 \leq D_\phi(x, y) \leq \frac{\beta}{2}\|x - y\|^2,$$

where β is the smoothness parameter of ϕ.

Mirror descent update. The MD algorithm can be written in proximal form as

$$x_{k+1} = \Pi_C^{D_\phi}\left(\operatorname*{arg\,min}_{x\in\mathbb{R}^n}\left\{f(x_k) + \nabla f(x_k)^\top (x - x_k) + \frac{1}{\eta_k} D_\phi(x, x_k)\right\}\right),$$

where $\Pi_C^{D_\phi}(y) = \arg\min_{x\in C} D_\phi(x, y)$ projects onto the feasible set C using the Bregman divergence. For an in-depth analysis of the MD algorithm and its extensions, we refer the reader to Ref. [74, 77, 78].

Example B.4.2. Consider the following optimization over the unit simplex:

$$f(x) \to \min_{x \in S}, \quad \text{s.t. } S = \{x \in \mathbb{R}^n : x^\top \mathbf{1} = 1, x \geq 0\}.$$

Using the negative entropy function $\phi(x) = \sum_{i=1}^n x_i \ln x_i$ as the mirror map, the MD update becomes

$$x_{k+1} = \frac{(x_k)_i \exp(-\eta_k \nabla f(x_k)_i)}{\sum_{j=1}^n (x_k)_j \exp(-\eta_k \nabla f(x_k)_j)},$$

where the update is a renormalization that ensures x remains in the simplex.

Continuous-time analysis. In the case of β-smooth convex functions, the continuous-time dynamics of MD are best described using the Lyapunov function in the dual space:

$$V(Z(t)) = D_{\phi^*}(Z(t), z^*), \quad Z(t) = \nabla\phi(X(t)),$$

where ϕ is a strongly convex distance-generating function. Differentiating the Lyapunov function yields

$$\frac{d}{dt}V(Z(t)) = -\nabla f(X(t))^\top (X(t) - x^*) \leq -(f(X(t)) - f^*).$$

Integrating both sides and applying Jensen's inequality leads to the convergence rate:

$$f\left(\frac{1}{t}\int_0^t X(\tau)\, d\tau\right) - f^* \leq \frac{V(Z(0))}{t}.$$

Thus, the MD algorithm achieves a convergence rate of $O(1/k)$ for smooth functions. For non-smooth convex functions, MD exhibits a convergence rate of $O(1/\sqrt{k})$, similar to the behavior of GD in non-smooth settings.

References

[1] M. A. Lavrentiev and B. V. Shabat, *Methods of the Theory of Functions of a Complex Variable.* Nauka (in Russian), English translation by AMS, Providence, RI, 1973.

[2] J. B. Conway, *Functions of One Complex Variable.* Springer, New York, NY, 1978.

[3] L. V. Ahlfors, *Complex Analysis: An Introduction to the Theory of Analytic Functions of One Complex Variable.* McGraw-Hill, New York, NY, 1979.

[4] M. Tabor, *Principles and Methods of Applied Mathematics.* University of Arizona Press, Tucson, AZ, 1999.

[5] R. N. Bracewell, *The Fourier Transform and Its Applications.* McGraw-Hill, New York, NY, 1999.

[6] M. J. Lighthill, *Introduction to Fourier Analysis and Generalised Functions.* Cambridge University Press, Cambridge, UK, 1958.

[7] V. I. Arnold, *Ordinary Differential Equations.* Springer, Berlin, Germany, 1992.

[8] E. A. Coddington and N. Levinson, *Theory of Ordinary Differential Equations.* McGraw-Hill, New York, NY, 1955.

[9] D. Zwillinger, *Handbook of Differential Equations.* Academic Press, San Diego, CA, 1997.

[10] P. Hartman, *Ordinary Differential Equations.* SIAM, Philadelphia, PA, 2002.

[11] V. Arnold, *Ordinary Differential Equations.* The MIT Press, Cambridge, MA, 1973.

[12] V. I. Smirnov, *A Course of Higher Mathematics: Volume IV, Part II: Partial Differential Equations.* Pergamon Press, Oxford, UK, 1964.

[13] G. B. Folland, *Introduction to Partial Differential Equations.* Princeton University Press, Princeton, NJ, 1995.

[14] O. A. Ladyzhenskaya, *The Boundary Value Problems of Mathematical Physics.* Springer, Berlin, Germany, 1985.

[15] I. Stakgold and M. Holst, *Green's Functions and Boundary Value Problems.* Wiley, Hoboken, NJ, 2011.

[16] R. Courant and D. Hilbert, *Methods of Mathematical Physics, Vol. 2.* Wiley, New York, NY, 1989.

[17] G. Falkovich, V. Lvov, and V. Zakharov, *Kolmogorov Spectra of Turbulence I: Wave Turbulence.* Springer-Verlag, Berlin, Germany, 1992.

[18] E. Hopf, "The partial differential equation $u_t + uu_x = \mu u_{xx}$," *Communications on Pure and Applied Mathematics*, **3**(3), pp. 201–230, 1950.

[19] J. D. Cole, "On a quasi-linear parabolic equation occurring in aerodynamics," *Quarterly of Applied Mathematics*, **9**(3), pp. 225–236, 1951.

[20] I. M. Gelfand and S. V. Fomin, *Calculus of Variations.* Dover, Mineola, NY, 2000.

[21] M. Giaquinta and S. Hildebrandt, *Calculus of Variations I.* Springer, Berlin, Germany, 1996.

[22] V. I. Arnold, *Mathematical Methods of Classical Mechanics.* Springer, Berlin, Germany, 1989.

[23] L. I. Rudin, S. Osher, and E. Fatemi, "Nonlinear total variation based noise removal algorithms," *Physica D: Nonlinear Phenomena*, **60**(1), pp. 259–268, 1992.

[24] J. Calder, "The calculus of variations (lecture notes)." 2019. http://www-users.math.umn.edu/~jwcalder/CalculusOfVariations.pdf.

[25] B. Polyak, "Some methods of speeding up the convergence of iteration methods," *USSR Computational Mathematics and Mathematical Physics*, **4**(5), pp. 1–17, 1964.

[26] Y. E. Nesterov, "A method for solving the convex programming problem with convergence rate $o(1/k^2)$," *Dokl. Akad. Nauk SSSR*, **269**, pp. 543–547, 1983.

[27] W. Su, S. Boyd, and E. J. Candes, "A Differential Equation for Modeling Nesterov's Accelerated Gradient Method: Theory and Insights," *arXiv:1503.01243*, 2015.

[28] A. C. Wilson, B. Recht, and M. I. Jordan, "A Lyapunov Analysis of Momentum Methods in Optimization," *arXiv:1611.02635*, 2016.

[29] M. Levi, *Classical Mechanics with Calculus of Variations and Optimal Control: An Intuitive Introduction.* AMS, Providence, RI, 2014.

[30] H. Touchette, "Legendre-fenchel transforms in a nutshell," 2014.

[31] A. Chambolle, "An algorithm for total variation minimization and applications," *Journal of Mathematical Imaging and Vision*, **20** pp. 89–97, 2004.

[32] R. K. P. Zia, E. F. Redish, and S. R. McKay, "Making sense of the legendre transform," *American Journal of Physics*, **77**(7), pp. 614–622, Jul 2009. Available: http://dx.doi.org/10.1119/1.3119512.

[33] L. S. Pontryagin *et al.*, *The Mathematical Theory of Optimal Processes*. Interscience Publishers, New York, NY, 1962.

[34] R. Bellman, *Dynamic Programming*. Dover, Mineola, NY, 2003.

[35] D. P. Bertsekas, *Dynamic Programming and Optimal Control*. Athena Scientific, Belmont, MA, 2017.

[36] V. G. Boltyanskii, *Mathematical Methods of Optimal Control*. Holt, Rinehart & Winston, New York, NY, 1971.

[37] L. Pontryagin, V. Boltayanskii, R. Gamkrelidze, and E. Mishchenko, *The Mathematical Theory of Optimal Processes (translated from Russian in 1962)*. Wiley, New York, NY, 1956.

[38] A. T. Fuller, "Bibliography of pontryagm's maximum principle," *Journal of Electronics and Control*, **15**(5), pp. 513–517, 1963.

[39] R. Bellman, "On the theory of dynamic programming," *PNAS*, **38**(8), p. 716, 1952.

[40] C. Moore and S. Mertens, *The Nature of Computation*. New York, USA: Oxford University Press, New York, NY, 2011.

[41] E. T. Jaynes, *Probability Theory: The Logic of Science*. Cambridge University Press, Cambridge, UK, 2003.

[42] R. Durrett, *Probability: Theory and Examples*. Cambridge University Press, Cambridge, UK, 2010.

[43] D. J. C. Mackay, *Information Theory, Inference, and Learning Algorithms*. Cambridge University Press, Cambridge, UK, 2003.

[44] A. Sinclair, "Uc berkley, cs271 "randomness & computation" course," 2020. Available: https://people.eecs.berkeley.edu/~sinclair/cs271/n13.pdf.

[45] C. E. Shannon, "Prediction and entropy of printed english," *The Bell System Technical Journal*, **30**(1), pp. 50–64, 1951.

[46] J. L. Doob, *Stochastic Processes*. John Wiley & Sons, New York, NY, 1953.

[47] C. W. Gardiner, *Handbook of Stochastic Methods*. Springer, Berlin, Germany, 1985.

[48] G. Grimmett and D. Stirzaker, *Probability and Random Processes*. Oxford University Press, Oxford, UK, 2001.

[49] B. Oksendal, *Stochastic Differential Equations: An Introduction with Applications*. Springer, Berlin, Germany, 2013.

[50] R. S. Sutton and A. G. Barto, *Reinforcement Learning: An Introduction*, 2nd ed. The MIT Press, Cambridge, MA, 2018.

[51] F. Kelly and E. Yudovina, *Stochastic Networks*, ser. Institute of Mathematical Statistics Textbooks. Cambridge University Press, Cambridge, UK, 2014.

[52] V. Y. Chernyak, M. Chertkov, D. A. Goldberg, and K. Turitsyn, "Non-equilibrium statistical physics of currents in Queuing networks," *Journal of Statistical Physics*, **140**(5), pp. 819–845, Sep. 2010. Available: https://doi.org/10.1007/s10955-010-0018-5.

[53] S. Asmussen and P. W. Glynn, *Stochastic Simulation: Algorithms and Analysis*, ser. Stochastic Modelling and Applied Probability. **57**. New York, NY: Springer, 2007.

[54] C. M. Bishop, *Pattern Recognition and Machine Learning*. Springer, New York, NY, 2006.

[55] M. J. Wainwright and M. I. Jordan, "Graphical models, exponential families, and variational inference," *Foundations and Trends in Machine Learning*, **1**(1), pp. 1–305, 2008.

[56] M. Chertkov, *INFERLO: Inference, Learning & Optimization with Graphical Models*. 2004. https://sites.google.com/site/mchertkov/research/living-books.

[57] D. B. Wilson, "Perfectly random sampling with markov chains," 1996. http://www.dbwilson.com/exact/.

[58] T. Richardson and R. Urbanke, *Modern Coding Theory*. Cambridge, UK: Cambridge University Press, 2008.

[59] J. S. Yedidia, W. T. Freeman, and Y. Weiss, "Constructing free-energy approximations and generalized belief propagation algorithms," *IEEE Transactions on Information Theory*, **51**(7), pp. 2282–2312, 2005.

[60] V. Likhosherstov, Y. Maximov, and M. Chertkov, "Inference and sampling of k_{33}-free ising models," in *Proceedings of the 36th International Conference on Machine Learning*, ser. Proceedings of Machine Learning Research, K. Chaudhuri and R. Salakhutdinov, Eds., **97**. PMLR, 09–15 Jun 2019, pp. 3963–3972. Available: http://proceedings.mlr.press/v97/likhosherstov19a.html.

[61] A. Y. Lokhov, M. Vuffray, S. Misra, and M. Chertkov, "Optimal structure and parameter learning of ising models," *Science Advances*, **4**(3), 2018. Available: https://advances.sciencemag.org/content/4/3/e1700791.

[62] C. Chow and C. Liu, "Approximating discrete probability distributions with dependence trees," *IEEE Transactions on Information Theory*, **14**(3), pp. 462–467, 1968.

[63] J. B. Kruskal, "On the shortest spanning subtree of a graph and the traveling salesman problem," *Proceedings of the American Mathematical Society*, **7**(1), pp. 48–50, 1956.

[64] G. Strang, *Linear Algebra and Learning from Data.* Wellesley-Cambridge Press, Wellesley, MA, 2019.

[65] G. Cybenko, "Approximation by superpositions of a sigmoidal function," *Mathematics of Control, Signals, and Systems,* **2**(4), pp. 303–314, 1989.

[66] K. Hornik, "Approximation capabilities of multilayer feedforward networks," *Neural Networks,* **4**(2), pp. 251–257, 1991.

[67] A. Pinkus, "Approximation theory of the mlp model in neural networks," *Acta Numerica,* **8**, pp. 143–195, 1999.

[68] P. Kidger and T. Lyons, "Universal approximation with deep narrow networks," in *Proceedings of Thirty Third Conference on Learning Theory,* ser. Proceedings of Machine Learning Research, J. Abernethy and S. Agarwal, Eds., **125**. PMLR, 09–12 Jul 2020, pp. 2306–2327. Available: https://proceedings.mlr.press/v125/kidger20a.html.

[69] S. Levine, "Deep reinforcement learning." 2021. https://rail.eecs.berkeley.edu/deeprlcourse/,

[70] E. Brunskill, "Reinforcement learning." 2022. https://web.stanford.edu/class/cs234/modules.html.

[71] S. Agrawal, "Reinforcement learning." 2019. https://ieor8100.github.io/rl/.

[72] C. J. Watkins and P. Dayan, "Q-learning," *Machine Learning,* **8**(3-4), pp. 279–292, 1992.

[73] S. Boyd and L. Vandenberghe, *Convex Optimization.* New York, NY: Cambridge University Press, 2004.

[74] A. Ben-Tal and A. Nemirovski, *Lectures on Modern Convex Optimization: Analysis, Algorithms, and Engineering Applications,* ser. MOS-SIAM Series on Optimization. Philadelphia, PA: Society for Industrial and Applied Mathematics (SIAM), 2001.

[75] B. Polyak, *Introduction to Optimization.* Optimization Software, New York, NY, 1987.

[76] M. Jaggi, "Revisiting frank-wolfe: Projection-free sparse convex optimization," in *International Conference on Machine Learning (ICML).* PMLR, 2013, pp. 427–435.

[77] A. Beck and M. Teboulle, "Mirror descent and nonlinear projected subgradient methods for convex optimization," *Operations Research Letters,* **31**(3), pp. 167–175, 2003.

[78] Y. Nesterov, "Primal-dual subgradient methods for convex problems," *Mathematical Programming,* **120**(1), pp. 221–259, 2009.

Index

D

E

F

G

M

O

P

Q

R

S

V

W

Z

www.ingramcontent.com/pod-product-compliance
Lightning Source LLC
Chambersburg PA
CBHW062138271225
36892CB00004B/74
* 9 7 8 9 8 1 9 8 0 8 2 4 3 *